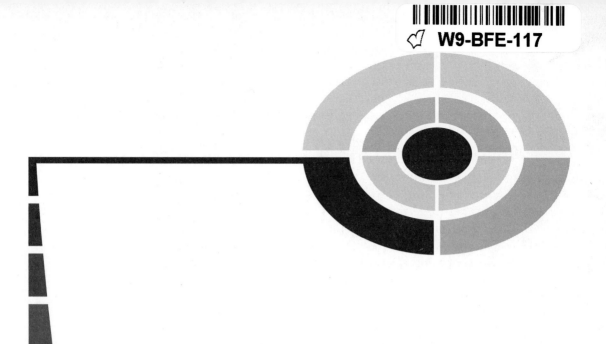

Organic Chemistry Demystified

Demystified Series

Accounting Demystified
Advanced Statistics Demystified
Algebra Demystified
Anatomy Demystified
asp.net 2.0 Demystified
Astronomy Demystified
Biology Demystified
Biotechnology Demystified
Business Calculus Demystified
Business Math Demystified
Business Statistics Demystified
C++ Demystified
Calculus Demystified
Chemistry Demystified
College Algebra Demystified
Corporate Finance Demystified
Databases Demystified
Data Structures Demystified
Differential Equations Demystified
Digital Electronics Demystified
Earth Science Demystified
Electricity Demystified
Electronics Demystified
Environmental Science Demystified
Everyday Math Demystified
Forensics Demystified
Genetics Demystified
Geometry Demystified
Home Networking Demystified
Investing Demystified
Java Demystified
JavaScript Demystified
Linear Algebra Demystified
Macroeconomics Demystified

Management Accounting Demystified
Math Proofs Demystified
Math Word Problems Demystified
Medical Terminology Demystified
Meteorology Demystified
Microbiology Demystified
Microeconomics Demystified
Nanotechnology Demystified
OOP Demystified
Options Demystified
Organic Chemistry Demystified
Personal Computing Demystified
Pharmacology Demystified
Physics Demystified
Physiology Demystified
Pre-Algebra Demystified
Precalculus Demystified
Probability Demystified
Project Management Demystified
Psychology Demystified
Quality Management Demystified
Quantum Mechanics Demystified
Relativity Demystified
Robotics Demystified
Six Sigma Demystified
sql Demystified
Statistics Demystified
Technical Math Demystified
Trigonometry Demystified
uml Demystified
Visual Basic 2005 Demystified
Visual C# 2005 Demystified
xml Demystified

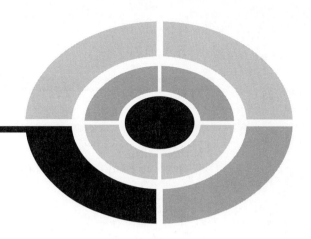

Organic Chemistry
Demystified

DANIEL R. BLOCH

McGRAW-HILL
New York   Chicago   San Francisco   Lisbon   London
Madrid   Mexico City   Milan   New Delhi
San Juan   Seoul   Singapore   Sydney   Toronto

The McGraw·Hill Companies

Library of Congress Cataloging-in-Publication Data

Bloch, Daniel, Date.
    Organic chemistry demystified / Daniel Bloch.
        p.   cm.
    Includes index.
    ISBN 0-07-145920-0
        1. Chemistry, Organic—Textbooks.   I. Title.

    QD253.2.B56   2005
    547—dc22   2005054685

5 6 7 8 9 0   DOC/DOC   0 1 0 9 8

ISBN 0-07-145920-0

The sponsoring editor for this book was Judy Bass and the production supervisor was Pamela A. Pelton. It was set in Times Roman by TechBooks. The art director for the cover was Margaret Webster-Shapiro.

Printed and bound by RR Donnelley.

McGraw-Hill books are available at special quantity discounts to use as premiums and sales promotions, or for use in corporate training programs. For more information, please write to the Director of Special Sales, McGraw-Hill Professional, Two Penn Plaza, New York, NY 10121-2298. Or contact your local bookstore.

To Nan for her assistance, patience, and helpful comments.

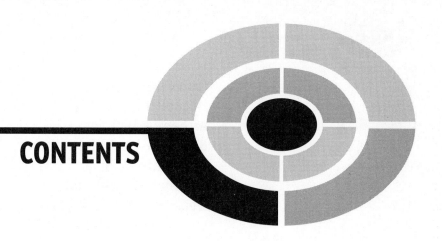

CONTENTS

CONTENTS

CONTENTS

PREFACE

Organic chemistry is the chemistry of carbon-containing compounds. Every living organism, plant and animal, is composed of organic compounds. Anyone with an interest in life and living things needs to have a basic understanding of organic chemistry.

Articles continue to appear in newspapers and magazines describing the development of new medicines and diagnostic tests. These new products and technologies are a result of a better understanding of the structure and function of DNA, proteins, and other organic biological molecules. The reactions and interactions of these complex molecules are the same reactions and interactions that occur in more simple organic molecules.

This text was written to help those who are intimidated by the words *organic chemistry*. Those who have never had a formal course in organic chemistry and students currently taking or planning to take a formal course will find this text an easy-to-read introduction and supplement to other texts.

The chapters are written in the same general order as found in most college textbooks. It would be helpful, but not necessary, if the reader had a course in introductory chemistry. The first three chapters cover the background material typically covered in general chemistry courses. It is not necessary that chapters be read sequentially, but since material tends to build on previous concepts it will be easier to understand the material if the chapters are read in sequential order.

Key terms and concepts are italicized. Be sure you understand these concepts as they will continue to appear in other sections of this book. Questions (and answers) are given within each chapter to help you measure your understanding. Each chapter ends with a quiz covering the material presented. Use each quiz to check your comprehension and progress. The answers to quizzes are given in the back of the text. Review those problems (immediately) you did not get correct. Be sure you understand the concepts before going to the next chapter as new material often builds upon previous concepts.

As you read each chapter, take frequent breaks (you can munch on the extra gum drops used to make models in Chapter 5). The book contains a lot of figures and diagrams. Follow these as you read the text. It is often easier to understand a reaction mechanism in a diagram than to describe it in words.

Yes, there is some memorization. New terms will appear that you probably have never heard before. For a series of terms I recommend making a mnemonic and I suggested a few. Reaction mechanisms are not as difficult as they may appear. You can predict most reactions in that negative species will be attracted to positive species (opposites attract). Atoms with electrons to share will be attracted to species that want more electrons—it is just that simple.

There is a multiple-choice final exam at the end of the text. The final exam has more general, but similar, questions than those in the quizzes. Answers are given in the back of the book. If you are able to answer 80% of the final exam questions correctly (the first time), you will have a good understanding of the material.

I hope you will enjoy reading about organic chemistry as much as I have enjoyed writing about it.

Daniel R. Bloch

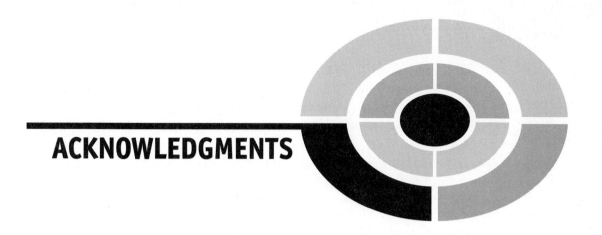

ACKNOWLEDGMENTS

The following individuals were kind enough to review various chapters in this book:

Vaughn Ausman, Marquette University
Kate Bichler, University of Wisconsin Center—Manitowoc
Peter Conigliaro (retired), S.C. Johnson
Sheldon Cramer (emeritus), Marquette University
Timothy Eckert, Carthage College
Sharbil Firsan, Sigma-Aldrich
Kevin Glaeske, Wisconsin Lutheran College
Bruce Holman
Shashi Jasti, Sigma-Aldrich
Steven Levsen, Mount Mary College
Julie Lukesh, University of Wisconsin—Green Bay
Kevin Morris, Carthage College
Pat Nylen, University of Wisconsin—Milwaukee
Stephen Templin, Cardinal Stritch University

A special thanks to Priyanka Negi and Judy Bass who assisted with the technical editing of this book.

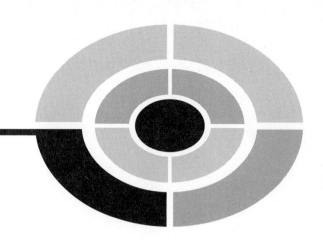

Organic Chemistry
Demystified

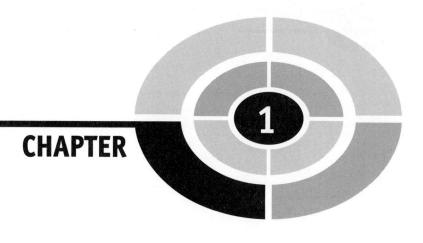

CHAPTER 1

Structure and Bonding

Introduction

The study of organic chemistry involves the reactions and interactions of molecules. Since molecules are composed of atoms, it is necessary to review the structure of atoms and how they contribute to the properties of molecules.

Atomic Structure

Atoms are composed of a *nucleus* surrounded by *electrons*, as shown in Fig. 1-1. The nucleus consists of positively charged *protons* and neutral *neutrons*. Although the nucleus consists of other subatomic particles, the proton, neutron, and electron are the only subatomic particles that will be discussed in this text.

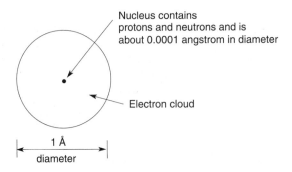

Fig. 1-1. Structure of an atom.

The atom is extremely small. It has a diameter of about 10^{-10} m (0.000,000,000,1 m or 0.000,000,004 in.). These small dimensions are usually expressed in angstroms (Å), where 1 Å equals 1×10^{-10} m, or pm where 1 pm equals 1×10^{-12} m. The nucleus is about 1/10,000th the diameter of the atom, or about 10^{-4} Å. A *key point*: *most of the volume of an atom is occupied by the electrons.* To put this in terms that are easier to understand, if the atom was magnified so that the nucleus was the size of a marble, the area occupied by the electrons would be the size of a football stadium. Take a minute to visualize that. The area occupied by electrons is huge relative to that of the nucleus. The area occupied by electrons is referred to as the *electron cloud*.

MASSES OF ATOMS

The mass of an atom is concentrated in the nucleus. A proton and a neutron each have a mass of about 1.66×10^{-24} g. An electron has a mass of 1/1800th that of a proton. Since these are such very small numbers, it is more convenient to give the mass of a proton and a neutron in *atomic mass units (amu)*. One amu is equal to 1.66×10^{-24} g. The mass of individual atoms is also given in a.m.u. The mass of 1 mole of atoms (a mole is a specific number, approximately 6.022×10^{23}) is the *atomic mass*, which we usually call the *atomic weight*, of an element. The atomic weight is expressed in grams/mol.

ELECTRON CLOUDS

Structures of molecules are usually written as shown in Fig. 1-2. Structure 1-2a implies that atoms are quite far apart, relative to their size. This is certainly true for the nuclei, but *not* for the electron clouds. The distance between a hydrogen nucleus and a carbon nucleus in a carbon-hydrogen bond is about 1.14 Å. The

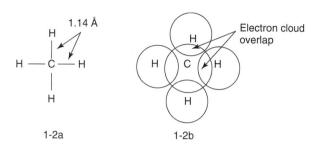

Fig. 1-2. Bond formation resulting from electron cloud overlap.

radius of the electron cloud of an isolated hydrogen atom is calculated to be 0.79 Å and the radius of an isolated carbon atom is calculated to be 0.91 Å. Since the sum of the radii of the two atoms is 1.70 Å and the actual bond length is only 1.14 Å, the electron clouds must overlap to form the C−H bond. Generally, the greater the electron cloud overlap, the greater the electron density in the bond and the stronger the bond. The area occupied by the electrons, the electron cloud, is much greater than implied in the structures typically drawn (such as Structure 1-2a) in this book and other organic chemistry textbooks. The area occupied by electrons in a molecule is more accurately represented by Structure 1-2b.

Why do we need to be so concerned with electrons and electron clouds? Organic chemistry involves physical interactions and chemical reactions between molecules. Electrons are primarily responsible for these interactions and reactions.

QUESTION 1-1
Atoms consist of which three subatomic particles?

ANSWER 1-1
Protons, neutrons, and electrons.

Atomic Number and Atomic Mass (Weight)

The *atomic number* (Z) for an element is equal to the number of protons in the nucleus of an atom of a given element. The sum of the number of protons and neutrons is the *mass number* (A). If the number of protons changes (a nuclear reaction), a new element results. There are no changes in the number of protons in an atom in chemical reactions.

An element is identified with a *symbol*. The symbol is an abbreviation for an element: H stands for hydrogen, C for carbon, He for helium, and Na for sodium. Symbols are not always the first letters of a current name as some symbols are derived from historical or non-English names. A symbol is often shown with a superscript indicating the atomic weight and a subscript indicating the atomic number, e.g., ^4_ZHe or ^4_2He.

ISOTOPES

The number of neutrons may vary for a particular element. About 99% of carbon atoms have six neutrons, about 1% have seven neutrons, and a very small percentage contain eight neutrons. Atoms with the same *atomic number* (and thus the same number of protons) but different mass numbers (the sum of protons and neutrons) are called *isotopes*. The average mass of carbon is 12.0107 g/mol, the element's atomic weight. Note that the atomic weights (we should really say atomic masses, but organic chemists usually use the term weight) of elements in the periodic table (Appendix A) are not whole numbers as they represent the average of the isotopic composition. The number of electrons in a neutral atom, one without a charge, equals the number of protons. Electrons also contribute to an atom's molecular weight, but an electron's total weight is about 1/2000th that of a proton and their weight contribution is usually ignored.

QUESTION 1-2
A sample consists of three atoms of chlorine: one has a mass of 35.0 amu and two have masses of 36.0 amu. How many protons, neutrons, and electrons are in each atom? What is the average mass of the sample?

ANSWER 1-2
The atom of 35.0 amu has 17 electrons, 17 protons, and 18 neutrons. The other atoms have 17 electrons, 17 protons, and 19 neutrons. The average mass is $(1 \times 35.0 + 2 \times 36.0)/3 = 35.7$ amu.

Electron Energy Levels

Electrons occupy concentric *shells* and *subshells* around a nucleus. The shells are given numbers called *principle quantum numbers* of 1, 2, 3, etc., to identify the levels. The energy of each shell and distance between the electrons in a shell and the nucleus increases with increasing principle quantum number. Level 1 is the lowest energy level and the electrons in that shell are nearest to the nucleus. Level 2 is higher in energy and the electrons in this level are found further from the nucleus than are the electrons in Level 1. Shells are composed of subshells. Subshells have designations s, p, d, and f. The energy of the shells and subshells increases as shown in Fig. 1-3. The electron configurations for hydrogen, helium,

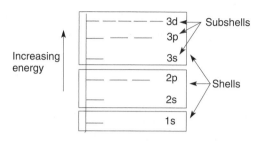

Fig. 1-3. Energy levels of shells and subshells.

carbon, nitrogen, and oxygen atoms are shown in Fig. 1-4. Electrons prefer to occupy the lowest energy levels available to them. This represents their most stable state called their *ground state*.

AUFBAU PRINCIPLE

Figure 1-4 is a more concise method showing how electrons fill the subshells as the atomic number of the element increases. Each additional electron goes into the lowest energy subshell available to it. This is called the *aufbau* (building up) principle. Figure 1-4 shows the lowest energy electron configuration of six common elements. Each s subshell consists of one orbital. Each p subshell consists of three orbitals. Note the term *orbital*, not orbit, is used. An orbital is defined in a following section.

Each orbital can hold a maximum of two electrons. When the orbitals in a subshell are filled, electrons go into the next higher-energy subshell. Each principle shell has only one s orbital: 1s, 2s, 3s, etc. Each principle shell of Level 2 and higher has three p orbitals, p_x, p_y, and p_z. All p orbitals in the same subshell ($2p_x$, $2p_y$, and $2p_z$) are of equal energy. Orbitals of equal energy are

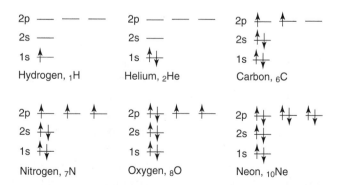

Fig. 1-4. Electron configuration of elements.

called *degenerate orbitals*. The maximum number of electrons in a main shell is $2n^2$, where n is the principle quantum number, 1, 2, 3, etc.

PAULI EXCLUSION PRINCIPLE

Since electrons have negative charges, there is some resistance for two electrons to occupy the same orbital, that is, to pair up. Species of like charge (two negative charges) repel each other. The helium atom has two electrons to be placed in orbitals. (See the electron configuration of helium in Fig. 1-4.) One electron can be put into the lowest energy orbital, the 1s orbital. The second electron can go into the 1s orbital or the 2s orbital. The energy required to put the second electron into the higher energy 2s orbital is greater than the energy required (electron-electron repulsion) to pair the electrons in the 1s orbital. Therefore the second electron goes into the 1s orbital. Each electron is said to have a spin, like a top, and the spin can be clockwise or counterclockwise. The spin direction is indicated by an arrow pointing up or down. Two electrons in the same orbital must have opposite spins (*Pauli exclusion principle*). Helium's two electrons are shown with opposite spins ($\uparrow\downarrow$) in Fig. 1-4.

HUND'S RULE

Consider the carbon atom with six electrons. The electron configuration is shown in Fig. 1-4. Using the aufbau principle, the first two electrons go into the 1s orbital. The next two electrons go into the next higher energy 2s orbital. Then the last two electrons go into the higher energy 2p orbitals. The last two electrons could go into one p orbital or each could go into two different p orbitals. For degenerate (equal energy) orbitals, it is more energy efficient for electrons to go into different degenerate orbitals until they must pair up (*Hund's rule*).

Now consider oxygen with eight electrons. When seven electrons are added by the aufbau principle, the electron configuration will be the same as shown for nitrogen (see Fig. 1-4). The last electron added pairs with an electron already in a 2p orbital. Their spins must be opposite (Pauli exclusion principle) as shown in Fig. 1-4.

ELECTRON CONFIGURATIONS

A short-hand method for writing the electron configurations of atoms is shown for oxygen as $1s^2 2s^2 2p^4$. Verbally one would say one s two, two s two, two p four. The prefix indicates the principle energy level (1, 2, 3, etc.), the letter (s, p, d, or f) gives the type of orbital, and the exponent is the number of electrons in that orbital.

QUESTION 1-3
Draw the short-hand electron configuration for the sodium atom.

ANSWER 1-3
The sodium atom has 11 electrons: $1s^2 2s^2 2p^6 3s^1$.

VALENCE ELECTRONS

The electrons in the *outermost* shell are called the *valence electrons*. Elements in the first row (period) in the periodic table, hydrogen and helium, have only a 1s orbital. The maximum number of electrons these two elements can accommodate is 2. A 2-electron configuration will be called a *duet*. When hydrogen has 2 valence electrons in its 1s orbital it will be called *duet happy*. Elements in the second row (period) in the periodic table, from lithium to neon, can hold a maximum of 10 electrons. The outermost shell, the valence shell, has a principle quantum number of 2 and can hold a maximum of 8 electrons, $2s^2$, $2p^6$. When the valence shell orbitals are filled, the atom will be called *octet happy*. The number of *valence electrons* in the elements in the first three rows of the periodic table is equal to their group number (see the periodic table in Appendix A). Hydrogen in Group IA has 1 valence electron, carbon in Group IVB has 4 valence electrons, and fluorine in Group VIIB has 7 valence electrons. An atom can gain valence electrons from, or loose electrons to, other atoms. Valence electrons are important since they are involved in forming chemical bonds.

QUESTION 1-4
How many electrons does a nitrogen atom have? How many valence electrons does it have?

ANSWER 1-4
It has seven electrons and five valence electrons.

The Octet Rule

Neon, argon, and the other elements in column VIIIB in the periodic table are called the *noble gases*. They have eight electrons in their valence shell. Helium is an exception since its valence shell (1s) can hold only two electrons. Noble gases are so called because they are, of course, gases and tend to be unreactive or inert. There is a special stability associated with atoms with eight electrons in their valence shell (except for the elements in row 1). The *octet rule* states that elements will gain, lose, or share electrons to achieve eight electrons in their outermost (valence) shell. An explanation for this special stability is beyond the scope of this book.

There are some exceptions to the octet rule. Third row elements (such as sulfur and phosphorus) can hold up to 18 electrons in their outermost valence shell (3s, 3p, and 3d orbitals). Beryllium and boron atoms can have less than 8 electrons in their valence shells. An example of a boron compound will be discussed in a following section.

Valences

The *bonding capacity* or the number of bonds to an atom is called its *valence.* (It would be helpful to look at the periodic table in Appendix A as you read this paragraph.) The valence of atoms in Groups IA to IVA is the same as the group number. Lithium (Group IA) has a valence of one and will have a single bond to another atom. Carbon (Group IVB) has a valence of four and there are four bonds to each carbon atom. Carbon is called tetravalent. The valence of elements in Groups VB to VIIB is 3, 2, and 1 (or eight minus the group number) respectively. Elements in Groups VB to VIIB can have multiple positive valences, but those situations will not be discussed here.

QUESTION 1-5
What valence does oxygen have? What is the bonding capacity for oxygen? What is the bonding capacity for boron?

ANSWER 1-5
Oxygen has a valence and bonding capacity of 2. Boron has a bonding capacity of 3.

Atoms in the second row tend to gain or lose electrons to achieve the electron structure of the nearest noble gas. They want to be duet or octet happy as are the noble gases. Thus, lithium tends to lose one electron to achieve the helium electron configuration. Fluorine tends to gain an electron to achieve the neon electron configuration.

QUESTION 1-6
How many electrons must oxygen gain to become octet happy?

ANSWER 1-6
Oxygen will gain 2 electrons to have a full octet in its valence shell.

BOND FORMATION

Atoms form bonds by transferring or sharing electrons with other atoms. An atom that loses an electron has a positive charge and is called a *cation*. If an atom gains an electron, it has a negative charge and is called an *anion*. Atoms of elements in Groups IA and IIA tend to transfer electrons to elements in Groups VIB and VIIB. The resulting cation from Group IA or IIA forms an ionic bond with the resulting anion from Group VIB or VIIB. Elements in Groups VIB and VIIB tend to share electrons if they react with elements in Groups IVB to VIIB. Sharing electrons result in covalent bonds.

Carbon, in Group IVB, tends to form bonds with many other elements. One reason there are so many organic compounds is that carbon atoms can form bonds with other carbon atoms, resulting in a large number of compounds.

Lewis Structures

Lewis structures are a convenient way of showing an atom's valence electrons. Dots are used to indicate the valence electrons. The inner electrons, the *core electrons*, are not shown. Lewis structures for carbon, nitrogen, and fluorine atoms are shown in Fig. 1-5.

LEWIS STRUCTURE FOR CH$_4$

The Lewis structures of some compounds are shown in Fig. 1-6. Bonds in compounds are shown by a pair of dots or a solid line representing two electrons. How does one know where to put the electrons? First consider methane, CH$_4$. There are four simple rules to follow:

1. Sum the valence electrons of all the atoms in a molecule.
    CH$_4$ has eight valence electrons, four from the carbon atom and four from the hydrogen atoms (one from each).
2. Show the structure of the compound by connecting the atoms with a single (two electron) bond. You may have to be told how the atoms are connected if they can be connected in more than one way. Methane is

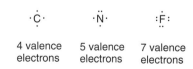

·Ċ·	·N̈·	:F̈:
4 valence electrons	5 valence electrons	7 valence electrons

Fig. 1-5. Lewis structures.

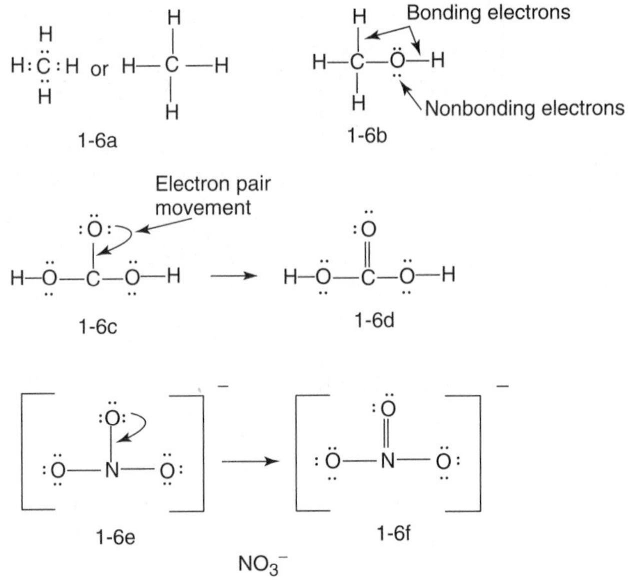

H:C:H or H—C—H

1-6a

H—C—O—H   Bonding electrons ... Nonbonding electrons

1-6b

Electron pair movement

H—O—C—O—H → H—O—C—O—H

1-6c   1-6d

1-6e   →   1-6f

NO_3^-

Fig. 1-6. Lewis structures for molecules and ions.

shown as Structure 1-6a. Each hydrogen atom is bonded to the carbon atom with a single bond.

3. Each bond consists of two electrons. Subtract the number of bonding electrons from the total number of valence electrons. There are eight electrons in the bonds and eight valence electrons. In this case, all valence electrons are assigned to the four C—H bonds in CH_4.

4. If there are additional unassigned electrons, place them on the second row elements to give full octets. Pairs of electrons not involved in bonding are called nonbonding electrons, as shown in Structures 1-6b–1-6f. (In CH_4 there are no nonbonding electrons and all atoms are duet or octet happy.)

5. Move electrons in pairs (shown by the curved arrow in Structures 1-6c and 1-6e) to make all atoms duet or octet happy, if possible.

LEWIS STRUCTURE FOR CH_3OH

Three examples using CH_3OH, H_2CO_3, and NO_3^- will help explain how to draw Lewis structures. Atom connectivity is shown in Fig. 1-6. First consider CH_3OH. There are 14 valence electrons: 4 from the carbon atom, 4 from the four hydrogen atoms, and 6 from the oxygen atom. The atoms are connected as shown in Structure 1-6b. The five bonds use 10 electrons (2 electrons in each bond). There are 4 valence electrons left to assign. Put these electrons on the oxygen atom to make it octet happy. All atoms are now duet or octet happy. The Lewis structure is shown as Structure 1-6b.

   A key point: the total number of valence electrons for an atom in a compound is the sum of all bonding and nonbonding electrons. Both electrons in a bond are counted as valence electrons for *each atom* connected by that bond. In a C–H bond the two electrons are counted as valence electrons for H *and* for C. Thus electrons in bonds are double counted as valence electrons.

LEWIS STRUCTURE FOR H_2CO_3

Now consider H_2CO_3, Structures 1-6c and 1-6d. There are 24 valence electrons, 2 from the two hydrogen atoms, 4 from the carbon atom, and 18 from the three oxygen atoms. Ten electrons are used in the five bonds connecting the atoms. The remaining 14 electrons are put on the oxygen atoms. Four electrons are put on the two oxygen atoms bonded to H and C. Six electrons are put on the third oxygen atom bonded only to C, as shown in Structure 1-6c. All atoms are octet happy except the carbon atom. Move one electron pair, as shown by the curved arrow in Structure 1-6c, to be shared by the oxygen atom and the carbon atom. This results in two bonds (a double bond) between the oxygen and carbon atoms as shown in Structure 1-6d. Now every atom is duet or octet happy.

LEWIS STRUCTURE FOR $NO_3{}^-$

There is an additional step to consider for ions and ionic compounds that have a net charge. If the ion has a negative charge, an additional valence electron needs to be added for each negative charge. If the ion has a positive charge, one valence electron has to be removed for each positive charge.

   The nitrate anion, $NO_3{}^-$, has a net negative charge. There are 24 valence electrons, 5 from the nitrogen atom, 18 from the three oxygen atoms, and an additional electron due to the negative charge. The atoms are connected as shown in Structure 1-6e. Six electrons are used in the three bonds. The remaining 18 electrons are put on the oxygen atoms, 6 electrons on each. All atoms are octet happy, except nitrogen. Move an electron pair, as shown by the curved arrow in Structure 1-6e, between any one of the three oxygen atoms and the nitrogen atom. This results in two bonds between one oxygen atom and the nitrogen atom (Structure 1-6f). Now each atom is octet happy.

QUESTION 1-7
Draw the Lewis structures for CH_3F, ICl, H_2O, HCN, and $CH_3CO_2{}^-$.

ANSWER 1-7

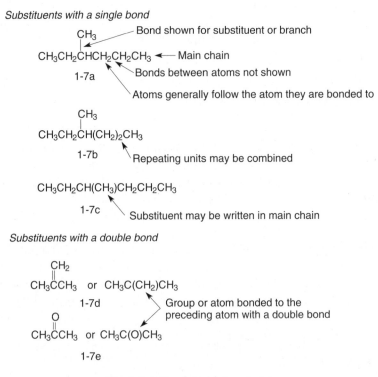

KEKULÉ STRUCTURES

Kekulé structures are similar to the Lewis structures but exclude the nonbonding electrons. All bonds are shown as lines and not dot pairs.

CONDENSED STRUCTURES

Condensed structures are another way of drawing chemical structures. Follow along in Fig. 1-7 as you read this paragraph. Structures 1-7a–c represent the same compound written in different ways. The carbon atoms are usually (but

Fig. 1-7. Condensed structures.

not always) written on a horizontal line and called the main or parent chain since the carbon atoms are connected to each other as in a chain.

Dashes (–) are not used to indicate bonds unless an atom or group is written on the line above or below the atoms in the horizontal line. An atom or group that appears above or below the main chain is called a *substituent, side group, or a branch*, as shown in Structures 1-7a and 1-7b. A substituent (branch) may also be shown in parentheses between the atoms on the horizontal line (in the main chain) as in Structure 1-7c. The substituent is bonded to the carbon atom preceding it in the horizontal line. Additional atoms connected to a carbon atom in the main chain are usually shown on the same horizontal line directly following that carbon. If a group appears consecutively, it may be shown in parentheses, with a subscript indicating the number of repeating groups. That is, CH_2CH_2 may be shown a $(CH_2)_2$ as in Structure 1-7b.

If a substituent is bonded to the preceding carbon atom with a double bond, as in $=CH_2$ or $=O$, it is shown in the condensed structure without the double bond. Examples are Structures 1-7d and 1-7e.

LINE-BOND STRUCTURES

The *line-bond* (also called *bond-line, line, line-angle, skeleton,* or *stick-structure*) method is another way to draw chemical structures. The rules for drawing structures are given below. Follow along in Fig. 1-8 as you read the rules. Lewis and line-bond structures are shown for comparison. The application for each rule has the corresponding letter (a, b, c, etc.) in the structure in Fig. 1-8. Each arrow in the structure is associated with a letter.

(a) A line is drawn showing the bond between two carbon atoms.
(b) The intersection (angle) where two lines meet represents a carbon atom.
(c) The symbol C is not shown at these intersections or at the end of any line.

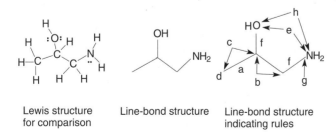

Lewis structure   Line-bond structure   Line-bond structure
for comparison   indicating rules

Fig. 1-8.   Line-bond structures.

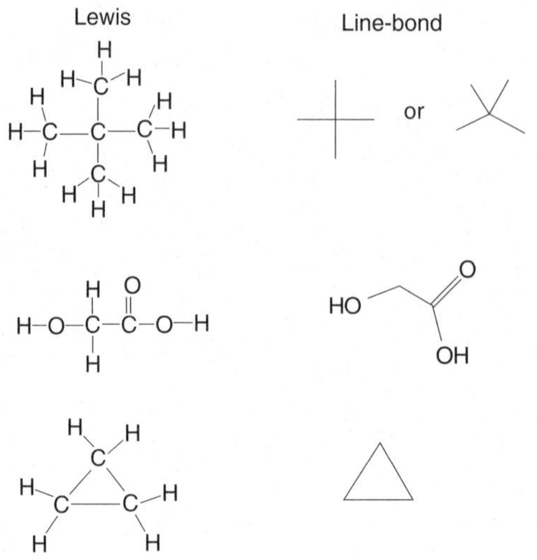

Lewis   Line-bond

Fig. 1-9.   Examples of Lewis and line-bond structures.

(d) Hydrogen atoms attached to carbon atoms are not shown. The number of hydrogen atoms attached to each carbon atom is such that there is a total of four bonds to each carbon atom.

(e) If a *heteroatom* (noncarbon atoms such as N, O, Cl, S, etc.) is present, that atom is shown. Hydrogen atoms attached to heteroatoms are shown.

(f) A line indicating a bond is drawn to each heteroatom.

(g) There is *no* carbon atom at the end of the line bonding the heteroatom.

(h) Nonbonding electron pairs are not shown.

A few examples will help you become more proficient in drawing line-bond structures. Figure 1-9 shows several Lewis structures and the corresponding line-bond structures.

QUESTION 1-8
Write the line-bond structures for $(CH_3)_2CHCH_3$, $CH_3(CH_2)_4CH(CH_3)CO_2H$, and $NH_2CH(CH_3)C(O)CH_3$.

ANSWER 1-8

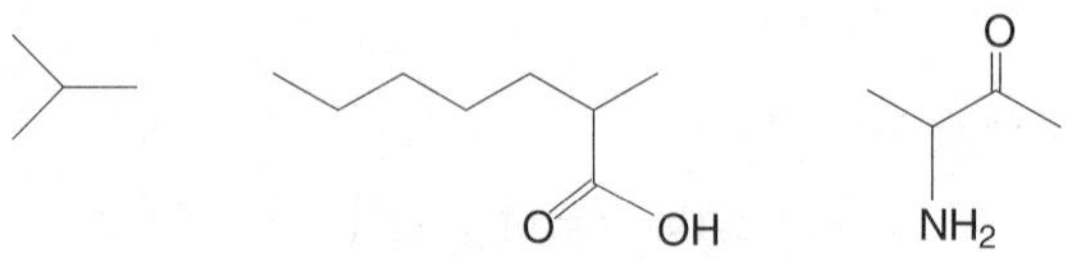

Orbital Shapes

Subshells s, p, and d are orbitals with different three-dimensional (3-D) structures. Since first row elements contain only the s subshell and the second row elements (Li through F) contain only s and p subshells, this discussion will be limited to these two subshells. What is an orbital? How does an orbital differ from an orbit? If we think of a satellite circulating the earth, gravity is attracting it toward the earth and its momentum is propelling it toward outer space. A balance of these two forces keeps the satellite in an *orbit* around the earth. It is logical to think an electron is in an orbit for similar reasons. The negatively charged electron is attracted to the positively charged nucleus and the electron's momentum propels it away from the nucleus. But that is not a satisfactory explanation for the energy or the area (electron cloud) occupied by an electron.

Quantum Mechanics

We are most familiar with describing the velocity and position of matter that has an easily measurable mass. A bouncing ball would be one example. Classical physics can be used to describe where the ball is at any instant in time. However, light rays consist of photons that are massless packets of energy. Light is usually described in terms of an oscillating wave, such as waves on a body of water, as shown in Structure 1-10a in Fig. 1-10. Electrons have a very small mass and have properties of *both* matter and waves. A mathematical approach describing the wave nature of electrons is the best model we have to predict the energy and *most probable* area occupied by electrons. This approach is known as *quantum mechanics* or *wave mechanics*.

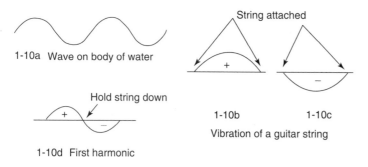

1-10a Wave on body of water

String attached

1-10b   1-10c

Vibration of a guitar string

Hold string down

1-10d First harmonic

Fig. 1-10. Oscillating wave.

THE NATURE OF WAVES

First consider the properties of *standing waves*. The best analogy is a guitar string. It is fixed at both ends. Strumming it produces a vibration that can be described as a wave that extends first above and then below the plane defined by points of attachment of the string. Waves 1-10b and 1-10c in Fig. 1-10 are examples of a guitar string standing wave. This is similar to the wave on a body of water. Mathematically, we can give the wave a plus (+) sign when it is above a defined plane and a negative (−) sign when it is below this plane. Although this is a 2-D description, the quantum mechanical approach for describing an electron is a 3-D description.

If we hold down the guitar sting at its center and strum it again, we would get a wave shown in Structure 1-10d, called the first harmonic. The point where the string crosses the plane, or goes from a + to a − sign, is called a *node*. Remember, the + and − signs show regions of space relative to some fixed coordinate system. What does this wave system have to do with electrons?

QUESTION 1-9
Draw the structure of a wave with two nodes.

ANSWER 1-9

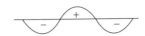

WAVE EQUATIONS

Several brilliant scientists (Schrödinger, Dirac, and Heisenberg) developed a rather complex *wave equation* to describe the properties of the electron in a hydrogen atom. This equation is based on the properties of waves, like the vibrations of a guitar string. Solutions to this equation are called *wave functions*, given the symbol ψ. Wave functions are mathematical descriptions of the energy, shape, and 3-D character of the various atomic orbitals. The wave equation has several solutions that describe the various orbitals (s, p, d, and f). We are most accustomed to working with equations that have one solution. But consider the equation for a straight line, $Y = mX + b$. Different values of Y and X satisfy this equation for specific values of m and b. Similarly, different values of ψ satisfy the wave equation.

Although ψ has no physical meaning (this is a rather difficult concept to comprehend), its square (ψ^2) is the probability of finding an electron at some

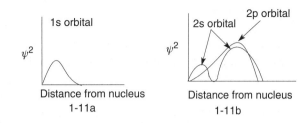

Fig. 1-11. Probability of finding an electron in atomic orbitals.

point in space. Different solutions to the wave equation describe the 3-D shapes of the various subshells, the 1s, 2s, and 2p orbitals.

Electron density probability

Consider the 1s orbital to consist of a series of thin shells of increasing diameter as an onion consists of layers (thin shells) of increasing diameter. Graph 1-11a in Fig. 1-11 is a plot of the probability (ψ^2) of finding a 1s electron in a thin spherical shell at some distance from the nucleus. The probability of finding an electron is zero at the nucleus, increases, and then decreases as a function of distance in any direction from the nucleus. Graph 1-11b shows probabilities of finding an electron in a thin layer some distance from the nucleus for the 2s and 2p orbitals. Note in Level 2, the electrons are found, on an average, further from the nucleus than are electrons in Level 1. But remember, we are just using a mathematical model to describe the wave properties of an electron. This seems to be the best model for describing the properties of electrons, at least until a better model is developed.

Atomic orbitals

The s and p orbitals localized on atoms are called *atomic orbitals* (AOs). The shapes of the 1s, 2s, and 2p AOs described by the quantum mechanical approach are shown in Fig. 1-12. The 1s and 2s orbitals are *spherically symmetrical*. The 2s orbital has one *node* (the dashed circle), a region of zero probability of finding an electron. This is also seen in Graph 1-11b where the probability of finding a 2s electron increases, goes to zero, and increases again. The 2p orbital has a *dumbbell*, or perhaps more accurately a *doorknob*, shape. There are three 2p orbitals, each perpendicular (*orthogonal*) to each other, identified as p_x, p_y, and p_z, shown in Fig. 1-12. These orbitals all have the same energy (they are degenerate), and have directional character (the x, y, and z directions). There are also one 3s and three 3p orbitals for third row elements.

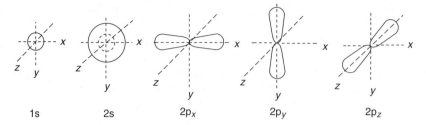

Fig. 1-12. Shapes of s and p orbitals.

QUESTION 1-10
Draw the structure superimposing all three p orbitals.

ANSWER 1-10

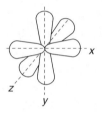

Bond Formation

The atomic 1s orbital of a hydrogen atom contains one electron ($1s^1$). If the 1s orbitals of two hydrogen atoms overlap, they can share their electrons and form a two-electron bond between the atoms. The amount of orbital overlap is a measure of *bond strength. A key point: generally the greater the orbital overlap, the stronger the bond.*

The overlap of two s orbitals is called *head-to-head* or *end-on* overlap. The area of overlap is cylindrically symmetrical along a line connecting the two nuclei. This type of overlap, shown in Fig. 1-13, results in a *sigma (σ) bond*.

BOND LENGTH AND STRENGTH

Why do orbitals overlap? Chemical reactions occur because the resulting species (the products) are usually more stable than the starting materials (the reactants). In an atom, we know the negatively charged electrons are attracted to positively

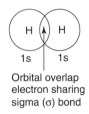

Orbital overlap
electron sharing
sigma (σ) bond

Fig. 1-13.   Sigma bond between two hydrogen atoms.

charged nuclei. As two hydrogen atoms approach each other from an infinite distance, the electrons on one atom become attracted to the nucleus of the other atom. This mutual attraction increases as the distance between atoms decreases. As the atoms get closer and closer, the positively charged nuclei began to repel each other.

A diagram of energy versus the distance between two hydrogen atoms is shown in Fig. 1-14. There is an optimum distance between two bonded atoms at which the energy is at a minimum. This distance is called the *bond length*: 0.74 Å for the hydrogen molecule. The energy required to break the bond is called the *bond strength*. This energy value is 435 kJ/mol (104 kcal/mol) for a hydrogen-hydrogen bond. Bond energies have historically been reported in kilocalories per mole. More recent texts give bond energies in kilojoules per mole. One kcal = 4.184 kJ.

The two separate hydrogen atoms are higher in energy (less stable) than is the hydrogen molecule. Atoms and molecules like to be in the lowest energy (most stable) state. Thus when a bond is formed, energy is given off and the resulting molecule has a lower potential energy (is more stable) than the starting individual species has. So a hydrogen molecule, H_2, is more stable than two separate hydrogen atoms. It follows then that energy must be added to the hydrogen molecule to break the bond, giving the original separate hydrogen atoms.

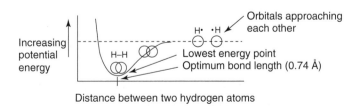

Fig. 1-14.   Potential energy as a function of distance of two hydrogen atoms.

QUESTION 1-11
Would you expect a stronger bond for H_2 if the bond length was 0.50 Å rather than the observed 0.74 Å?

ANSWER 1-11
No, repulsion between nuclei would increase the energy of the H_2 molecule, decreasing the energy required to break the bond.

Valence Bond Theory

Valence bond theory describes bond formation from overlapping valence AOs. In the above example of the hydrogen molecule, electrons in the sigma bond are shared and each atom can be considered to have two valence electrons, at least part of the time. Each hydrogen atom is now duet happy.

Consider an element in the second row of the periodic table that has electrons in p orbitals. The electron configuration for carbon $(1s^2 2s^2 2p_x{}^1 p_y{}^1 p_z{}^0)$ shows one electron in the $2p_x$ orbital and one electron in the $2p_y$ orbital. (See Fig. 1-4 for a review of the aufbau principle.) The p orbitals on an atom are perpendicular (orthogonal) to each other. One of the p orbitals, say $2p_x$, on one carbon atom can overlap in a head-to-head manner with the $2p_x$ orbital on another carbon atom, giving a sigma (head-to-head, cylindrically symmetrical) bond, as shown in Fig. 1-15. The $2p_y$ orbitals, one on each carbon atom, will have maximum overlap if they are in the same plane. The $2p_y$ orbitals can overlap in a *side-to-side* manner as shown in Fig. 1-15. The resulting bond is called a pi (π) bond. The orbital overlap in a π bond is not as great as the overlap in a sigma bond, and thus a π bond is usually weaker than a sigma bond.

Most drawings easily show the head-to-head p orbital overlap forming a sigma bond. In these drawings, it appears that the p orbitals are too far apart

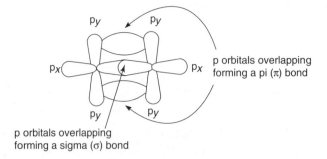

Fig. 1-15. Overlapping p orbitals of two carbon atoms.

for any significant side-to-side overlap. This is just a limitation of the drawings. Remember the previous example of the nucleus being the size of a marble while the electrons occupy the area of a football field? The electron clouds are much larger than depicted on most drawings, including the drawing in Fig. 1-15.

QUESTION 1-12
Would you expect a greater bond strength between two p orbitals if they were in the same plane or perpendicular to each other?

ANSWER 1-12
They must be in the same plane for greatest overlap and greatest bond strength.

Molecular Orbitals

Overlapping AOs remain localized on each atom. The AO model gives more information (energy and bond direction) than does a Lewis structure (which just shows valence electrons), but it lacks information about *excited states* of molecules. Excited states exist when all electrons are not in the lowest possible energy levels, the *ground state*. Molecular orbital (MO) theory explains electron excited states.

MO theory describes AOs combining (mixing) to form *new* orbitals, MOs. AOs combine mathematically in two ways: by adding the wave functions of two AOs and by subtracting one wave function from the other. The former is called *constructive overlap* and the latter *destructive overlap*. Remember, wave functions are just mathematical models describing 3-D shapes (spheres, dumbbells, etc.) of orbitals. Combining orbitals in this way is called the *linear combination of atomic orbitals* (*LCAO*). AOs cease to exist after they mix to form new MOs. The combining (mixing) of orbitals conserves the number of orbitals. That is, *two* AOs combine to give *two* MOs. The energy diagram for mixing two 1s orbitals is shown in Fig. 1-16. MOs have a defined energy, size, and shape. The electrons are now delocalized throughout the entire MO.

Bonding and Antibonding MOs

One of the new MOs formed is more stable (of lower energy) than the originating AOs. This MO is called a *bonding orbital*. The other MO is less stable than the originating AOs and it is called an *antibonding MO*. Antibonding

1-16a 1s Atomic oribtals and ground state for molecular orbitals for H· and H$_2$

σ* MO after mixing

Energy

1s AO
H·
before mixing

1s AO
H·
before mixing

H$_2$ σ MO after mixing
More stable than starting AO

1-16b 1s Atomic orbitals and excited state for molecular orbitals for H· and H$_2$

σ* MO after mixing

1s AO
H·
before mixing

1s AO
H·
before mixing

Energy

H$_2$ σ MO after mixing

1-16c 2p Atomic orbitals and ground state for molecular orbitals for N· and N$_2$

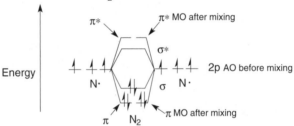

π*

π* MO after mixing

σ*

Energy

2p AO before mixing

N·

σ

N·

π

N$_2$

π MO after mixing

Fig. 1-16. Molecular orbital diagram of s and p orbitals.

MOs are indicated by an asterisk (*). There are bonding sigma (σ) MOs and antibonding sigma (σ*) MOs, as shown in Diagram 1-16a and 1-16b in Fig 1-16. There are also bonding pi (π) and antibonding pi (π*) MOs shown in Diagram 1-16c in Fig. 1-16. If an electron in a bonding MO is promoted to a higher energy MO, the new electron configuration is called an *excited state*. Diagram 1-16b shows an example of an excited state. Lewis structures and localized AOs do not include the concept of an excited (*) state. A *key point: when MOs form, the AOs (used to make the new MOs) no longer exist,* even though AOs and MOs are both shown in diagrams as in Fig. 1-16.

Diagrams 1-16a and 1-16b show the MOs for 1s orbital mixing and Diagram 1-16c shows the mixing of p orbitals. MOs are usually shown only for the valence shell orbitals and not for the core orbitals.

QUESTION 1-13
If two AOs overlap to form MOs, will the two MOs be degenerate?

ANSWER 1-13
No, one will be a bonding MO (lower energy) and one will be an antibonding MO (higher energy).

Bonding and 3-D Molecular Shape

The discussion so far has been about the energy, size, and shapes of orbitals on individual atoms. How does this relate to the 3-D structure of molecules? First consider the 3-D structure of methane, CH_4. Four hydrogen atoms are bonded to one carbon atom. The electron configuration of carbon (see Fig. 1-4) is $1s^2 2s^2 2p_x^1 2p_y^1 2p_z^0$. According to Hund's rule, two of the p orbitals each contain one electron. The third p orbital is empty. Two of the four hydrogen atoms could form sigma bonds with the two p orbitals on the carbon atom that already contain one electron. This would put two electrons into each of the two sigma bonds formed by the overlap of a 1s hydrogen and a 2p carbon orbital. One might propose the last two hydrogen atoms could form two sigma bonds by bonding to opposite sides of the empty carbon p orbital. Such an arrangement, shown in Fig. 1-17, would make all H—C—H bond angles 90°.

It has been experimentally determined that all H—C—H bond angles in methane are 109.5° not 90°. Well, 90° is close to 109.5°—but not close enough. Another theory is needed to explain the observed bond angles.

A convention was developed to represent a 3-D structure in two dimensional space (such as this page). A solid line (—) indicates a bond in the plane of the paper. A solid wedge (——) indicates a bond coming out the paper toward you. A dashed wedge (......) indicates a bond going behind the plane of the paper. An example of methane (CH_4) is shown in Fig. 1-18. It is very useful to have molecular models to view and understand these structures.

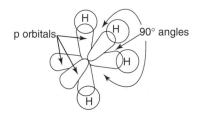

Fig. 1-17. Overlapping of unhybridized p and s orbitals in CH_4.

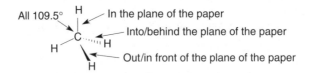

Fig. 1-18. 3-D structure of methane.

HYBRID ORBITALS

How are the 109.5° bond angles for H—C—H in methane explained? The theory of combining (mixing) AOs can explain the observed angles. This theory combines orbitals on the same atom. (MOs, described previously, result from combining/mixing orbitals of different atoms.) Do we know for a fact that AOs mix? No, but this theory does explain the correct bond angles in many molecules. Atoms other than carbon form hybridized orbitals but only carbon hybridization will be discussed.

sp³ ORBITALS ON CARBON ATOMS

A single 2s AO is combined (mathematically) with three 2p AOs on the same carbon atom to form four new *sp³ hybrid orbitals*. The total number of orbitals is conserved (stays the same). Four AOs (one 2s and three 2p) combined to give four new sp³ hybrid orbitals. These new hybrid orbitals are still localized AOs; that is, they are orbitals on a single atom. They are not MOs. A *key point: the AOs that undergo mixing cease to exist after the new hybrid orbitals are formed.* Orbital formation is a mathematical process of adding (mixing) the wave functions (orbitals) of one 2s and three 2p orbitals. The resulting hybrid orbitals have two lobes and one node, shown in Fig. 1-19. Since one lobe is much smaller (has a much lower electron density) than the other lobe, only the larger lobe is usually shown as the new sp³ orbital.

QUESTION 1-14
The sp³ orbital in Fig. 1-19 has a large lobe with a + sign and a smaller lobe with a − sign. Does this indicate the electron has a positive charge in one lobe and a negative charge in the other lobe?

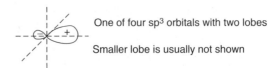

Fig. 1-19. sp³ Hybrid orbital.

ANSWER 1-14
No, + and − are mathematical descriptions. The electron always has a negative charge.

VSEPR THEORY

The *valence-shell electron-pair repulsion* (*VSEPR*) theory states that valence electrons, in both bonding and nonbonding electron pairs, want to be as far from each other as possible. The four bonding electron pairs in methane can achieve this arrangement by assuming a tetrahedral structure. All bond angles in a tetrahedral structure are 109.5°, which is experimentally observed for methane. All bond lengths and strengths are the same for each of the four C—H bonds. A 3-D structure of methane is shown in Fig. 1-18.

Methane has four sigma bonds and no nonbonding electron pairs on the central carbon atom. Each bond to a hydrogen atom contains two electrons and there are four carbon-hydrogen bonds for a total of eight valence electrons. The carbon atom has eight valence electrons and is octet happy. Each hydrogen atom has two valence electrons in its bond to the carbon atom, and each hydrogen atom is duet happy. The molecule is *symmetrical* and has a tetrahedral *electron-domain and molecular geometry*.

Electron-domain geometry

An electron domain is an area of electron density. A nonbonding electron pair is one electron domain. A single bond is one electron domain. A double or triple bond is considered *one* electron domain. An sp^3 hybridized carbon atom has four electron domains and a tetrahedral electron-domain geometry.

The structures of ammonia and water molecules are shown in Fig. 1-20. The nitrogen atom in ammonia has eight valence electrons, six from the three bonds to hydrogen atoms and one pair of nonbonding electrons. The oxygen atom in water also has eight valence electrons, four from the two bonds to hydrogen

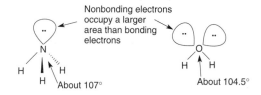

Fig. 1-20.   3-D structures of ammonia and water.

atoms and two pairs of nonbonding electrons. Both oxygen and nitrogen have eight valence electrons and are octet happy. A tetrahedral *electron-domain geometry* results when the sum of sigma bonds and nonbonding electron pairs equals four.

Water and ammonia do not have perfect electron-domain tetrahedral structures. A nonbonding electron pair occupies a larger spatial area than does a bonding electron pair. The nonbonding electrons in ammonia force the hydrogen atoms closer together than the ideal H—N—H angle of 109.5°. An H—N—H bond angle of about 107° is observed. Similarly, the two nonbonding electron pairs in water force the two hydrogen atoms even closer together, resulting in an H—O—H bond angle of about 104.5°. The electron-domain geometry is a distorted tetrahedral structure for both molecules.

Molecular geometry

The *molecular geometry* is based on the arrangement of atoms in a molecule and does not consider the nonbonding electrons. The molecular geometry is *trigonal pyramidal* for ammonia, *bent* for water, and *tetrahedral* for methane. The electron-domain geometry takes into account the 3-D arrangement of a molecule including all electron pairs (bonding and nonbonding electrons). See Table 1-1 for examples of electron-domain and molecular geometries. A *key point: the molecular and electron domain geometries are not always the same.*

QUESTION 1-15
What is the electron-domain 3-D geometry about the carbon and oxygen atoms in CH_3OH?

ANSWER 1-15
The electron-domain geometry about C and O is tetrahedral since both are sp^3 hybridized.

Table 1-1. Hybridization chart.

# of σ bonds	# of lone pairs	Total pairs	Hybridization	Electron-domian geometry	Molecular geometry	Example
4	0	4	sp^3	Tetrahedral	Tetrahedral	CH_4
3	1	4	sp^3	Tetrahedral	Trigonal pyramidal	NH_3
2	2	4	sp^3	Tetrahedral	Bent	H_2O
3	0	3	sp^2	Trigonal planar	Trigonal planar	BF_3
2	0	2	sp	Linear	Linear	C_2H_2

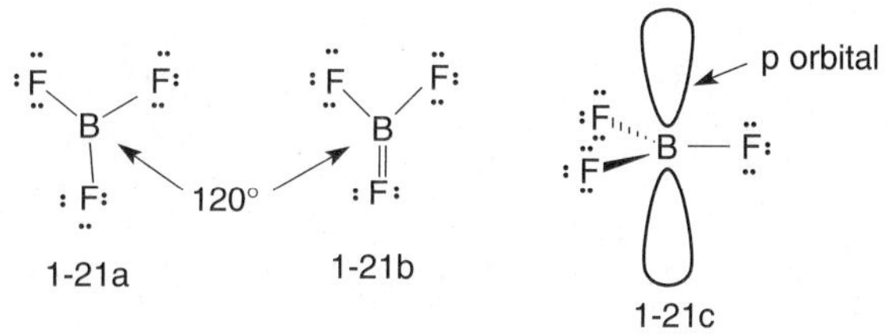

Fig. 1-21. Stucture of BF_3

sp^2 HYBRIDIZATION OF BF$_3$

The Lewis structure of boron trifluoride (BF_3) is shown in Fig. 1-21. There are 24 valence electrons, 3 from boron and 21 from the three fluorine atoms. Six electrons are used to form the bonds between boron and the three fluorine atoms. The remaining 18 valence electrons are placed around the three fluorine atoms, as shown in Structure 1-21a. The three fluorine atoms are all octet happy, with 8 valence electrons each. Boron, a second row element, has only 6 valence electrons. If one of the fluorine atoms was willing to share an electron pair (see Structure 1-21b), boron would be octet happy. As will be discussed below, fluorine has a great attraction for electrons and does *not* want to share electrons with other atoms and therefore, the best Lewis structure is 1-21a. Studies have shown BF_3 to have a *trigonal planar structure* with F—B—F bond angles of 120°. Boron trifluoride is an example of an exception to the octet rule.

The observed bond angles of 120° for BF_3 can be explained by the combination (hybridization) of the 2s and two 2p orbitals, giving three sp^2 orbitals. Three AOs combine to give three hybrid AOs. An sp^2 hybrid orbital has one large and one small lobe and has a shape similar to an sp^3 orbital (Fig. 1-19). The small lobe is again ignored since orbital overlap (bonding) of this small lobe with orbitals of other atoms is negligible. The three sp^2 orbitals are furthest from each other with a planar structure and bond angles of 120°, as predicted by VSEPR theory.

When the new hybrid sp^2 orbitals form, the original atomic 2s and two 2p orbitals disappear. But there is still one p orbital that was not hybridized. What happened to this p orbital? Nothing! It remains an unhybridized 2p orbital. No electrons are in this orbital in BF_3 and the p orbital will be centered on boron and perpendicular to the plane of the three hybrid sp^2 orbitals as shown in Structure 1-21c.

sp^2 HYBRIDIZATION OF ETHYLENE

Ethylene has the formula C_2H_4. Lewis structures for ethylene are shown in Fig. 1-22. There are 12 valence electrons, 8 from the two carbon atoms and

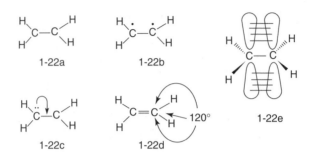

Fig. 1-22.   Structure of ethylene (ethene).

4 from the four hydrogen atoms. Connecting the atoms with five bonds uses 10 valence electrons (Structure 1-22a). We could put one of the 2 remaining electrons on each carbon atom (Structure 1-22b) but carbon would still not have a full octet (not be octet happy). Also, compounds with unpaired electrons tend to be unstable. Another possibility is to put the remaining 2 electrons on one of the carbon atoms (Structure 1-22c). One carbon atom is now octet happy, but the other carbon atom only has 6 valence electrons. In Structure 1-22c, a curved arrow shows the movement of the nonbonding electrons on the carbon atom to form a second bond between the two carbon atoms giving Structure 1-22d. Now both carbon atoms have full valence electron octets.

PI (π) BONDS

What is the hybridization of the identical carbon atoms in ethylene? Studies have shown the molecule to be planar with H−C−H and H−C−C bond angles of about 120°. These angles would be consistent with sp^2 hybridization as discussed for BF_3. The unhybridized 2p orbital on each carbon atom is perpendicular to the three planar sp^2 hybridized orbitals. The p orbital consists of two lobes of equal size, one above and one below the plane of the sp^2 orbitals. Rotation can occur around the C−C single bond until both lobes of each p orbital overlap with each other (see Structure 1-22e), forming *one* π *bond* containing two electrons. Side-to-side overlap of p orbitals on adjacent carbon atoms (a π bond) is not as great as head-to-head overlap of sp^2 orbitals (a σ bond) on adjacent carbon atoms. Less orbital overlap results in a π bond being weaker than a σ bond.

Lewis drawings imply that the p orbitals are too far apart to overlap, but this is just a limitation of the drawing. Remember the analogy of the marble (nucleus) in the football stadium (electron cloud). The electrons occupy a huge volume relative to the size of the nucleus. The Lewis and line-bond structures do not accurately convey the areas electrons occupy. The Lewis structure shows the two bonds, called a *double bond*, as identical solid lines (see Structure 1-22d). The bonds are not identical, one is a σ bond and one is a π bond.

sp HYBRIDIZATION

Acetylene has the formula C_2H_2. Lewis structures (Fig. 1-23) are drawn using 10 valence electrons, 8 from the two carbon atoms and 2 from the two hydrogen atoms. Connecting the atoms with three bonds uses 6 valence electrons (Structure 1-23a), leaving 4 valence electrons unassigned. One electron pair can be put on each carbon atom (Structure 1-23b), but each carbon atom would still have only 6 valence electrons. The curved arrows show electron pair movement in Structure 1-23b forming two identical bonds between the two carbon atoms, giving Structure 1-23c. Each bond holds 2 electrons. Studies have shown that acetylene is planar and linear. This can be explained if the 2s and one 2p orbital combine to form two sp hybrid orbitals. The hybrid orbital will have one large lobe and a small lobe (which is ignored). The two sp hybrid orbitals are furthest apart when they are 180° from each other. One sp orbital on each carbon atom overlaps head-to-head to form a σ C—C bond. The other sp orbital overlaps with the 1s orbital of a hydrogen atom, forming a σ H—C bond on each carbon atom (Structure 1-23d).

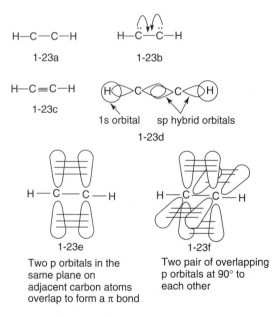

Fig. 1-23.   Structure of acetylene (ethyne).

Two p orbitals on each carbon atom did not hybridize. Two p orbitals, one on each carbon atom, are parallel to each other and in a common vertical plane. These p orbitals can overlap to form a π bond, shown as Structure 1-23e. The two remaining p orbitals, one on each carbon atom, are parallel to each other and in a common horizontal plane. These two p orbitals overlap to form a second π bond, shown in Structure 1-23f. The pair of electrons on each carbon atom, as shown in Structure 1-23b, are put into these π orbitals.

Lewis structure 1-23c implies three equivalent bonds represented by identical solid lines. The bonds are not all equivalent. A triple bond consists of one σ bond and two π bonds. Again, the σ bond is stronger than each π bond.

How can you determine the hybridization of an atom? The total number of sigma bonds and lone pairs can be used to predict the hybridization as shown in Table 1-1.

QUESTION 1-16
What is the hybridization of each carbon atom in the molecule below? (Subscripts are used to identify carbon atoms.)

$$H-\underset{\underset{H}{|}}{\overset{\overset{H}{|}}{C_1}}-\underset{\underset{Br}{|}}{\overset{\overset{Br}{|}}{C_2}}-C_3\equiv C_4-\overset{\overset{H}{|}}{C_5}=C_6=\overset{\overset{H}{|}}{C_7}-\underset{\underset{H}{|}}{\overset{\overset{H}{|}}{C_8}}-H$$

ANSWER 1-16
C_1, C_2, and C_8 are sp^3, C_3, C_4, and C_6 are sp, and C_5 and C_7 are sp^2. Sum the number of σ bonds and lone pairs and use the data in Table 1-1 to predict hybridization.

Curved Arrows

Curved arrows show electron movement. You need to understand the meaning of curved arrows to avoid "o-chem frustration." The curved arrow indicates the direction of movement of electrons from the base of the arrow to the head ($\rightarrow$) of the arrow. We will *always* show electrons moving toward an electron-deficient site and *never* show an electron-deficient site approaching an electron pair. This is simply the convention that is used. Electron movement does not necessarily mean the electrons are totally transferred and in most cases they are shared in a new covalent bond.

Electronegativity and Bond Polarity

Lewis structures show bonding electron pairs as solid lines (—) in molecules. This structure may imply that the electrons are equally shared between the two bonded atoms. This is *usually not* the case. Atoms have different affinities for the electrons in a bond. Pauling developed a relative *electronegativity* (*EN*) scale, shown in Table 1-2. Electronegativity is a measure of an atom's attraction for the electrons in a bond to that atom. Fluorine is the most electronegative atom (4.0) and cesium is the least electronegative (0.7) as one might expect as these elements are at the opposite corners of the periodic table. Electronegativity increases going from left to right in a row and from bottom to top in a column in the periodic table.

IONIC BONDS

First consider the reaction between a lithium atom and a fluorine atom. Lithium is an *electropositive* atom. It would rather lose an electron than gain an electron. Losing an electron would give it the duet happy electron configuration of He. Fluorine has a large affinity for electrons. If fluorine gained an electron, it would have the octet happy electron configuration of noble gas Ne. The difference in EN values between Li and F is so great that a Li electron is essentially completely

Table 1-2.   Pauling electronegativity (EN) values.

IA	IIA		IIIB	IVB	VB	VIB	VIIB
H 2.1							
Li 1.0	Be 1.5		B 2.0	C 2.5	N 3.0	O 3.5	F 4.0
Na 0.9	Mg 1.2		AI 1.5	Si 1.8	P 2.2	S 2.5	CI 3.0
K 0.8							Br 2.8
Rb 0.8							I 2.5
Cs 0.7							

$$Li \quad \overset{\cdot\cdot}{\underset{\cdot\cdot}{F}}: \quad \longrightarrow \quad Li^+ : \overset{\cdot\cdot}{\underset{\cdot\cdot}{F}} :^{-}$$

Ionic bond

Fig. 1-24. Ionic bond.

transferred to the fluorine atom. The two charged species form an *ionic bond*, shown in Fig. 1-24.

COVALENT BONDS

A *covalent bond* is one in which the electron pair is shared between both atoms in the bond but not necessarily shared equally. When two identical atoms form a bond to each other, such as two hydrogen atoms or two carbon atoms, they share electrons equally, giving a *nonpolar covalent bond* (see Fig. 1-25). Since both identical atoms have identical EN values, an equal attraction for the electrons in their common bond is expected. The bond in H—Cl is an example of a *polar covalent bond*. The difference in EN values (ΔEN) is 0.9. ΔEN between hydrogen and carbon is 0.4 and this covalent bond is considered to be nonpolar. As an approximation, if the difference in the EN value is 0.4 or less, the bond is considered to be nonpolar covalent, even though it is slightly polar.

If the electrons are not shared equally in a bond and ΔEN for the atoms in a bond is between 2.0 and 0.4, the bond is called a polar covalent bond. If ΔEN is greater than 2.0, the bond is considered ionic. Remember, EN values are just an approximation for predicting bond polarity.

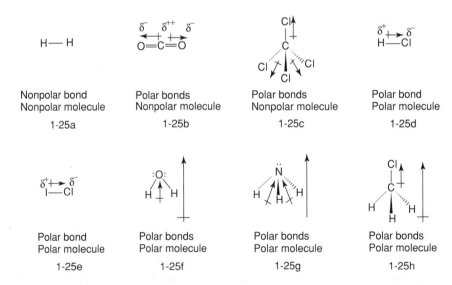

Fig. 1-25. Polarity in molecules.

Dipole Moments

The polarity in bonds is quantified by the molecule's dipole moment. A dipole refers to a molecule with a net positive charge at one place in the molecule and a net negative charge in another place in the molecule (see Fig. 1-25). The degree of polarity in different bonds can differ greatly. A dipole moment (μ) is equal to the charge (c) of one pole times the distance (d) between poles ($\mu = dc$). The larger the charge or the greater the distance between charges, the larger the dipole moment. The units of a dipole moment are debye units. Why is it necessary to consider the polarity of bonds and molecules? The polarity of bonds and molecules predicts how molecules interact with each other, with other molecules, and with an external electromagnetic field.

NONPOLAR MOLECULES

If a molecule consists of just two atoms, it is easy to predict the polarity of the molecule by looking at the Pauling EN values. If a molecule consists of three or more atoms, one must consider the 3-D structure of the molecule, including nonbonding electrons, to determine if the *molecule* is polar or nonpolar. It may help to draw a Lewis structure to determine the 3-D structure of a molecule, using Table 1-1 to determine hybridization and the molecular and electron-domain geometries. Symmetrical molecules tend to be nonpolar and nonsymmetrical molecules tend to be polar. (Symmetry is discussed in more detail in Chapter 5.)

Bond polarity and molecule polarity are shown for several molecules in Fig. 1-25. Hydrogen (H_2; Structure 1-25a), carbon dioxide (CO_2; Structure 1-25b), and carbon tetrachloride (CCl_4; Structure 1-25c) are nonpolar molecules. The carbon atom in CO_2 is sp hybridized, and the molecule is linear and symmetrical. The carbon atom in CCl_4 is sp^3 hybridized and the molecule is tetrahedral, giving a symmetrical structure. In CO_2 and CCl_4 the center of the positive and negative charges *coincide* (are on top of each other). Although the bonds in a molecule may be polar, one needs to determine the *vector (magnitude and direction) sum* for the centers of the partial positive and negative charges. If the centers of the net positive and net negative charges co-incide, the molecule is nonpolar. If the term vector sums is confusing, think of the charges as having positive and negative weights and determine where the average weight (charge) is in the molecule. Highly symmetrical molecules tend to be nonpolar.

The electrons in the carbon-oxygen bond in CO_2 (Structure 1-25b) are not shared equally. An oxygen atom is more electronegative than a carbon atom, and thus one would expect the electrons to spend more time closer to the oxygen atom. The electrons are still shared (as compared to an ionic bond where electrons are transferred), but not equally. This is an example of a polar covalent bond. The polarity of a covalent bond is shown by an arrow pointing in the direction of greater electron density. The base of the arrow has a cross, indicating the positive end of the arrow. The partial positive and partial negative charges on atoms can also be indicated by δ^+ and δ^-. The partial positive charge in CO_2 is centered on the carbon atom. The average of the two partial negative charges is also centered on the carbon atom. The average δ^+ and δ^- charges coincide and the molecule is nonpolar. A similar analysis explains why CCl_4 (Structure 1-25c) is nonpolar.

POLAR MOLECULES

Structures of polar molecules are also shown in Fig. 1-25. Molecules such as HCl (Structure 1-25d) and ICl (Structure 1-25e) are polar since they are linear molecules and their differences in EN values are greater than 0.4. Water (Structure 1-25f) and ammonia (Structure 1-25g) are polar because of their polar bonds and their 3-D geometry. H_3CCl (Structure 1-25h) is also a polar molecule. Although it has a tetrahedral geometry, there is a net dipole when one considers that the C—Cl bond is polar and the H—C bonds are considered nonpolar. Molecules 1-25f, 1-25g, and 1-25h are polar as their centers of positive and negative charges do not coincide. The arrows in these structures show the direction of polarity.

QUESTION 1-17
Predict which molecules are polar: CH_3OCH_3, CH_3OH, CH_2Cl_2, and CF_4.

ANSWER 1-17
All are polar except CF_4.

Formal Charges

Formal charge (FC) is a method of assigning a charge to an atom in a Lewis structure. Formal charges are an *indication* of bond polarity and do not represent actual charges on atoms in a molecule. To determine a formal charge, first write a valid Lewis structure. More than one Lewis structure is possible for many molecules and the charge on the same atom may differ in different structures. The formal charge is calculated by subtracting the valence electrons *assigned to an atom* in a compound from the valence electrons of that atom in its elemental state. The number of assigned electrons to an atom in a compound equals the sum of the nonbonding electrons plus one half of the number of bonding electrons

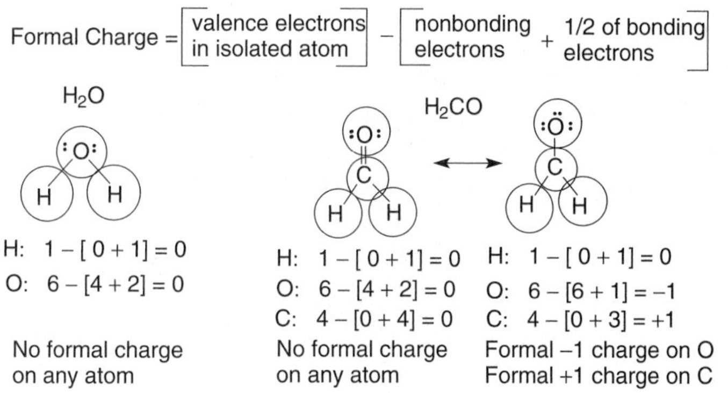

$$\text{Formal Charge} = \boxed{\begin{array}{c}\text{valence electrons}\\\text{in isolated atom}\end{array}} - \boxed{\begin{array}{c}\text{nonbonding}\\\text{electrons}\end{array}} + \boxed{\begin{array}{c}\text{1/2 of bonding}\\\text{electrons}\end{array}}$$

H_2O

H: $1 - [0 + 1] = 0$
O: $6 - [4 + 2] = 0$

No formal charge
on any atom

H_2CO

H: $1 - [0 + 1] = 0$
O: $6 - [4 + 2] = 0$
C: $4 - [0 + 4] = 0$

No formal charge
on any atom

H: $1 - [0 + 1] = 0$
O: $6 - [6 + 1] = -1$
C: $4 - [0 + 3] = +1$

Formal -1 charge on O
Formal $+1$ charge on C

Fig. 1-26. Formal charges.

(those in bonds to that atom). Examples of assigning formal charges are given in Fig. 1-26.

FC = valence electrons in an isolated atom − assigned valence electrons
in a Lewis structure

A key point: *when determining formal charges each valence electron in the molecule is counted only once.* When assigning electrons in an O—H bond, one electron is assigned to (owned by) H and the other electron in the bond is assigned to O. Note this is different from determining the total number of valence electrons owned or shared by an atom in a molecule, where electrons in bonds are double counted. That is, the two electrons in an H—O bond are counted as two valence electrons for H and as two valence electrons for O (see the section entitled *Valence Electrons* in this chapter).

QUESTION 1-18
What is the formal charge on each atom in

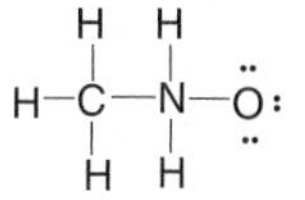

ANSWER 1-18
All H and C = 0, N = +1, and O = −1.

Resonance Structures

Structures of some molecules can be represented by more than one valid Lewis structure. The actual molecule is *not* accurately represented by any one Lewis structure, but more accurately by some combination of two or more Lewis structures. Lewis structures that differ only in the location of non-bonding electron pairs or π bond electron pairs are called *resonance structures*. A combination of these structures is called a *resonance hybrid* structure, a resonance hybrid, or a hybrid structure. Each individual resonance structure is fictitious/imaginary/artificial/theoretical, whichever term you prefer. Each does not represent the actual structure of the molecule. The actual structure is some combination (hybrid) of the individual resonance structures. It is difficult to draw the hybrid structure, so the actual structure is usually represented by the best Lewis resonance structure.

THE DOUBLE-HEADED ARROW

A *double-headed arrow* (↔) is used to indicate resonance structures. This double-headed arrow does *not* represent an equilibrium between resonance structures. Resonance structures do not shift back and forth as do different species in an equilibrium reaction.

There are a few rules to follow for drawing contributing resonance structures:

1.   Structures should be valid Lewis structures.
2.   The positions of the atoms in different Lewis structures do not change.
3.   The number of electron pairs stays the same in each structure.
4.   Not all structures contribute equally to the resonance hybrid. The most stable structures contribute the most.

A key point: the individual atoms keep exactly the same positions in all resonance structures; only the electron pairs move.

Examples of resonance structures for formaldehyde and the carbonate anion are shown in Fig. 1-27. Structures 1-27a and 1-27b differ by movement of an electron pair. The curved arrow shows electron movement. One pair of electrons in the carbon-oxygen double bond in formaldehyde moves to the oxygen atom. All atoms in Structure 1-27a are octet/duet happy. Resonance structure 1-27b is less stable (and less important) since the carbon atom is not octet happy. The actual molecule looks more like Structure 1-27a than Structure 1-27b. Three resonance structures, 1-27c, 1-27d, and 1-27e, are shown for the carbonate anion. All three are equivalent and of equal stability. The actual molecule is an

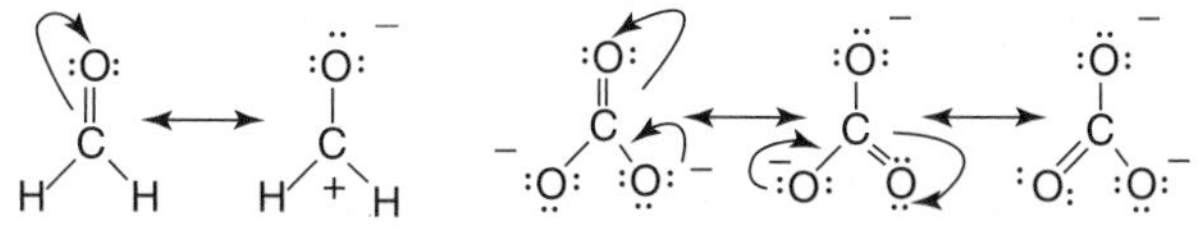

1-27a   1-27b

Formaldehyde

All atoms in 1-27a
are octet happy, C
in 1-27b does not
have a full octet

1-27c   1-27d   1-27e

Carbonate anion

All atoms are octet happy in each
resonance form and all structures
are of equal importance

Fig. 1-27.   Resonance forms.

average of the three structures shown. Each C—O bond is of equal energy and
equal bond length.

Resonance can also be described as electron pair delocalization. That is, some
electron pairs are not localized on one atom or between two atoms but spend
time on several different atoms. Delocalization of electron pairs, just like delocalization of a positive or negative charge, increases the stability of that species.

QUESTION 1-19
Which is the most stable resonance structure in each pair?

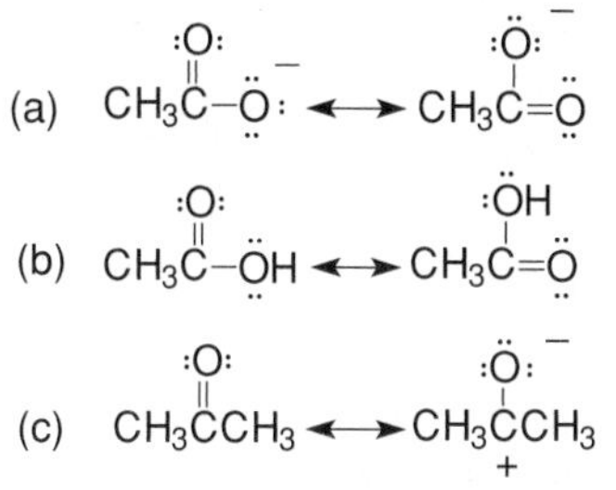

(a) $CH_3C-\ddot{O}: \longleftrightarrow CH_3C=\ddot{O}$

(b) $CH_3C-\ddot{O}H \longleftrightarrow CH_3C=\ddot{O}$

(c) $CH_3CCH_3 \longleftrightarrow CH_3CCH_3$

ANSWER 1-19
(a) Both are equivalent. (b) They are not resonance forms since the position of
H is changed. (c) The first structure is better since the carbon atom is not octet
happy in the second structure.

Intermolecular Forces

Forces between molecules are responsible for the magnitude of the melting and
boiling temperatures and for solubility characteristics of molecules. The greater
the attraction between molecules of a specific compound, the higher the melting
and boiling points are likely to be. Solubility characteristics use the classic

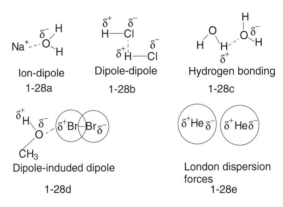

Fig. 1-28. Intermolecular forces ("-----" indicates interaction).

saying, *like dissolves like*. Polar molecules are most soluble in polar solvents and nonpolar molecules are most soluble in nonpolar solvents. There are five categories of interactions. Examples of each are shown in Fig. 1-28.

ION-DIPOLE INTERACTIONS

Ion-dipole interactions describe the interaction between an ion and the partial charge (δ) of a polar molecule. Cations (positively charged ions) are attracted to the negative pole of a dipole, and anions (negatively charged ions) are attracted to the positive pole of a dipole. This interaction is particularly important for solutions of ionic compounds in polar solvents, such as an ionic Na^+ ion interacting with a polar water molecule, shown in Structure 1-28a.

DIPOLE-DIPOLE INTERACTIONS

Dipoles result from unequal sharing of electrons in bonds. If molecules are close to each other, the negative pole of one molecule is attracted to the positive pole of another molecule as shown in Structure 1-28b. These interactions are generally weaker than ion-dipole interactions.

HYDROGEN BONDS

This is a special type of dipole-dipole interaction. It does not refer to an actual bond, but a strong interaction between a covalently bonded hydrogen atom and a molecule containing an atom with nonbonding electrons, such as oxygen, nitrogen, and the halogens. The hydrogen atom undergoing hydrogen bonding must be covalently bonded to an oxygen, a nitrogen, or a fluorine atom, resulting in a highly polar covalent bond. This puts a large partial positive charge (δ^+) on the covalently bonded hydrogen atom and it seeks an electron pair on another atom. Hydrogen bonds are stronger than most dipole-dipole interactions. An example is shown for water in Structure 1-28c.

DIPOLE-INDUCED DIPOLE INTERACTIONS

If a polar molecule is close to a nonpolar molecule, it can influence the electron cloud of the nonpolar molecule, making the latter somewhat polar. This results in an attraction between dipoles of unlike charges on two different molecules, as shown in Structure 1-28d.

LONDON DISPERSION FORCES

At any given instant, the electrons surrounding an atom or molecule are not uniformly distributed; that is, one side of the atom may have a greater electron density than the other side. This results in a momentary dipole within the atom. The dipole on one atom may induce a dipole on another atom. The net result is an attraction between atoms. This explains the interaction between helium atoms (Structure 1-28e), that are nonpolar, yet they must have attraction for each other since they form a liquid when cooled sufficiently. All molecules exhibit dispersion forces.

Quiz

1. Atoms consist of how many types of subatomic particles?
    (a) One
    (b) Two
    (c) Three
    (d) Four

2. The nucleus of an atom is much larger than its electron cloud.
    (a) True
    (b) False
    (c) They are the same size

3. The atomic number for second row elements is smaller than the atomic weight (mass).
    (a) True
    (b) False
    (c) They are the same

4. Isotopes of an element
   (a) all have the same weight
   (b) have the same number of protons
   (c) have the same number of neutrons
   (d) consist of ice crystals

5. Electrons have
   (a) a positive charge
   (b) a negative charge
   (c) no charge

6. The ____ orbitals are degenerate.
   (a) 1s and 2s
   (b) 2s and 2p
   (c) $2p_x$ and $2p_y$
   (d) 1s and 2p

7. Each orbital can hold a maximum of ____ electrons.
   (a) one
   (b) two
   (c) three
   (d) four

8. A carbon atom has a total of ____ electrons.
   (a) Two
   (b) Four
   (c) Six
   (d) Eight

   $1s^2\ 2s^2 2p^2$

   $-\overset{\displaystyle .}{\underset{\displaystyle .}{C}}\cdot$

9. An octet happy nitrogen atom in a compound has ____ valence electrons.
   (a) two
   (b) four
   (c) six
   (d) eight

10. Atoms in Group 1A (alkali metals) tend to
    (a) lose electrons
    (b) gain electrons
    (c) keep the same number of electrons
    (d) do not have electrons

11. A Lewis structure shows
    (a) all electrons
    (b) valence electrons

(c) core electrons

(d) no electrons

12. The ____ orbital has the same shape as the 1s orbital.
    (a) 2s
    (b) $2p_x$
    (c) $2p_y$
    (d) $2p_z$

13. The orbital in the following series with the highest energy is a
    (a) 1s orbital
    (b) 2s orbital
    (c) 2p orbital
    (d) 3s orbital

14. The greater the orbital overlap on adjacent atoms,
    (a) the stronger the bond
    (b) the weaker the bond
    (c) makes no difference in bond strength
    (d) orbitals never overlap

15. Molecular orbitals (MOs) are located on
    (a) a single atom — AO
    (b) two or more atoms ← MO
    (c) only hydrogen atoms
    (d) vocal cows

16. A hybrid orbital is
    (a) a combination of only s orbitals
    (b) a combination of only p orbitals
    (c) a combination of s and p orbitals
    (d) a combination of a horse and a donkey

17. In the 3-D structure of a molecule, the bonds to atoms coming out of the plane of the paper are shown as
    (a) solid lines
    (b) solid wedges
    (c) hatched wedges
    (d) not shown

18. The angle between sp^2 orbitals is
    (a) 90°
    (b) 109.5°
    (c) 120°
    (d) 180°

19. VSEPR theory states that
     (a) electron clouds attract each other
     (b) electron clouds repel each other
     (c) electron clouds ignore each other
     (d) one should talk quietly

20. A curved arrow shows
     (a) an equilibrium
     (b) a resonance structure
     (c) electron pair movement
     (d) where to insert an apostrophe

21. A polar bond
     (a) shares electrons equally
     (b) shares electrons unequally
     (c) is ionic
     (d) is a very cold bond

22. A formal charge
     (a) is always positive
     (b) is always negative
     (c) is the apparent charge on an atom
     (d) means wearing a tux

23. Each resonance structure
     (a) represents the true structure of a molecule
     (b) represents some percentage of the true structure of the molecule
     (c) shows single electron movement
     (d) contributes equally to the true structure of the molecule

24. London dispersion forces exist
     (a) between all molecules
     (b) only in polar compounds
     (c) only in ionic compounds
     (d) in England

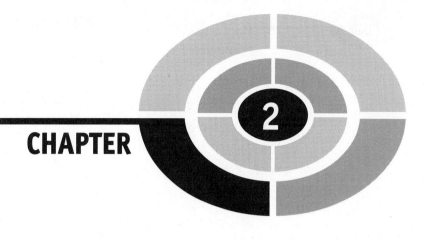

2

Families and Functional Groups

Organic compounds are classified into families that contain functional groups. This chapter gives the name, generic structure, and a specific example of the major functional groups. You will need to learn the names and structures of the families and functional groups, since these terms are commonly used to describe molecules throughout this text. Each group is discussed in more detail in the corresponding chapters. Use this chapter as a reference if you do not remember the structure of a functional group as you read the various chapters.

Family	Functional group	Example	Chapter reference
Alcohol	$\ce{>C-OH}$	CH_3OH	13
Aldehyde	$\ce{>C-\overset{O}{\overset{\|}{C}}-H}$	$CH_3\overset{O}{\overset{\|}{C}}H$	19
Alkane	none	$CH_3CH_2CH_3$	4
Alkene	$\ce{>C=C<}$	$H_2C{=}CH_2$	6 and 8
Alkyne	$-C{\equiv}C-$	$HC{\equiv}CH$	9
Amide	$\ce{-C-\overset{O}{\overset{\|}{C}}-N<}$	$CH_3\overset{O}{\overset{\|}{C}}NH_2,\ \ CH_3\overset{O}{\overset{\|}{C}}NHCH_3$	21
Arene	⬡	⬡—CH_3	17 and 18
Carboxylic acid	$\ce{>C-\overset{O}{\overset{\|}{C}}-O-H}$	$CH_3\overset{O}{\overset{\|}{C}}OH$	9
Carboxylic acid anhydride	$\ce{>C-\overset{O}{\overset{\|}{C}}-O-\overset{O}{\overset{\|}{C}}-C<}$	$CH_3\overset{O}{\overset{\|}{C}}O\overset{O}{\overset{\|}{C}}CH_3$	21
Carboxylic acid ester	$\ce{>C-\overset{O}{\overset{\|}{C}}-O-C<}$	$CH_3\overset{O}{\overset{\|}{C}}OCH_3$	21
Carboxylic acid halide	$\ce{>C-\overset{O}{\overset{\|}{C}}-X}$	$CH_3\overset{O}{\overset{\|}{C}}Cl$	21
Ether	$\ce{>C-O-C<}$	CH_3OCH_3	14
Haloalkane (alkyl halide)	$\ce{>C-X}$	CH_3Cl	11
Ketone	$\ce{>C-\overset{O}{\overset{\|}{C}}-C<}$	$CH_3\overset{O}{\overset{\|}{C}}CH_3$	19

(continued)

Family	Functional group	Example	Chapter reference
Phenol			17
Thiol	$\ce{>C-S-H}$	CH_3SH	15
Sulfide	$\ce{>C-S-C<}$	CH_3SCH_3	15
Amine	$\ce{>C-N<}$	CH_3NH_2, $(CH_3)_2NH$, $(CH_3)_3N$	

Acids and Bases

Introduction

It may seem strange to talk about acids and bases in organic chemistry as these terms generally make one think of inorganic compounds such as hydrochloric acid (HCl) and sodium hydroxide (NaOH). There are families of organic acids such as carboxylic acids (acetic acid or vinegar) and organic bases including amine (nitrogen-containing) compounds. Organic compounds other than carboxylic acids and amines also have properties of acids and bases, depending upon the reaction conditions. Many organic reactions are acid-base reactions although they are not referred to by this term. A review of acid-base chemistry is necessary to better understand organic reactions.

Arrhenius Definition

Arrhenius acids are materials that produce protons (H^+) in water and Arrhenius bases are materials that produce hydroxide anions (^-OH) in water. This definition is limited to aqueous systems.

Brønsted-Lowry Definition

The *Brønsted-Lowry* concept defines acids as proton donors and bases as proton acceptors. This concept broadens the definition of Arrhenius acids and bases. Acids include, of course, carboxylic acids, but can also include alcohols and amines that contain a hydrogen atom that can be donated as a proton under the appropriate conditions. Compounds that contain atoms with nonbonding electron pairs, such as the oxygen atom in water (H_2O), alcohols (ROH), and carboxylic acids (RCO_2H), can act as bases under the appropriate conditions. Wow, a carboxylic acid acting as a base and an amine acting as an acid? This apparent contradictory behavior will be explained.

QUESTION 3-1
Write the balanced equation for the ionization of an organic acid in water.

ANSWER 3-1
$$RCO_2H + H_2O \rightarrow H_3O^+ + RCO_2^-$$

Conjugate Acids and Bases

A *conjugate base* is a species that results when an acid loses a proton. The species that results when a base accepts a proton from an acid is called the *conjugate acid*. This equilibrium reaction is shown in Fig. 3-1. *In each acid-base reaction an acid reacts with a base to give a conjugate base and a conjugate acid.*

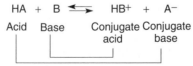

An acid is associated with its conjugate base
and a base is associated with its conjugated acid

Fig. 3-1.   Acids, bases, and their conjugates.

QUESTION 3-2
Write the conjugate base of HCl.

ANSWER 3-2
Cl^-.

K_a AND pK_a

To what extent does an acid donate a proton to a base? That depends on the base it reacts with. Most tables list acid strength in water (the base). Acid strength is a measure of a material's willingness to donate a proton to water. An equilibrium expression (Eq. 3-2a) for generic acid HA is shown in Fig. 3-2. The equilibrium constant, K_{eq} (Eq. 3-2b), and acid-dissociation constant, K_a (Eq. 3-2c), for HA are also shown in Fig. 3-2. Since the concentration of water in dilute aqueous solutions is approximately constant (55.5 mol/L), it is multiplied by K_{eq} to yield K_a (Eq. 3-2d). The value of K_a is a measure of the hydronium ion (H_3O^+) concentration, that is, the acidity. *Larger values of K_a indicate stronger acids.* A proton (H^+) is often written in its hydrated form as H_3O^+.

Approximate K_a values of several acids are listed in Table 3-1. Since the range of K_a values is so large, they are often expressed as pK_a, where p$K_a = -\log K_a$. *A key point: there is an inverse relationship between K_a and pK_a.* The larger the value of K_a, the smaller the value of pK_a. Acid strength increases as pK_a values decrease. Negative values of pK_a indicate very strong acids.

QUESTION 3-3
Which is the stronger acid, ROH_2^+ or ROH?

ANSWER 3-3
ROH_2^+ (p$K_a = -3$) is stronger than ROH (p$K_a = 16$–17).

$$\text{(a)} \quad HA + H_2O \rightleftharpoons H_3O^+ + A^-$$

$$\text{(b)} \quad K_{eq} = \frac{[H_3O^+][A^-]}{[HA][H_2O]}$$

$$\text{(c)} \quad K_a = \frac{[H_3O^+][A^-]}{[HA]}$$

$$\text{(d)} \quad K_a = K_{eq}[H_2O]$$

Fig. 3-2. Equilibrium and acidity expressions.

Table 3-1.   Acid dissociation constants.

Acid	K_a	pK_a	Conjugate base
Inorganic acids (HCl and HI)	$10^{+7} - 10^{+10}$	-7 to -10	Cl^-, I^-
Protonated carboxylic acid ($RCO_2H_2^+$)	10^{+6}	-6	RCO_2H
Protonated alcohol (ROH_2^+)	10^{+3}	-3	ROH
Hydronium ions (H_3O^+)	55	-1.7	H_2O
Carboxylic acids (RCO_2H)	$10^{-4} - 10^{-5}$	$+4$ to $+5$	RCO_2^-
Water (H_2O)	10^{-16}	$+16$	HO^-
Alcohols (ROH)	$10^{-16} - 10^{-17}$	$+16$ to $+17$	RO^-
Amines (RNH_2)	10^{-38}	$+38$	RHN^-
Alkanes (C_nH_{2n+2})	10^{-50}	$+50$	$C_nH_{2n+1}^-$

Equilibrium Reactions

We can predict the relative equilibrium concentrations of acid and conjugate acid by looking at the pK_a values of the acid and the conjugate acid. In acid-base reactions, the rule is *survival of the weakest*. This is just the opposite of the "rule of the jungle." In acid-base reactions, the equilibrium is shifted toward the weaker acid (or weaker base). The acid and conjugate acid will be on opposite sides of the equilibrium arrows, as shown in Fig. 3-1.

WEAK ACIDS

Consider two cases: water reacting with acetic acid and water reacting with ammonia, as shown in Fig. 3-3. In Reaction (a), the pK_a values for acetic acid and hydronium ion (smaller pK_a values mean stronger acids) show that the hydronium ion is a stronger acid than acetic acid, and the equilibrium is

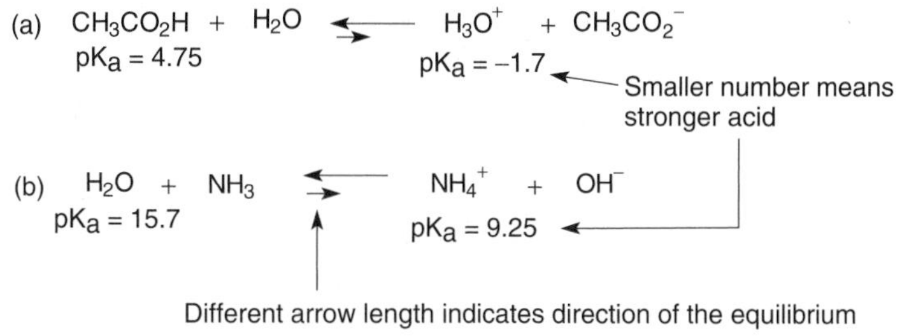

(a) CH_3CO_2H + H_2O ⇌ H_3O^+ + $CH_3CO_2^-$
       $pK_a = 4.75$ $pK_a = -1.7$ ← Smaller number means stronger acid

(b) H_2O + NH_3 ⇌ NH_4^+ + OH^-
       $pK_a = 15.7$ $pK_a = 9.25$

Different arrow length indicates direction of the equilibrium

Fig. 3-3.   Survival of the weakest acid.

shifted toward acetic acid, the weaker acid. The direction of the equilibrium is indicated by the relative size of the two equilibrium arrows. In Reaction (b), the pK_a value for the ammonium ion is larger than the pK_a value for water. In this expression, water is the acid and the ammonium ion is the conjugate acid. The equilibrium is shifted to the left toward the weaker acid, which is water in this example. The difference in magnitude of the two pK_a values is indicative of the extent to which the equilibrium is shifted in the direction of the weaker acid (or base).

STRONG ACIDS

Most organic acids fall into the category of weak acids. There are two acids in Table 3-1 that may appear unfamiliar, the protonated carboxylic acid $(RCO_2H_2^+)$ and the protonated alcohol (ROH_2^+). Both are fairly strong acids. These acids will be seen again in reactions where a strong acid such as HCl is used as a catalyst, resulting in protonation and activation of a carboxylic acid or alcohol in a chemical reaction. Figure 3-4 shows an equilibrium expression where a protonated alcohol (ROH_2^+) is the weaker acid and the equilibrium is shifted in the direction of the weaker acid. Both acids in this equilibrium are very strong acids.

BASES

Bases are classified as weak and strong. Weak bases are stable species and do not readily share their electrons to form bonds with protons. Strong bases want to share their electrons to form bonds with protons. Strong acids disassociate to form weak (stable) conjugate bases. The more stable the conjugate base, the further the equilibrium will be shifted toward that conjugate base (see Fig. 3-4). *A key point: strong acids have weak conjugate bases and weak acids have strong conjugate bases.*

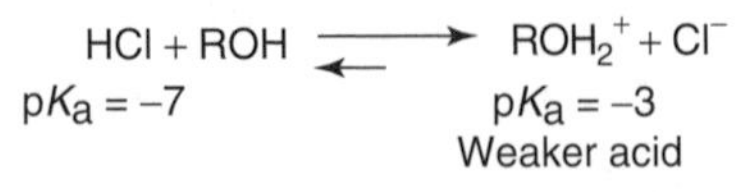

$$HCl + ROH \; \rightleftharpoons \; ROH_2^+ + Cl^-$$
$$pK_a = -7 \qquad\qquad pK_a = -3$$
$$\text{Weaker acid}$$

Fig. 3-4.   An equilibrium expression for protonated alcohols.

RELATIVE STRENGTH OF BASES

Electronegativity

What factors make bases strong or weak bases? In a bond between two different atoms, one of the atoms usually has a greater attraction for the bonding electrons. This degree of attraction is called the atom's *electronegativity*. The larger the electronegativity value for an atom, the more it wants the bonding electrons. Consider four atoms *in the same row* in the periodic table: carbon (C), nitrogen (N), oxygen (O), and fluorine (F). Electronegativity increases going from C to F. Fluorine has the greatest attraction for electrons, holds on to them very strongly, and does not want to share them with a proton. The fluoride anion is the weakest, most stable base in this series.

Consider the acidity of the series CH_4, NH_3, H_2O, and HF (protonated forms of C, N, O, and F) and the stability of each corresponding conjugate base. In this series, HF is the strongest acid since F^- is the weakest conjugate base. CH_4 is the weakest acid since H_3C^- is the strongest conjugate base. The H_3C^- anion wants to share its nonbonding electron pair with a proton to a greater extent than does F^-.

Electron density

If the electron density (charge per unit area) in a base is decreased, the base will have less attraction for a proton. Consider the hydrogen halide series HF, HCl, HBr, and HI and their conjugate bases F^-, Cl^-, Br^-, and I^-. The iodide anion has a much larger electron cloud than does the fluoride anion. The iodide anion can disperse its negative charge over a larger area (its entire electron cloud). The negative charge can be considered more "dilute" and it will have less attraction for a positively charged proton. The iodide anion is the most stable, weakest base in this series and HI is the strongest acid. Relative sizes of halide anions are shown in Fig. 3-5.

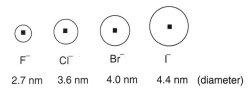

| F^- | Cl^- | Br^- | I^- |
| 2.7 nm | 3.6 nm | 4.0 nm | 4.4 nm (diameter) |

Fig. 3-5. Relative sizes of halogen anions. The larger the diameter, the more "dilute" the electron density.

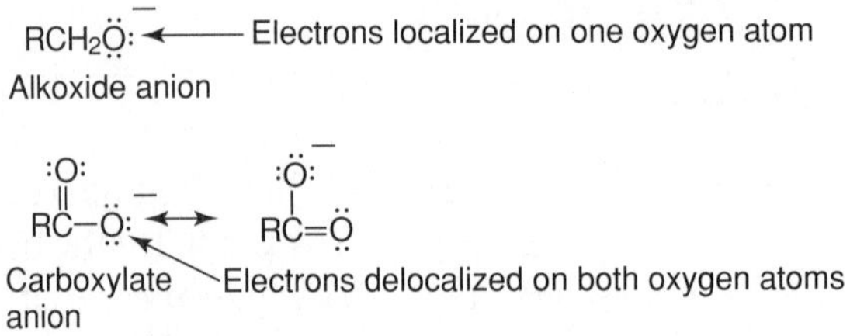

RCH$_2\ddot{\text{O}}$: $\overset{-}{}$ ⟵───── Electrons localized on one oxygen atom

Alkoxide anion

:O:
‖
RC–$\ddot{\text{O}}$: $\overset{-}{}$ ⟷ $\ddot{\text{O}}$: $\overset{-}{}$ RC=$\ddot{\text{O}}$

Carboxylate ⟍Electrons delocalized on both oxygen atoms
anion

Fig. 3-6.   Delocalization of electrons.

QUESTION 3-4

Which is the stronger base, F$^-$ or I$^-$?

ANSWER 3-4

F$^-$; it has the weaker conjugate acid.

ELECTRON DELOCALIZATION

Why is a carboxylic acid a stronger acid than an alcohol? In both cases, the hydrogen atom is bonded to an oxygen atom. Consider the stability of the conjugate bases shown in Fig. 3-6. The electrons on the *alkoxide anion* (an alcohol without the hydrogen atom bonded to the oxygen atom) are *localized* on the oxygen atom. The electrons on the *carboxylate anion* can be *delocalized* on both oxygen atoms as shown in the two resonance structures in Fig. 3-6. This delocalization reduces the electron density per oxygen atom and thus the carboxylate anion has less attraction for a proton when compared to the alkoxide anion, where the electrons are localized on one oxygen atom. A carboxylate anion is a weaker, more stable conjugate base than is an alkoxide anion and a carboxylic acid is a stronger acid than is an alcohol.

INDUCTIVE STABILIZATION

Ions and molecules can be stabilized by an *inductive effect*. An inductive effect results from the differences in electronegativity between sigma-bonded atoms. This is a *short-range effect* in that the atom attracting the electrons must be close (one or two bonds) to the atom or group being stabilized. The effect of chlorine content on the acidity of a series of chlorinated acetic acids is shown by the magnitude of pK_a values given in Fig. 3-7. Increasing the chlorine content increases the acidity since chlorine atoms pull the bonding electrons toward themselves, reducing the electron density on the oxygen atoms in the carboxylate anion (see Fig. 3-7). The lower the electron density of the oxygen

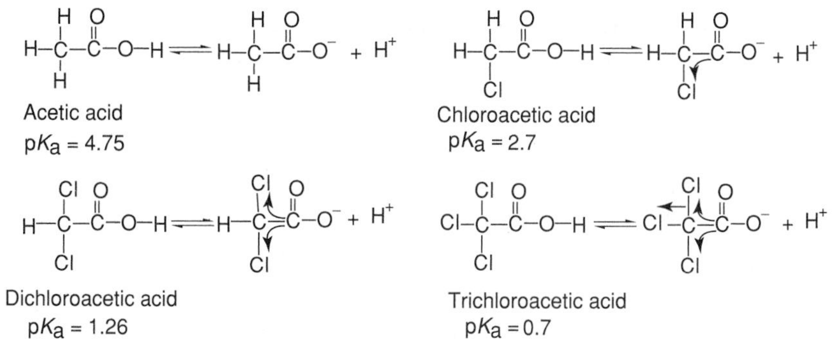

Fig. 3-7.   pK_a values of chlorinated acetic acid.

atoms in the conjugate base, the less attraction they have for protons and the stronger is the corresponding conjugate acid.

QUESTION 3-5
Which is the stronger acid, CH_3CO_2H or F_3CCO_2H?

ANSWER 3-5
F_3CCO_2H; the three fluorine atoms withdraw electrons from the carboxylate anion and hence stabilize it.

Weak Hydrocarbon Acids

Alkanes, alkenes, and alkynes differ in acid strength. All are very weak proton donors but they can still act as acids when a strong base is present. Alkanes have pK_a values of about 50, alkenes have pK_a values of about 40, and terminal alkynes have pK_a values of about 25. Thus alkynes are much more acidic than alkanes. Why? Consider the hybridization of the carbon atom in each case. An alkane carbon atom is sp^3 hybridized, an alkene carbon atom is sp^2 hybridized, and an alkyne carbon atom is sp hybridized. The electrons in a 2s orbital of an unhybridized carbon atom are held closer to the nucleus than are the electrons in 2p orbitals. The closer the electrons are to the nucleus, the more tightly they are held and the less available they are to form a bond with a proton.

An sp orbital has more "*s character*" (50%) than does an sp^2 orbital (33% s character), which has more s character than does an sp^3 orbital (25% s character). The electrons in an sp orbital are held closer to the nucleus than are the electrons in an sp^3 orbital. The electrons in the sp orbital of an *acetylide* anion (RC≡C$^-$, a terminal alkyne without a proton) are held more tightly than are the electrons

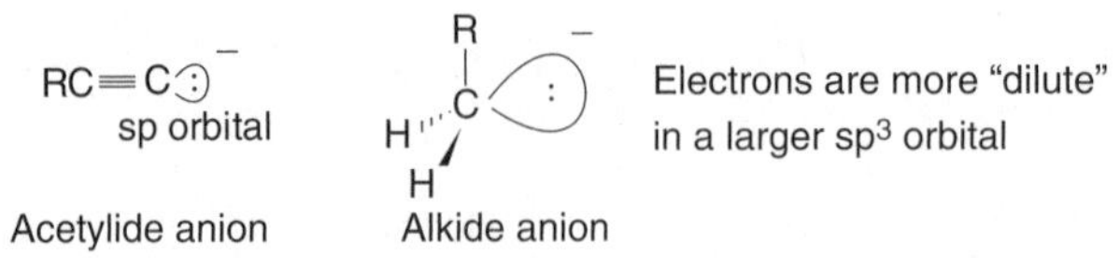

Fig. 3-8. Electron density in sp and sp^3 orbitals.

in an sp^3 orbital of an alkide anion (RCH$_2^-$, an alkane without a proton), as shown in Fig. 3-8. The acetylide anions are weaker bases than are alkide anions since the former do not want to share electrons with a proton as readily. Weak bases have strong conjugate acids; so relative to each other, alkynes are stronger acids than are alkenes and alkenes are stronger acids than are alkanes.

QUESTION 3-6
Which is a stronger acid, CH$_3$CH=CH$_2$ or CH$_3$C≡CH?

ANSWER 3-6
CH$_3$C≡CH; terminal alkynes are stronger acids than are alkenes.

Lewis Acids and Bases

There is a third definition of acids and bases that includes and extends the Brønsted-Lowry definition. These are *Lewis acids* and *Lewis bases*. Lewis bases are electron pair donors and Lewis acids are electron pair acceptors. Three Lewis acid-base reactions are shown in Fig. 3-9. In Reaction (a), $^-$OH (the base) donates (shares) an electron pair and H$^+$ (an acid) accepts an electron pair. In Reaction (b), NH$_3$ donates an electron pair and H$^+$ accepts an electron pair.

Borane (BH$_3$) is not octet happy. It has an empty p orbital. Borane can accept an electron pair from ammonia to form a B—N bond (Reaction (c)) and become octet happy. This is an acid-base reaction without H$^+$ or $^-$OH ions being involved. Lewis acids tend to be electron deficient. Examples include cations of

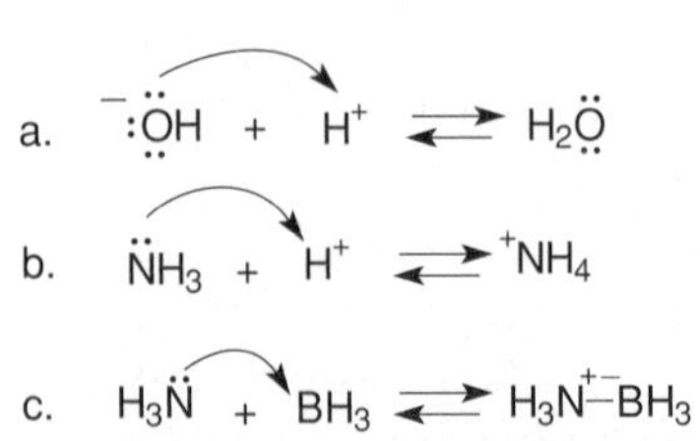

Fig. 3-9. Lewis acid-base reactions.

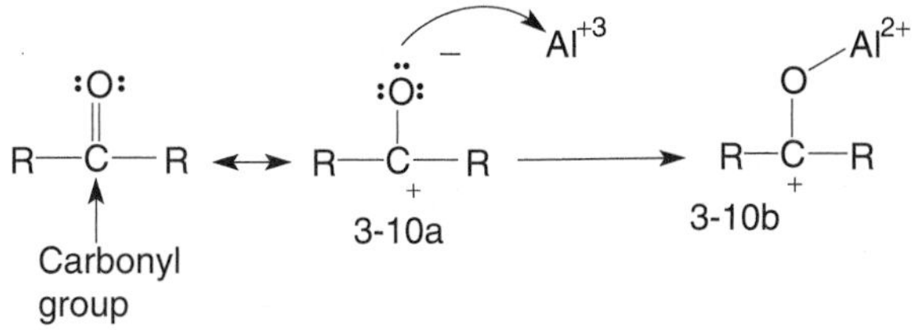

Fig. 3-10. Lewis base reaction between the carbonyl group and Al^{3+}.

metals, such as Al^{3+} and Fe^{3+}, and electron-deficient species such as carbocations (carbon atoms with a formal positive charge, R_3C^+). Lewis bases have nonbonding or π electrons they can share.

Consider the reaction of an aluminum ion with a carbonyl group as shown in Fig. 3-10. This is a Lewis acid-base reaction. One resonance form of the carbonyl group, Structure 3-10a, has a positive charge on the carbonyl carbon atom. This carbon atom is electron deficient and can act as a Lewis acid. The oxygen atom is electron rich and shares its electrons with the electron-poor aluminum ion. The oxygen atom acts as a Lewis base.

The electrons in pi bonds of alkenes and alkynes are more loosely held than are the electrons in sigma bonds. Alkenes and alkynes can act as Lewis bases since their loosely held pi electrons can react with a proton or other electron-deficient species. An example is given in Fig. 3-11, Reaction (a).

NUCLEOPHILIC AND ELECTROPHILIC REACTIONS

Organic chemists call many acid-base reactions by other names. In the case of a proton reacting with π electrons of an alkene, the proton, an electrophilic (electron-loving) species, forms a bond with an alkene carbon atom. These

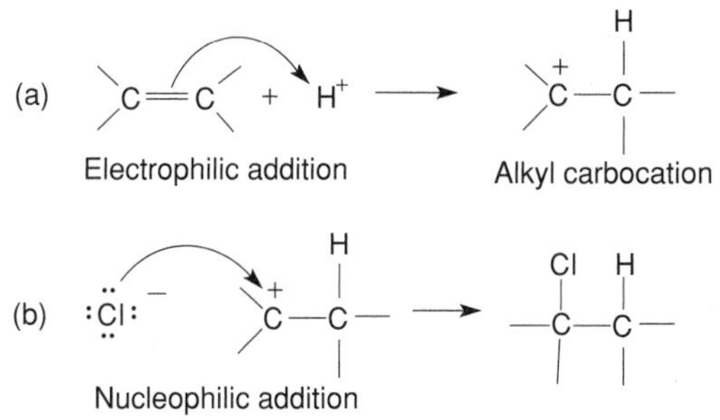

Fig. 3-11. Electrophilic and nucleophilic addition reactions.

reactions are called *electrophilic addition reactions*. After the proton reacts with the carbon atom (Reaction (a)) in Fig. 3-11, a *carbocation* is formed. A carbocation is an electron-deficient carbon atom. A species with electrons to share (a Lewis base) can share its electrons with an electron-deficient species, such as the carbocation. Since Lewis bases are attracted to electron-deficient species, this reaction is called a *nucleophilic (nucleus-loving) addition reaction*. The reaction of a chloride anion with a carbocation (Reaction (b)) in Fig. 3-11 is called a *nucleophilic addition reaction*.

QUESTION 3-7
Write the structure of a nucleophile.

ANSWER 3-7
Any neutral or negative species with a nonbonding electron pair or pi electrons it is willing to share, e.g.,

$$R\ddot{O}H, \quad R\ddot{O}:^-, \quad H_2C=CH_2, \text{ and } HC\equiv CH$$

Quiz

1. An acid is a material that
    (a) accepts a proton
    (b) donates a proton
    (c) donates an electron
    (d) accepts a neutron

2. A conjugate base results from
    (a) an acid losing a proton
    (b) a base accepting a proton
    (c) an acid gaining a proton
    (d) a base accepting an electron pair

3. K_a is an expression for
    (a) potassium
    (b) a universal constant
    (c) an equilibrium
    (d) a rate of reaction

4. pK_a is
   (a) a symbol for an element
   (b) always large
   (c) an abbreviation for *putting potassium K away*
   (d) equal to $-\log K_a$

5. Survival of the weakest refers to
   (a) a rugby match
   (b) the law of the jungle
   (c) the rate of a reaction
   (d) an acid-base equilibrium expression

6. Weak acids have
   (a) weak conjugate bases
   (b) strong conjugate bases
   (c) large K_a values
   (d) small pK_a values

7. The weakest acid is
   (a) CH_4
   (b) NH_4^+
   (c) H_2O
   (d) HF

8. The weakest base is
   (a) F^-
   (b) Cl^-
   (c) Br^-
   (d) I^-

9. The strongest acid is
   (a) CH_3CO_2H
   (b) $ClCH_2CO_2H$
   (c) Cl_2CHCO_2H
   (d) Cl_3CCO_2H

10. The strongest acid is
    (a) CH_4
    (b) $H_2C=CH_2$
    (c) $HC\equiv CH$
    (d) H_3CCH_3

11. Lewis acids
    (a) donate electrons

    (b) accept electrons ⟵
    (c) are electron rich
    (d) are proton donors

12. Nucleophiles
    (a) like electrons
    (b) like carbocations
    (c) hate neutrons
    (d) like neutrons

13. Electrophiles
    (a) like electrons
    (b) like carbocations
    (c) hate neutrons
    (d) like neutrons

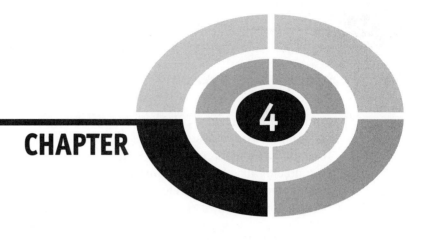

CHAPTER 4

Alkanes and Cycloalkanes

Introduction

Alkanes are one family of organic compounds. These materials contain only carbon and hydrogen atoms. Alkanes are not considered a functional group as they are relatively unreactive and do not undergo many chemical reactions. All bonds between carbon atoms and between carbon and hydrogen atoms in alkanes are single, sigma (σ) bonds. If an element is mentioned in this text, we are referring to an atom of that element. For example, the statement "hydrogen is bonded to carbon" means a hydrogen atom is bonded to a carbon atom.

Sources of Alkanes

Liquid petroleum, also called crude oil, is the main source of alkanes. Petroleum is pumped from wells containing the remains of prehistoric plants. Wells also contain gases that are the source of low molecular weight alkanes. Coal is also a source of alkanes. Petroleum is a mixture of thousands of compounds. These compounds can be separated into fractions of increasing molecular weight by distillation, though it is difficult to obtain pure, individual compounds by this method. The main distillation fractions are known as gases, gasoline, greases, and tars. Higher molecular weight compounds are degraded into useful smaller molecules by heating at relatively high temperatures.

The primary use of alkanes is as a fuel for heating, generating electricity, and transportation (internal combustion engines). The main reaction of these materials is combustion, the reaction with oxygen.

QUESTION 4-1
Where do you use alkanes in everyday life?

ANSWER 4-1
Methane to heat homes, propane in grills/stoves, butane in lighters, gasoline in autos, and oil in auto engines.

Acyclic and Cyclic Alkanes

Most alkanes are *acyclic* (noncyclic or not in a ring). Acyclic materials are classified into two subgroups: straight or linear chains and branched chains. A chain refers to the carbon atoms being connected to each other. Usually acyclic alkanes are just called alkanes. Branched chain alkanes have side groups. These side groups are called by various names: *substituents*, *side chains*, *pendent groups*, or just *branches*. This text will primarily use the term substituent. Alkanes may also be *cyclic*; that is, the carbon atoms are connected to form a ring. These materials are called *cycloalkanes*. Examples are shown in Fig. 4-1. Alkanes are occasionally called *aliphatic* (Greek, meaning fatty) compounds or *paraffins* (Latin, meaning slight affinity).

Alkanes, like people, come in all shapes and sizes. Real molecules are three dimensional (3-D). They can only be shown in two dimensions on paper, and therefore you will have to learn how to mentally visualize their 3-D structures. Many examples will be given to help you to visualize structures.

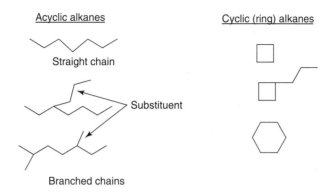

Fig. 4-1.   Shapes of alkanes.

Nomenclature

Nomenclature is a system for naming chemical compounds. Compounds have both formal and informal names. The formal system is called the IUPAC system. The structures and names of 20 acyclic alkanes are given in Table 4-1. This series of alkanes increases from 1 to 20 carbon atoms. There are times when you will have to memorize names and terms that are new to you. This is really not as difficult as it may appear. Look at all the names in Table 4-1. They all end in -*ane*. OK, that's what one might expect for an alk*ane*. Alkanes containing 5 to 19 carbon atoms have Greek prefixes that may already be familiar to you. Alkanes with 1 to 4 carbon atoms might be expected to be called unane, diane, triane, and tetraane—a reasonable assumption. But in many cases, molecules were named by individuals who could choose whatever name they felt was appropriate. These names are called *common* or *trivial names*.

The first four alkanes shown in Table 4-1 are methane, ethane, propane, and butane. A helpful hint is to make a mnemonic for memorizing a series of unfamiliar terms. In this case you might use *M*ike *E*ats *P*eanut *B*utter for the first four alkanes. You can probably make a better mnemonic. Note all names end in -ane. Compounds in families tend to have names with the same ending.

QUESTION 4-2
What is the name of an alkane with 15 carbon atoms?

ANSWER 4-2
Pentadecane.

Table 4-1. Formulas of alkanes.

Name	Molecular formula	Condensed formula
Methane	CH_4	CH_4
Ethane	C_2H_6	CH_3CH_3
Propane	C_3H_8	$CH_3CH_2CH_3$
Butane	C_4H_{10}	$CH_3CH_2CH_2CH_3$
Pentane	C_5H_{12}	$CH_3(CH_2)_3CH_3$
Hexane	C_6H_{14}	$CH_3(CH_2)_4CH_3$
Heptane	C_7H_{16}	$CH_3(CH_2)_5CH_3$
Octane	C_8H_{18}	$CH_3(CH_2)_6CH_3$
Nonane	C_9H_{20}	$CH_3(CH_2)_7CH_3$
Decane	$C_{10}H_{22}$	$CH_3(CH_2)_8CH_3$
Undecane	$C_{11}H_{24}$	$CH_3(CH_2)_9CH_3$
Dodecane	$C_{12}H_{26}$	$CH_3(CH_2)_{10}CH_3$
Tridecane	$C_{13}H_{28}$	$CH_3(CH_2)_{11}CH_3$
Tetradecane	$C_{14}H_{30}$	$CH_3(CH_2)_{12}CH_3$
Pentadecane	$C_{15}H_{32}$	$CH_3(CH_2)_{13}CH_3$
Hexadecane	$C_{16}H_{34}$	$CH_3(CH_2)_{14}CH_3$
Heptadecane	$C_{17}H_{36}$	$CH_3(CH_2)_{15}CH_3$
Octadecane	$C_{18}H_{38}$	$CH_3(CH_2)_{16}CH_3$
Nonadecane	$C_{19}H_{40}$	$CH_3(CH_2)_{17}CH_3$
Icosane	$C_{20}H_{42}$	$CH_3(CH_2)_{18}CH_3$

Formulas

Each successive molecular formula in Table 4-1 contains one additional carbon atom and two additional hydrogen atoms, a CH_2 unit. This list is known as a *homologous series* and the individual molecules are called *homologs*. The general formula for an acyclic alkane is C_nH_{2n+2}.

QUESTION 4-3
Derive a general formula for the acyclic alkanes in Table 4-1. Each homolog has two additional hydrogen atoms for each additional carbon atom.

ANSWER 4-3
$n(CH_2) + 2H = C_nH_{2n+2}$.

Structures

Alkanes are called *saturated* compounds because they contain the maximum number of hydrogen atoms that can be attached to the carbon atoms in a given compound. This will be demonstrated with structures shown below.

Sometimes, especially for large molecules, it is easier to visualize a molecule by drawing its structure. There are various ways of drawing structures. One method is to draw Lewis structures. Lewis structures are described in detail in the section entitled *Lewis Structures* in Chapter 1.

3-D STRUCTURES

The simplest alkane is methane, CH_4. The carbon atom is sp^3 hybridized (for a review of hybridized orbitals see the section entitled *Atomic Orbital Mixing* in Chapter 1) and all H—C—H bond angles are 109.5°. The solid lines (−) are in the plane of this page, the dashed wedge (⸺) indicates a bond going into (behind) the page, and the solid wedge (▬) indicates a bond coming out of the page toward you. It is highly recommended you get a molecular model kit to make 3-D structures for methane and other molecules. Basic kits are available at most college bookstores and can be purchased for $10–$20.

QUESTION 4-4
Using wedges and solid lines, draw the 3-D structure of ethane.

ANSWER 4-4

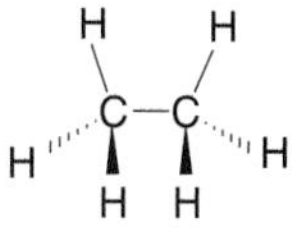

Now consider propane. Four methods for drawing structures are shown in Fig. 4-2: planar Lewis structures (Structure 4-2a), 3-D Lewis structures (Structure 4-2b), condensed structures (Structure 4-2c), and line-bond or skeletal structures (Structure 4-2d).

CONDENSED STRUCTURES

Condensed structures, such as Structure 4-2c in Fig. 4-2, show how atoms are connected but they do not show the 3-D orientation. Atoms connected to

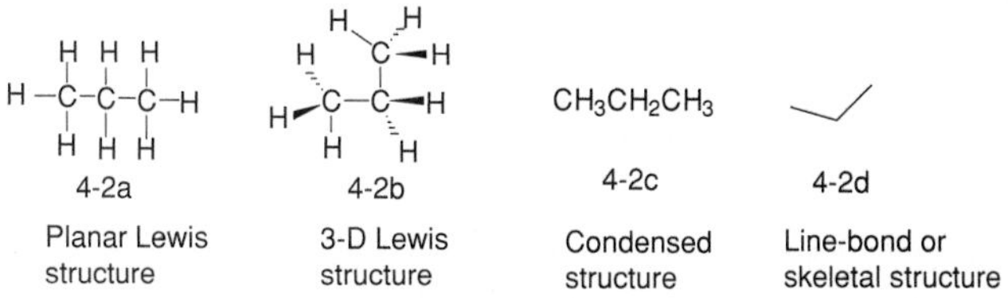

4-2a	4-2b	4-2c	4-2d
Planar Lewis structure	3-D Lewis structure	Condensed structure	Line-bond or skeletal structure

Fig. 4-2.   Four methods of drawing structures of propane.

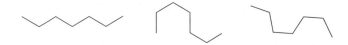

Fig. 4-3. Different orientations of *n*-heptane.

a particular carbon atom are usually written following that carbon atom. All uncharged carbon atoms will have four bonds to that carbon atom. For propane (Structure 4-2c), the first carbon atom is bonded to three hydrogen atoms and the following carbon atom, making four bonds to the first carbon atom. Remember, a hydrogen atom has only one bond to it. Occasionally hydrogen atoms bonded to the first carbon atom are written before that carbon atom, as in $H_3CCH_2CH_3$.

LINE-BOND STRUCTURES

Chemists like to simplify things—really. The *line-bond* (also called *line, line-angle*, and *skeletal*) structures are even easier and faster to draw. Alkanes can be represented by a series of connecting lines. The intersection or angle where two lines meet represents a carbon atom. The beginning and end of the connected lines (that is, the terminal positions) also represent carbon atoms. Line-bond structures are often drawn as zigzag structures showing bond angles. Ideal bond angles are 109.5°, but line-bond structures are usually drawn at other angles. Different orientations of line-bond structures of heptane are shown in Fig. 4-3.

QUESTION 4-5
Draw a line-bond structure for hexane.

ANSWER 4-5

One of several possible conformations (orientations).

Constitutional Isomers

There are several kinds of isomers (ice-o-mers). *Constitutional isomers* (also called structural isomers) are nonidentical compounds with the same molecular formula. They have the same number and kind of atoms but the atoms of each isomer are connected in a different order. We will refer to this order as their *connectivity*. As an example, suppose you have three cards that are the 5, 6, and 7 of spades. You can line them up on a table, or "connect" them, as follows: 5-6-7, 6-7-5, 6-5-7. You could also make 7-6-5, but that is really the same as 5-6-7 if you just rotate the three "connected" cards by 180° in the plane of the

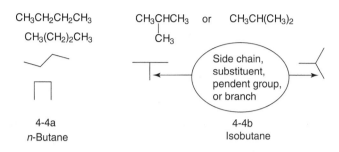

Fig. 4-4. Different representations of constitutional isomers of C_4H_{10}.

table. Similarly, 5-7-6 is the same as 6-7-5 and 7-5-6 is the same as 6-5-7. Thus there are three "constitutional isomers" of the three cards.

The atoms in methane, CH_4, ethane, CH_3CH_3, and propane, $CH_3CH_2CH_3$, can only be connected one way. Each has only one constitutional isomer. Butane, C_4H_{10}, can exist as two constitutional isomers. Different ways of drawing the condensed and line-bond structures of butane isomers are shown in Fig. 4-4. The two isomers are two different compounds, with different chemical and physical properties. They are called *n-butane* (Structure 4-4a) and *isobutane* (Structure 4-4b). The prefix n stands for "normal," meaning a linear, nonbranched alkane. The prefix n is usually not included when naming linear alkanes and alkyl groups. For example, butane and pentyl mean *n*-butane and *n*-pentyl. This text will use the n prefix for clarity purposes.

BRANCHED MOLECULES

Isobutane (Structure 4-4b) can be considered as one carbon atom attached to the center carbon atom of a linear three-carbon atom chain. Such materials are called branched molecules. The branch, a methyl group in this case, is called a substituent, a pendent group, a side chain, group, or a branch. These terms are used interchangeably.

DRAWING STRUCTURES

As the number of carbon atoms in an alkane increases, the number of constitutional isomers increases—exponentially. The number of constitutional isomers for the first 10 alkanes is shown in Table 4-2. Is there a formula that gives the number of isomers? Not that I know of. You just have to grind away and draw all the structures you can think of, eliminating duplicate structures.

Table 4-2. Number of isomers vs. number of carbon atoms.

Number of carbon atoms	Number of isomers
1	1
2	1
3	1
4	2
5	3
6	5
7	9
8	18
9	35
10	75

You need to be able to recognize the same isomer if it is drawn differently, as shown in Fig. 4-5. (Teachers get a certain pleasure out of drawing different orientations of the same isomer in exams to check a student's skills.) It may be difficult to recognize identical structures, but when you learn to name these compounds the task will become easier. Identical structures necessarily have identical names.

QUESTION 4-6
Draw all the isomers for pentane, C_5H_{12}.

ANSWER 4-6

Cycloalkanes

Cycloalkanes are compounds in which the carbon atoms are connected forming one or more rings. *Monocyclic* alkanes are molecules with just one ring. The

Fig. 4-5. Different orientations of the same alkane.

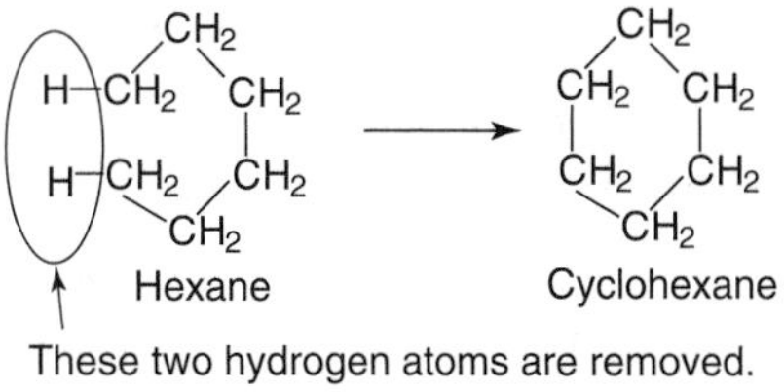

Fig. 4-6. Ring formation.

smallest ring is a three-membered ring. A ring could, theoretically, contain a large number of carbon atoms but generally rings do not contain more than about 10 carbon atoms.

In simple monocyclic alkanes, each carbon atom is bonded to two other carbon atoms and two hydrogen atoms. Each carbon atom can have no more than four bonds and no additional hydrogen atoms can be bonded to that carbon. These compounds are called *saturated cycloalkanes*. If the two ends of a linear chain are connected to form a ring, two hydrogen atoms must first be removed. An example of ring formation is shown in Fig. 4-6. The formula for a monocyclic alkane is C_nH_{2n}. This is different from the formula for a saturated acyclic alkane, C_nH_{2n+2}. Actually, the formula for a monocycloalk*ane* is the same as that for a monounsaturated alk*ene*. Alkenes are another family of organic compounds and one must be careful about identifying a family using only the formula.

Alkyl Groups

In the earlier example of isobutane (Structure 4-4b), the molecule can be described as a three-carbon linear chain with a CH_3- substituent group. This substituent is derived from methane (CH_4) by removing one hydrogen atom. Removing the hydrogen atom from an alkane gives an *alkyl* group, in this case a meth*yl* group. Note the *-yl* in both names. Similarly, removing one hydrogen atom from either carbon atom in ethane, CH_3CH_3, would give an eth*yl* group, CH_3CH_2-.

Consider propane shown in Fig. 4-7. Removing a hydrogen atom from either end of the molecule would give an *n-propyl* group (usually just called a propyl group). Removing a hydrogen atom from the central carbon atom gives a different alkyl group, an *isopropyl* group. Alkyl groups are not compounds. You cannot have a bottle of alkyl groups. The dash ($-$) to an alkyl group represents a bond to some other atom.

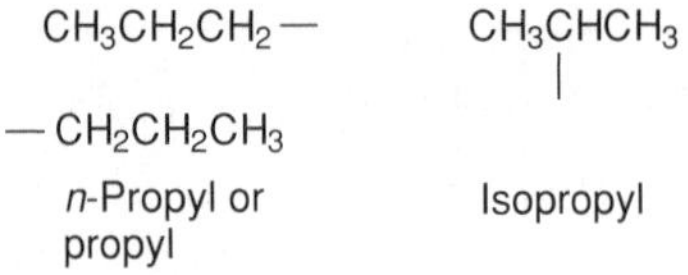

$$CH_3CH_2CH_2- \qquad CH_3CHCH_3$$

$$-CH_2CH_2CH_3$$

n-Propyl or propyl   Isopropyl

Fig. 4-7. Propyl and isopropyl groups.

ISOMERIC ALKYL GROUPS

Many alkyl groups, especially those containing three to five carbon atoms, have common names that need to be memorized. Many of the common names use the same prefixes: secondary- (sec-), normal- (n-), neo-, iso-, and tertiary- (tert- or t-). The acronym SNNIT can be used to remember these prefix names. Alkanes with five or more carbon atoms are best named by using the formal IUPAC naming system.

The common names for butyl groups (C_4H_9-) are shown in Fig. 4-8. If a hydrogen atom is removed from each carbon atom in *n*-butane, four alkyl structures result. But these represent only two different alkyl groups, *n*-butyl (Structure 4-8a) and *sec*-butyl (Structure 4-8b), since some of the four alkyl structures shown are identical structures. It is important to recognize identical structures that are formed by removing a hydrogen atom from different carbon atoms within the same molecule. Removing a hydrogen atom from each carbon atom of isobutane results in four alkyl structures. These four structures represent only two different alkyl groups, isobutyl (Structure 4-8c) and *tert*-butyl (Structure 4-8d) groups.

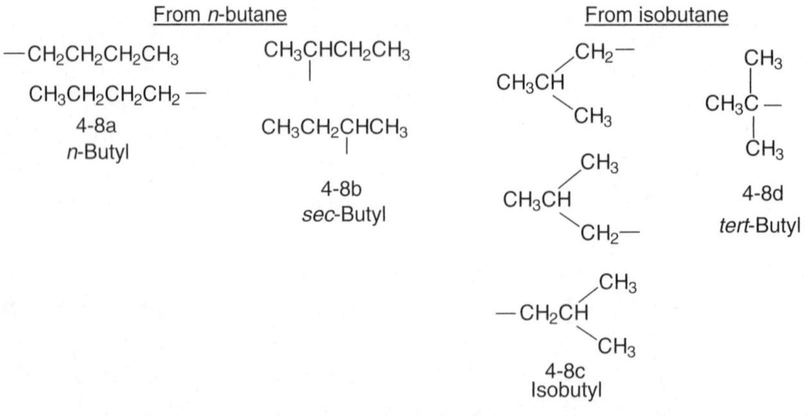

From *n*-butane

$$-CH_2CH_2CH_2CH_3 \qquad CH_3CHCH_2CH_3$$

$$CH_3CH_2CH_2CH_2-$$

4-8a
n-Butyl

$$CH_3CH_2CHCH_3$$

4-8b
sec-Butyl

From isobutane

$$CH_3CH \overset{\displaystyle CH_2-}{\underset{\displaystyle CH_3}{}}$$

$$CH_3C \overset{\displaystyle CH_3}{\underset{\displaystyle CH_3}{-}}$$

4-8d
tert-Butyl

$$CH_3CH \overset{\displaystyle CH_3}{\underset{\displaystyle CH_2-}{}}$$

$$-CH_2CH \overset{\displaystyle CH_3}{\underset{\displaystyle CH_3}{}}$$

4-8c
Isobutyl

Fig. 4-8. *n*-Butyl, *sec*-butyl, isobutyl, and *tert*-butyl groups.

QUESTION 4-7
How many alkyl groups can you draw for C_5H_{11}, derived from C_5H_{12}?

ANSWER 4-7
Eight, the alkyl groups are attached to the parent (or main chain).

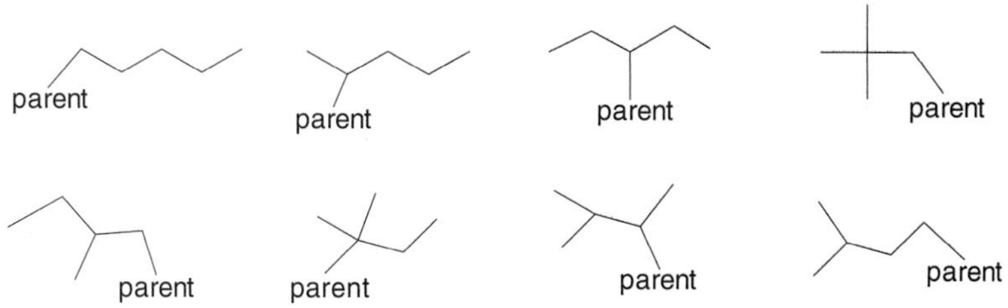

CLASSIFICATION OF CARBON ATOMS

A carbon atom is often classified by the number of other carbon atoms bonded to it. Primary (1°) carbon atoms are bonded to one other carbon atom, secondary (2°) carbon atoms are bonded to two other carbon atoms, tertiary (3°) carbon atoms have three other carbon atoms bonded to them, and quaternary (4°) carbon atoms are bonded to four other carbon atoms. Primary hydrogen atoms are bonded to a primary carbon atom, secondary hydrogen atoms are bonded to a secondary carbon atom, and tertiary hydrogen atoms are bonded to a tertiary carbon atom. The various types of hydrogen and carbon atoms are shown in Fig. 4-9.

QUESTION 4-8
Draw a structure that contains a quaternary hydrogen atom. [Do not work more than 3 min on this question.]

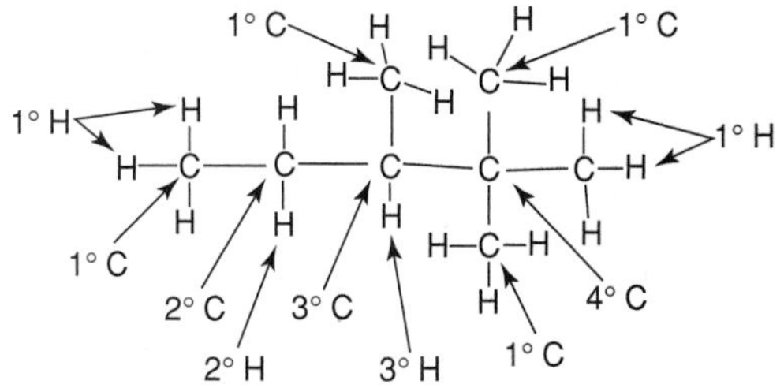

Fig. 4-9. Classification of carbon and hydrogen atoms.

$$-CH_2CH_2CH_2CH_2CH_3 \quad \text{1° pentyl group}$$

$$CH_3\overset{|}{C}HCH_2CH_2CH_3 \quad \text{2° pentyl group}$$

$$CH_3CH_2\overset{|}{C}HCH_2CH_3 \quad \text{2° pentyl group}$$

Fig. 4-10. Alkyl groups derived from *n*-pentane.

ANSWER 4-8

You cannot draw such a structure since a quaternary carbon atom has four carbon atoms and no hydrogen atoms bonded to it.

The alkyl groups derived from *n*-butane are *n*-butyl and *sec*-butyl. *tert*-Butyl and isobutyl groups are obtained from isobutane. These alkyl groups are shown in Fig. 4-8. You can see that *sec*- and *tert*-alkyl groups are so named because the carbon atom that is attached to the parent chain is a secondary or tertiary carbon atom respectively. The various alkyl groups for *n*-pentane are shown in Fig. 4-10. Two different alkyl groups have secondary carbon atoms that would be bonded to the parent chain. Which one will be called *sec*-pentyl? To avoid confusion, these substituents and those of higher molecular weight alkanes are best named by the IUPAC system.

Naming Compounds by the IUPAC System

Seeing how many isomers can be made from molecules containing a few carbon atoms, it is easy to understand why there are millions of different organic compounds. If each person who made a new material could choose any name for it, there would be millions of unrelated names to remember. There would probably be a lot of Smithanes, Jonesanes, Tomanes, etc. Around 1800, a worldwide system, the IUPAC system, was established to give formal names to all compounds using specific, uniform rules. Some exceptions to the IUPAC system are the use of common names, such as tert-butyl and isopropane, which were well established prior to the IUPAC system.

NOMENCLATURE RULES

Names consist of three parts; prefix, infix (or parent), and suffix. The prefix lists the substituents and their position, the infix (also called the parent, main chain, or backbone) gives the number of carbon atoms in the longest continuous carbon atom chain, and the suffix identifies the family. The suffix -ane refers to the alkane family.

Acyclic alkanes

As you read each rule, follow along in Fig. 4-11 for an example of that rule.

1. Select the longest continuous chain of carbon atoms. This is called the parent. The longest chain is not necessarily written horizontally. In fact, most of the time it isn't. The longest continuous carbon atom chain in this example consists of 10 carbon atoms.
2. Number each carbon atom in the parent chain. Begin from one end of the parent chain. If the parent chain contains a substituent(s), start numbering at the chain end closest to the first substituent. A methyl substituent is on the second carbon atom. Numbering the chain from the other end would put the first substituent on carbon atom number five. Substituents are given the lowest combination of numbers, e.g., a 2,2,5- is a lower combination than a 2,5,5- (see examples in Fig. 4-12).
3. Name each substituent and precede the name by the number of the carbon atom to which it is attached. The first substituent, a methyl group, is bonded to carbon atom number two (C-2), so it is named 2-methyl.

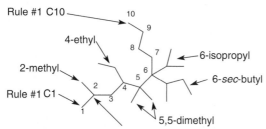

Rule #2: Methyl group on C2

Rule #3: (See placement on molecule).
   2-methyl   4-ethyl   5-methyl   5-methyl   6-isopropyl   6-sec-butyl

Rule #4: 2-Methyl, 5-methyl, and 5-methyl becomes 2,5,5-trimethyl
         Remember, each group requires a number.

Rule #5: Alphabetically, the first letter of each substituent is b, e, i, and m
         6-sec-butyl-4-ethyl-6-isopropyl-2,5,5-trimethyldecane.

Rule #6: Note sec is italicized and iso is not.

Rule #7: sec-Butyl could also be named as a complex substituent,
         6-(1-methylpropyl). Either name is acceptable.

Fig. 4-11.   Rules for naming acyclic alkanes.

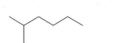
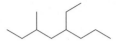

3-Methylheptane
not 2-ethylhexane
(heptane is the longest parent)

5-Ethyl-3-methyloctane
not 4-ethyl-6-methyloctane
(3,5 is less than 4,6)

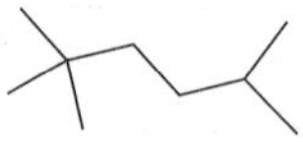

2,2,5-Trimethylhexane
not 2,5,5-trimethylhexane
(2,2,5 is less than 2,5,5)

Fig. 4-12. Naming acyclic alkanes.

4.   If the same substituent appears more than once, it is preceded by di-, tri- , tetra-, etc. It is alphabetized as the substituent, not by di, tri, or tetra. There is one methyl group on carbon atom number two (C-2) and two methyl groups on carbon atom number five (C5), giving 2,5,5-trimethyl. Each substituent requires a number and commas separate the numbers. Trimethyl is listed alphabetically by "m" and not "t."

5.   Write the full name. The substituents are listed alphabetically, not numerically. Hyphens separate numbers from names.

6.   Substituents that contain hyphens in their name, such as n-, sec-, and tert-, are named alphabetically by the parent substituent, not n, s, or t. Substituents containing prefixes iso and neo, with no following hyphen, are alphabetized by "i" and "n." This example contains a 6-*sec*-butyl group and a 6-isopropyl group. *sec*-Butyl alphabetically precedes isopropyl in the name.

7.   If the substituent is branched (a complex substituent) it can be given a common name, if it has one. If not, the substituent is named using Rules 1-4. The carbon atom of the substituent attached to the parent chain is *always* C-1. The complete complex substituent name is enclosed in parentheses and preceded by the number of the carbon atom on the parent chain to which the complex substituent is attached. The substitute on C-6 has the common name 6-*sec*-butyl or the formal name 6-(1-methylpropyl).

And this book is demystifying? Really, it is not as difficult as it may seem. Additional practice problems are shown in Fig. 4-12. It is equally important to be able to draw the structures of compounds when the names are given.

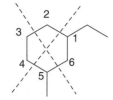

1-Ethyl-3-methylcyclohexane   1-Ethyl-5-methylcyclohexane

Rule #3: One could start numbering at the ethyl group or the methyl group. One starts with the ethyl group since it comes first alphabetically. Substitutents must have the lowest combination of numbers. A 1,3 combination is lower than a 1,5 combination.

Fig. 4-13. Naming cyclic alkanes.

Naming cyclic alkanes

The rules for naming cycloalkanes are essentially the same as those for acyclic compounds. Follow along in Fig. 4-13 as you go through the rules.

1. Count the number of carbon atoms in the parent ring and add cyclo to the alkane name. Since these are alkanes, the name ends in -ane. The example has six carbon atoms in the ring and therefore it is a cyclohexane.
2. A number is not needed if there is only one substituent.
3. If there is more than one substituent, the substituents are numbered so that they have the lowest possible numbers. One substituent will necessarily be on C-1. Substituents are named alphabetically. An ethyl group is on C-1 and a methyl group is on C-3.
4. If the substituent contains more carbon atoms than those on the ring, the substituent becomes the parent and the ring is the substituent.

Additional examples for naming cycloalkanes are given in Fig. 4-14.

1-Isopropyl-2-methylcyclohexane   1-Bromo-3-ethyl-2-methylcyclopentane
not 1-methyl-2-isopropylcyclohexane   not 1-bromo-2-methyl-3-ethylcyclopentane

Fig. 4-14. Naming cycloalkanes.

Physical Properties

All bonds in alkanes and cycloalkanes are C—C or C—H bonds. The C—C bond is nonpolar. The C—H bond is essentially nonpolar, and thus all bonds are nonpolar and therefore the molecules are nonpolar. Using the generalization *like dissolves like*, these materials are soluble in nonpolar solvents, such as carbon tetrachloride and other alkanes. They are *immiscible* (not soluble) in polar solvents like water.

DENSITIES

The density of alkanes is about 0.7 g/mL while the density of water is about 1.0 g/mL at room temperature. Since alkanes are less dense and immiscible with water, a mixture of the two materials forms two layers, with alkanes being the less dense upper layer. The most common example is oil floating on water. Alkanes are hydrophobic (water-fearing).

INTERMOLECULAR FORCES

Since alkanes are nonpolar, intermolecular attraction between molecules results from *London dispersion (LD)* forces. These forces are weak compared to ionic, hydrogen bonding, and highly polar intermolecular interactions. London dispersion forces result from momentary dipoles due to fluctuating electron densities. (For a review, see the section entitled *Intermolecular Forces* in Chapter 1.) The total effect of LD force interactions increases as the size of the molecule increases. The total intermolecular attraction forces of the lower molecular weight linear alkanes e.g., methane, propane, and butane, are small and these materials are gases at room temperature.

Linear alkanes

Larger molecules have a greater summation of LD forces, sufficiently strong to make them liquids. Pentane to heptadecane have sufficient LD forces to make them liquids at room temperature. Higher molecular weight alkanes have even greater total LD forces. As a result they are solids at room temperature. Increasing LD force interactions results in increasing boiling and melting point temperatures as the molecular weight increases in these linear molecules. Similar trends are seen for the cycloalkanes.

QUESTION 4-9
Alkanes can exist as liquids or solids at room temperature. What physical changes result if the temperature increases?

ANSWER 4-9
At higher temperatures sufficient energy may be present to vaporize liquids and melt solids.

QUESTION 4-10
What happens to the physical state of propane if pressure in a container of gaseous propane is increased? Think about the propane in a tank used for an outdoor grill.

ANSWER 4-10
Increasing the pressure will cause the molecules to bump into each other more often, increasing the chances they will "stick" together and form a liquid.

Branched alkanes

The physical state of a compound depends upon the attraction between molecules. Consider two molecules containing five carbon atoms, such as linear pentane and branched neopentane (2,2-dimethylpropane). As an analogy, consider pentane to be stretched out like a pencil. If one had a handful of pencils there would be significant contact between all the pencils. Neopentane is a branched, spherical molecule. If a pencil were compressed into a sphere it would have the size of a large marble. If you put a handful of marbles into a container and measured the total contact—sphere touching sphere—there is much less contact, or interaction, relative to the linear pencils. Less contact means less attraction between molecules. Generally, branched molecules boil and melt at lower temperatures owing to decreased LD force interactions. Less energy (lower temperatures) is thus required for a phase transition, solid to liquid or liquid to gas, to occur.

Chemical Properties

Alkanes tend to be chemically inert. This is an advantage if you have an application where you want the material to remain stable over a long period of time. An example would be the engine oil in your auto, which is a mixture of alkanes. The C–H bond is relatively strong. It is essentially nonpolar and nonpolar molecules tend to be less reactive than polar molecules. Reactivity factors will be discussed in greater detail in other sections of this book.

Alkanes readily react with oxygen (burn) when an ignition source is present. Methane is the primary component of "natural gas" that is burned to heat homes. Gasoline is primarily a mixture of liquid alkanes that is burned in internal combustion (auto) engines.

Conformations of Alkanes

BOND ROTATION

Molecules are like children with "ants in their pants." They wiggle, rotate, stretch, and bend. They just don't remain still. This section discusses rotation around single, sigma bonds. Sigma bonds result from head-to-head overlap of atomic orbitals. The strength of a bond is proportional to the total overlap of the bonding orbitals. Figure 4-15 shows rotation around the H—H bond axis in a hydrogen molecule (Structure 4-15a). Both hydrogen atoms rotate, or spin, but the total orbital overlap of the two 1s orbitals does not change, nor does the bond strength change. Consider the overlap of two sp^3 orbitals in a carbon-carbon bond as shown in Structure 4-15b. One or both of the carbon atoms can rotate around the bond axis. The orbital overlap, and thus the bond strength, does not change.

METHANE

From a bond rotation standpoint, methane is not too exciting. Although rotation occurs around the bond axis of each of the four H—C sigma bonds shown in the 3-D structure, the orbital overlap and bond strength remains the same. Bond rotation is shown in Fig. 4-16 for one of the four H—C bonds.

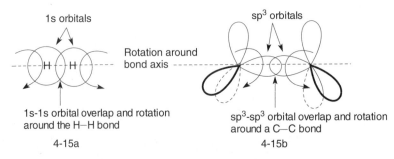

Fig. 4-15. Rotation around sigma bonds.

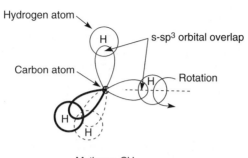

Methane, CH$_4$

Fig. 4-16. Bond rotation in methane.

ENERGY BARRIERS TO BOND ROTATION

Energy barriers to rotation around carbon-carbon bonds depend upon the size (volume) of the groups bonded to the two carbon atoms. Barriers to bond rotation are best assessed where solid lines, solid wedges, and hatched wedges are used to represent the 3-D structures of molecules. *Sawhorse structures* and *Newman projections* can be used to study energy barriers to rotation.

ETHANE

Sawhorse structures

Ethane can be drawn as a sawhorse structure (Fig. 4-17) or a Newman projection (Fig. 4-18). Sawhorse structures can be drawn with wedges (Structure 4-17a) but are usually drawn with solid lines (Structures 4.17b–d). It is important to remember that all bond angles in this molecule are 109.5°, even though it is not apparent in these drawings. In Fig. 4-17, note how the positions of hydrogen atom H_a changes relative to the hydrogen atoms on the back carbon atom when

| 4-17a | 4-17b | 4-17c | 4-17d |

3-D sawhorse　　　　Stick sawhorse with 120° rotations
　　　　　　　　　　giving three conformations

Fig. 4-17. Sawhorse structures of ethane.

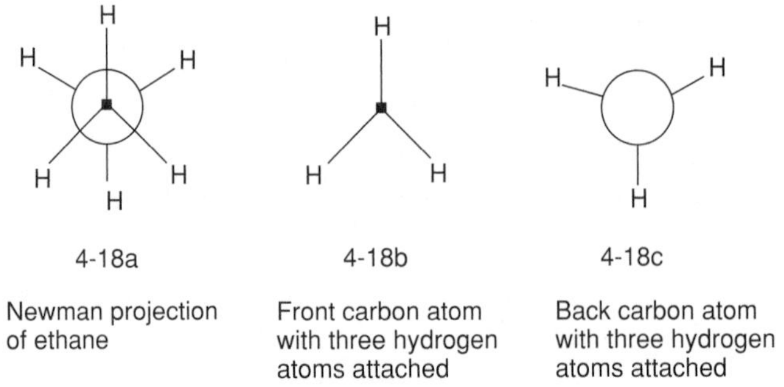

4-18a	4-18b	4-18c
Newman projection of ethane	Front carbon atom with three hydrogen atoms attached	Back carbon atom with three hydrogen atoms attached

Fig. 4-18. Newman projection of ethane.

the C—C bond is rotated. There is an infinite number of positions if the bond is rotated by a fraction of a degree. Structures 4-17b, 4-17c, and 4-17d show C—C bond rotations in 120° increments.

These different rotational positions are known as *conformations*. The study of the energy relationship of the various rotational positions is called *conformational analysis*.

NEWMAN PROJECTIONS

Newman projections are a more informative way to study conformations. A Newman projection for ethane is shown as Structure 4-18a in Fig. 4-18. In this structure you are looking straight down a C—C bond. The second carbon atom is directly behind the first carbon atom. The front carbon atom is represented by a dot as shown in Structure 4-18b. The back carbon atom is represented by a circle as shown in Structure 4-18c. Three hydrogen atoms radiate from each carbon atom. These planar structures show the H—C—H bond angles as 120°, while in reality they are 109.5°. Superimposing Structure 4-18b onto Structure 4-18c gives Structure 4-18a.

In the sawhorse and less so in the Newman structures, it appears that the atoms are quite far apart. The nuclei are quite far apart relative to their size. But the electron clouds, whose radii are about 2000 times larger than those of the nuclei, can overlap and invade each other's "personal space." Like the person looking over your shoulder helping you read your newspaper, invading one's space is even annoying at the molecular level.

Electron repulsions Don't electron repulsions contradict the previous discussion about molecules attracting each other because of London dispersion forces? Fluctuating electron densities create momentary dipole moments. These

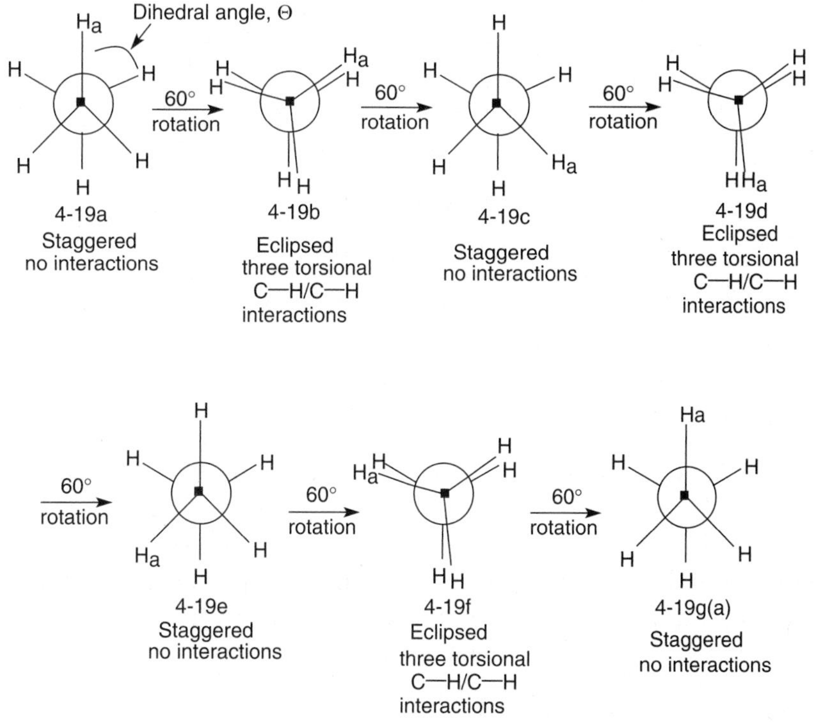

Fig. 4-19.   60° Rotational conformations of ethane.

dipole moments result in attraction between molecules—up to a point. Attraction continues until the negative electron clouds "touch." Energy is required to force the electron clouds to invade each other's space. Since molecules want to be in the lowest energy state possible, there are preferred conformations in molecules that minimize electron cloud overlapping. (For a review of electron repulsions, see the section entitled *VSEPR* in Chapter 1.)

Look at the seven conformations (Structures 4-19a–g) of ethane shown in Fig. 4-19. The last conformation is identical to the first conformation. A 360° clockwise rotation of the front carbon atom is done in 60° increments. The back carbon atom is not rotated. One hydrogen atom on the front carbon atom is designated H_a and its rotation will be followed. The angle between the H_a—C bond on the front carbon and any H—C bond on the back carbon is called the *dihedral angle*, noted as Θ in the first conformer shown in Fig. 4-19.

Staggered and eclipsed conformations The first conformation in Fig. 4-19, Structure 4.19a, shows all hydrogen atoms as far from each other as possible. This is called a *staggered conformation* and will represent the lowest energy

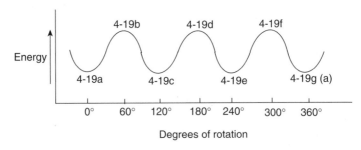

Fig. 4-20. Torsional strain in ethane as a function of C—C bond rotation.

conformation of this molecule. Structures 4-19a, 4-19c, 4-19e, and 4-19g are also staggered conformations. Rotating the front carbon atom of Structure 4-19a 60° clockwise gives Structure 4-19b in which each H on the front carbon atom overlaps with an H on the back carbon atom. This is called an *eclipsed conformation*. Structures 1-19b, 1-19d, and 1-19f are eclipsed conformations. The hydrogen atoms are not shown directly on top of each other for clarity purposes. In the eclipsed position, the electron clouds of the C—H bonds on the front and back carbon atoms overlap. Although there is some electron density around the hydrogen nucleus, most of the electron density, from the hydrogen atom, is in the H—C bond.

Torsional strain Electron-electron repulsion results when the C—H bonds overlap in the eclipsed position. This results in strain in the molecule and forces the molecule into a higher energy state. This bond-bond interaction is called *torsional strain* and is measured in kJ/mol. There are three torsional strain interactions (from the three bond-bond strain interactions) in each of the three eclipsed conformations of ethane. A plot of the relative energy of the molecule versus bond rotation is shown in Fig. 4-20.

The torsional strain energy in the eclipsed positions is about 4 kJ/mol (1 kcal/mol) for each C—H pair, or 12 kJ/mol (3 kcal/mol) per molecule. This is a very low energy compared to the strength of a C—H bond, of about 400 kJ/mol. Since the rotational energy barrier is so small, the C—C bond in ethane undergoes 360° rotation millions of times per second at room temperature. If the conformations of a mole of molecules could be frozen at any one instant, we would find more molecules in the staggered positions than in the higher energy eclipsed positions.

n-BUTANE

n-Butane can be considered an ethane molecule with one hydrogen atom on each carbon atom replaced by a methyl group. The four carbon atoms in butane

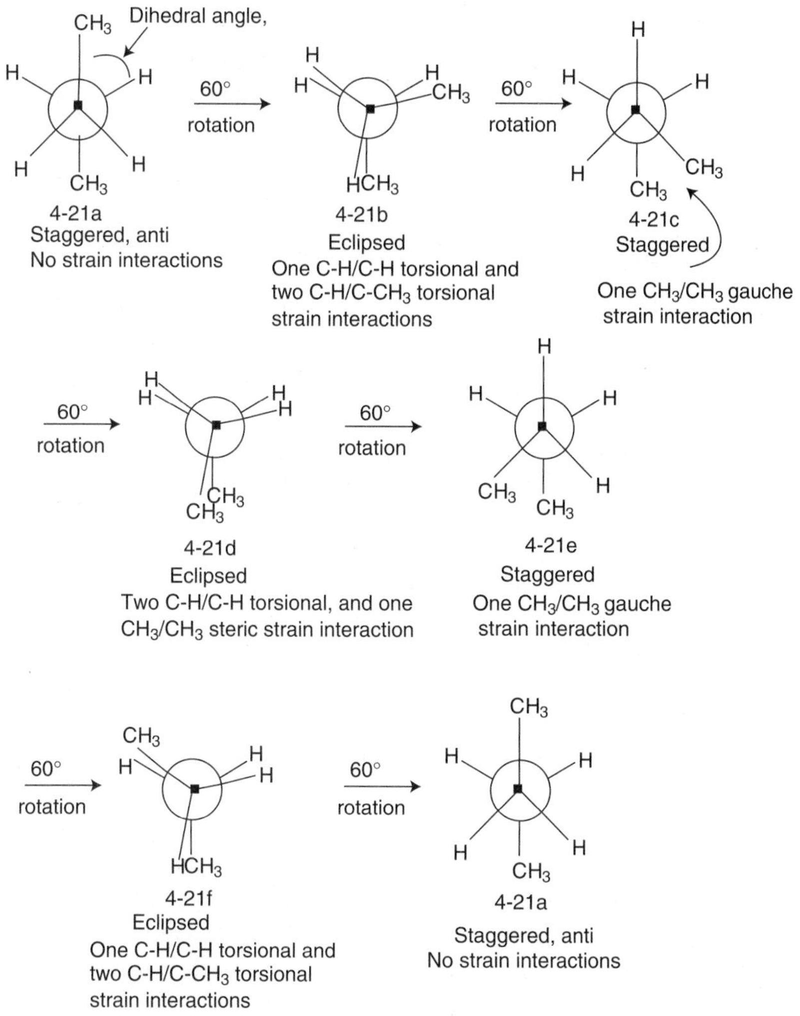

Fig. 4-21. 60° Rotational conformations of *n*-butane.

will be labeled C_1, C_2, C_3, and C_4. Conformational analysis can be done for rotation around the C_1–C_2 bond or the C_2–C_3 bond. An analysis for rotation around the C_2–C_3 bond axis is given here. Seven conformations of butane are shown in Fig. 4-21. Each structure differs by a 60° clockwise rotation of the front carbon atom. The last rotation gives a seventh conformation identical to the first conformation. It is easiest to follow this discussion by looking at the conformations in Fig. 4-21 as each is discussed.

Anti conformation

In the first staggered conformation (Structure 4-21a), the methyl groups are diagonally opposite (kitty cornered) to each other. This staggered position is called the *anti position*. This conformation has the lowest energy of all the conformations.

Gauche steric strain

Rotation by $60°$ gives an eclipsed structure (Structure 4-21b) in which there is torsional strain from the bonding electrons of one C—H/C—H pair and two C—H/C—CH_3 pairs. Rotating another $60°$ gives a staggered conformation (Structure 4-21c). There are now two CH_3 groups in *adjacent* positions, with a dihedral angle of $60°$. For groups other than hydrogen, this staggered position is called the *gauche (go shh)* position. The only strain here is a gauche CH_3/CH_3 *steric strain*. Steric strain results when the electron clouds around atoms or groups invade each other's personal space. Steric strain is also called *steric hindrance* or *van der Waals strain*. There is no H/CH_3 or H/H steric strain when H and CH_3 are in staggered positions as their electron clouds do not overlap.

Eclipsed steric strain

Rotating another $60°$ gives another eclipsed conformation (Structure 4-21d). The two methyl groups are eclipsed. This eclipsed steric strain is greater than the gauche steric strain. There are three strain interactions in Structure 4-21d, one CH_3/CH_3 steric strain interaction and two torsional strain C—H interactions. Rotating another $60°$ results in another staggered conformation (Structure 4-21e). The only strain here is a gauche CH_3/CH_3 steric strain. Rotating another $60°$ gives an eclipsed conformation (4-21f) with torsional strain from one C—H/C—H pair and two C—H/C—CH_3 pairs.

One more $60°$ rotation gives the initial anti conformation 4-21a. The total strain energy values for each rotation will not be discussed, but hopefully you now have an understanding of how to determine relative strain energies involved in each conformation. A more detailed discussion is available in any organic chemistry textbook.

Total steric strain

As the alkyl groups on adjacent carbon atoms become larger (methyl to ethyl to propyl, etc.) the steric interactions get larger, strain energies get greater, and bond rotation requires more energy. Numerical values for the strain energies discussed

Table 4-3.   Strain interaction energies.

Interaction	Strain	Energy barrier, ΔG
C—H/C—H	torsional - eclipsed	4.0 kJ/mol (1.0 kcal/mol)
C—H/C—CH$_3$	torsional - eclipsed	6.0 kJ/mol (1.4 kcal/mol)
HC—CH$_3$/C—CH$_3$	steric and torsional - eclipsed	11 kJ/mol (2.6 kcal/mol)
CH$_3$—CH$_3$	steric - gauche	3.2 kJ/mol (0.77 kcal/mol)

for ethane and butane are listed in Table 4-3. This table gives information about the magnitude of the different strain energies. More extensive tables can be found in standard organic chemistry textbooks.

QUESTION 4-11
Using Newman projections, draw the seven conformations of 60° rotations for propane. Which one or ones have the highest energy?

ANSWER 4-11
All eclipsed forms have higher energy than staggered forms have.

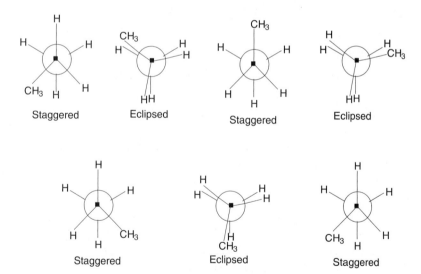

As a review, there are three types of strains in alkanes and cycloalkanes:

1. Torsional strain—eclipsing of bonding electrons in neighboring bonds.
2. Steric strain—repulsive interactions of electron clouds when atoms (other than hydrogen) approach each other too closely.
3. Angle strain—result of bonds being distorted from their ideal values (discussed in the following section).

| 60° | 90° | 108° | 120° |
| Cyclopropane | Cyclobutane | Cyclopentane | Cyclohexane |

Fig. 4-22. Internal bond angles of cycloalkanes.

Conformations of Cylcoalkanes

There are many natural materials, such as the various steroids, where the conformations of these molecules are related to their activity. The total strain energies determine the most favorable conformations and hence their biological activity. Since steroids contain six-membered rings (*cyclohexanes*), it is of interest to look briefly at the conformations of cycloalkanes and particularly cyclohexane.

Internal bond angles in the *planar polygons* cyclopropane, cyclobutane, cyclopentane, and cyclohexane are 60°, 90°, 108°, and 120°, respectively, as shown in Fig. 4-22. All carbon atoms are sp^3 hybridized and the ideal bond angles around a carbon atom are 109.5°. A bond angle different from 109.5° would be expected to contribute *angle strain* to a molecule.

ANGLE STRAIN

The greatest orbital overlap and hence the strongest bonds would be expected if sp^3 orbitals on adjacent atoms can undergo direct head-to-head overlap, giving an ideal bond angle of 109.5°. One would expect cyclopentane to be the most stable molecule in this series since it has bond angles (108°) closest to 109.5°.

Cyclopropane, with bond angles of 60°, is expected to be, and is, the least stable of the four cycloalkanes.

CYCLOPROPANE

Consider the cyclopropane molecule, Structure 4-23a in Fig. 4-23. In addition to bond angle strain, there is torsional strain from the eclipsing adjacent C—H bonds. The three C—H bond strain interactions on the "top side" of the ring are numbered and shown with a double-headed arrow (↔). Another three identical interactions exist on the "bottom side" of the molecule. However with just three carbons in the ring, the ring must remain planar.

RING PUCKERING

Cyclobutane (Structure 4-23b) also has angle and torsional strain. Strain can be reduced somewhat if the ring bends (puckers) from planarity. Cyclopentane

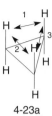

4-23a
Cyclopropane

The 3 double headed
arrows show the torsional
strain interactions on the
"top" of the molecule. An
additional 3 interactions occur
on the "bottom" side.

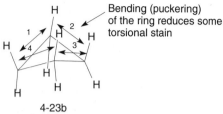

Bending (puckering)
of the ring reduces some
torsional stain

4-23b
Cyclobutane

The 4 double headed
arrows show the torsional
strain interactions on the
"top" of the molecule. An
additional 4 interactions occur
on the "bottom" side.

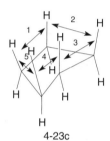

4-23c
Cyclopentane

The 5 double headed
arrows show the torsional
strain interactions on the
"top" of the molecule. An
additional 5 interactions occur
on the "bottom" side.

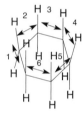

4-23d
Planar cyclohexane

The 6 double headed
arrows show the torsional
strain interactions on the
"top" of the molecule. An
additional 6 interactions occur
on the "bottom" side.

Fig. 4-23. Torsional strain in cylcoalkanes.

(Structure 4-23c) has little angle strain but there is significant torsional strain from the 10 pairs of C–H bond interactions. This ring also puckers somewhat, giving an "envelope" structure. In each of these structures, the torsional interactions are shown as numbered double-headed arrows for the "top side" of the molecule. An equal number of interactions are present on the "bottom side" of the molecule but are not shown in the structures.

QUESTION 4-12
Identify the 10 pairs of C–H torsional strain interactions in cyclopentane.

ANSWER 4-12
There are 5 torsional strain interactions between C–H bonds on the "top" of the ring and 5 torsional strain interactions between C–H bonds on the "bottom" of the ring.

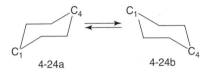

Fig. 4-24. Two chair conformations of cyclohexane.

CYCLOHEXANE

Cyclohexane is, surprisingly, strain free. If the ring were planar, there would angle strain and 12 torsional strain interactions as shown in Structure 4-23d. Instead of being planar, the ring puckers. It can form a conformation called a chair form, because it sort of looks like chair. Actually, this conformation can flip back and forth between two chair conformations, 4-24a and 4-24b shown in Fig. 4-24. In conformation 4-24a, one "end" carbon, C_1, is pointed down and the other "end" carbon, C_4, is pointed up. In chair conformation 4-24b, their positions are reversed, where C_1 is pointed up and C_4 is pointed down.

One chair conformation does not flip directly to the other chair conformation. Just for completeness, the flipping process goes from a chair to a half-chair to a twisted boat to a boat to a twisted boat to a half-chair and finally to the other chair. Since the present discussion deals exclusively with the chair forms, these other conformations are not discussed here, but are discussed in most elementary organic textbooks. You should practice drawing the chair forms.

Axial and equatorial positions

The hydrogen atoms attached to the cyclohexane ring are labeled H_a and H_e in Fig. 4-25. The H_a atoms point in the upward and downward (vertical) directions. These are called *axial* (a) positions. The H_e atoms point around the equator of the ring, and these are called *equatorial* (e) positions.

Axial positions   Equatorial positions   Axial and equatorial positions

Fig. 4-25. Axial and equatorial positions in cyclohexane.

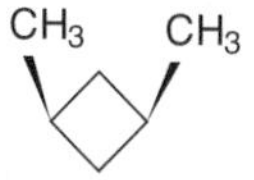

4-26a
cis-1,3-Dimethylcyclobutane

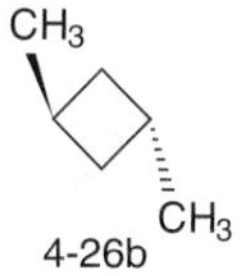

4-26b
trans-1,3-Dimethylcyclobutane

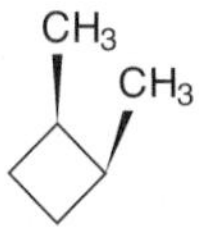

4-26c
cis-1,2-Dimethylcyclobutane

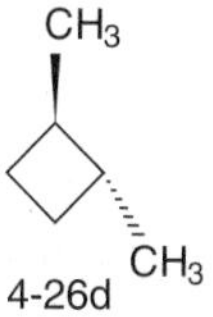

4-26d
trans-1,2-Dimethylcyclobutane

Fig. 4-26. Cis and trans dimethylcyclobutanes.

SUBSTITUTED CYCLOALKANES

Two substituents can be bonded to a cycloalkane in different ways. For example, the four dimethylcyclobutane isomers shown in Fig. 4-26 have a methyl group on carbon atoms C-1 and C-3 in Structures 4-26a and 4-26b and C-1 and C-2 in Structures 4-26c and 4-26d. From a stereochemistry (3-D) standpoint, the two methyl groups can be on "top" (above the plane of the ring) or one methyl group can be on the "top" and the other on the "bottom" (below the plane of the ring).

CIS AND TRANS ISOMERS

Cis and trans isomers are different compounds. If the methyl groups are on the "same side" of the molecule (Structures 4-26a and 4-26c), the groups are said to be cis to each other. To help remember the cis form, doesn't the arrangement of the methyl groups have the shape of a "C" tipped on its back? In Structures 4-26b and 4-26d the methyl groups are on opposite sides of the ring. This is the trans form. Trans means across from or on the other side, as in transcontinental. The terms "cis" and "trans" are italicized in a compound's name.

QUESTION 4-13

One isomer of 1,2-dimethylcyclopropane has both methyl groups on the "bottom" of the ring. Is this the cis or trans structure?

ANSWER 4-13

The structure is cis when both methyl groups are on the same side of the ring.

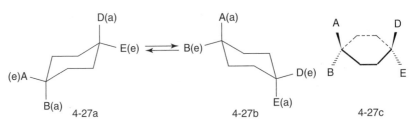

Fig. 4-27. Cis-trans and a-e conformations.

SUBSTITUTED CYCLOHEXANES

The chair structures in Fig. 4-27 have four *different* generic groups: A and B bonded to C-1 and D and E bonded to C-4 on cyclohexane. B and D are in axial positions and A and E are in equatorial positions in Structure 4-27a. Structure 4-27b results when the ring flips. Now B and D are equatorial and A and E are axial. *A key point: when a ring flips axial groups become equatorial and equatorial groups become axial—always.*

What is the cis-trans relationship of A and D and A and E in the first chair structure, 4-27a? It is often easier to determine this relationship by looking at the planar structure, 4-27c. It should be apparent that A and D are cis, even though they are in axial and equatorial positions in the chair conformations. A and E are trans to each other. What is the relationship when the ring flips? A and D are still cis and A and E are still trans. Cis-trans relationships do not change when the ring flips but a-e relationships do. *A key point: cis isomers remain cis and trans isomers remain trans in any conformational change while equatorial and axial positions are reversed on each carbon when a ring flips.*

Ring flipping is a conformational change. In the case of cyclohexane, there is restricted rotation around C—C bonds of just 60°, either clockwise or counterclockwise. No bonds are broken in this process. This is different from the case of ethane or butane discussed above, where there was 360° rotation around C—C bonds. The ring prevents free rotation around the C—C bonds.

STRAIN ENERGIES IN CYCLOHEXANES

A strain energy analysis can be done for the two chair conformations of cyclohexane. Bond angle strain is not significant since bond angles are about 109.5°. One need only consider the torsional and steric interactions as discussed

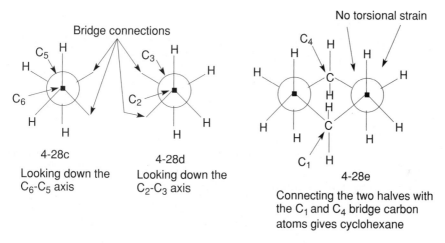

Fig. 4-28. Strain energy analysis for cyclohexane.

previously for ethane and butane. Figure 4-28 shows the two chair conformations of cyclohexane, 4-28a and 4-28b. You might predict that they are of equal stability since they look pretty much the same. Not only do they look the same, if you rotate conformer 4-28a 180° around the broken line axis shown, you get a structure identical to conformer 4-28b. The two conformations are identical spatially and of equal energy.

Newman projections

It is easier to do a strain energy analysis for cyclohexanes with Newman projections. This may get a little more challenging and molecular models would certainly help, but look at the structures in Fig. 4-28. Structure 4-28a is a typical chair form. The carbon atoms have been numbered to help follow the rotation. Rotate Structure 4-28a clockwise around the broken line axis so that you are looking directly down the C_6–C_5 bond and the C_2–C_3 bond. Looking down only the C_6–C_5 bond gives Newman projection 4-28c. Looking down only the C_2–C_3 bond gives Newman projection 4-28d. Structures 4-28c and 4-28d represent "two halves" of cyclohexane. These two structures should remind you of the Newman projection of ethane.

The two structures, 4-28c and 4-28d, can be connected with two bridges. In these two structures there are four bonds drawn without hydrogen atoms attached. When these four bonds are attached to two CH_2 groups, two bridges are formed as shown in Structure 4-28e. The two bonds on Structures 4-28c and 4-28d that are pointed downward are connected at C-1. The two bonds on Structures 4-28c and 4-28d pointed upward are connected at C-4. The C-1 and C-4 bridges connect the two halves. The result is Structure 4-28e, a Newman structure of cyclohexane. The torsional and steric interactions can now be evaluated in the Newman structure of cyclohexane just as they were for the ethane and butane molecules.

Strain analysis

All the hydrogen atoms in the chair form of cyclohexane are in staggered positions and there is no need to consider torsional C–H interactions. The methylene (–CH_2–) bridges, that is, C-1 and C-4, are staggered relative to each other and there is no torsional or angle strain interaction between them. There is a gauche steric strain interaction between C-1 and C-4. Since this strain is of equal magnitude in both chair forms it need not be considered when calculating the *difference in strain energies* of both chair conformations. The bridge carbon atoms, C-1 and C-6, reverse their "up" and "down" positions when the ring flips. Both chair conformations (4-28a and 4-28b) are of equal energy and one would expect to find equal amounts of both conformations at any given time.

Ring flipping

There is, however, an energy price to pay in going from one chair conformation to the other. This is because the ring-flipping process goes through the half-chair/twisted boat/boat conformations mentioned above that have higher strain energies than those of the chair form. The energy required in ring flipping for cyclohexane is about 41.8 kJ/mol (10 kcal/mol). As in the case for ethane and butane, this is a low energy barrier and the cyclohexane ring flips

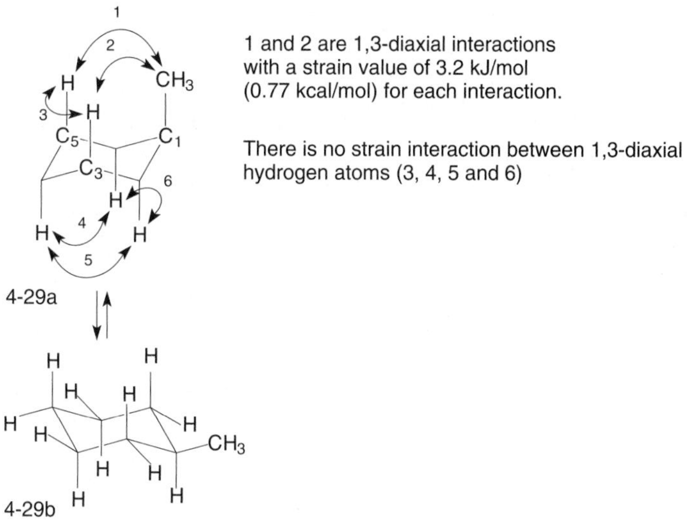

1 and 2 are 1,3-diaxial interactions with a strain value of 3.2 kJ/mol (0.77 kcal/mol) for each interaction.

There is no strain interaction between 1,3-diaxial hydrogen atoms (3, 4, 5 and 6)

Fig. 4-29.   1,3-Diaxial strain interactions in methylcyclohexane.

between both chair conformations rapidly (about 1000 times/sec) at room temperature.

METHYLCYCLOHEXANES

The strain energy picture changes if cyclohexane contains one or more substituents. The methyl group in methylcyclohexane is in an axial position in one chair conformation and in an equatorial position in the other chair conformation. Lewis structures for the two chair conformations are shown in Fig. 4-29. When the methyl group is axial (Structure 4-29a) there are two *1,3-diaxial* steric interactions between the methyl group and the axial hydrogen atoms on carbon atoms C-3 and C-5.

The hydrogen atoms on C-3 and C-5 are 1,3-diaxial to each other. There is no 1,3-diaxial steric interaction between these two hydrogen atoms as their van der Waals radii (electron clouds) are much smaller than that of a methyl group. There are also three potential 1,3-diaxial interactions on the "bottom" side of the ring. It is easy to forget that you have to consider both "sides" of the ring. All atoms in 1,3-diaxial positions on the bottom of the ring are hydrogen atoms and there are no strain interactions.

QUESTION 4-14
Do you have to specify if the methyl group is cis or trans in methylcyclohexane?

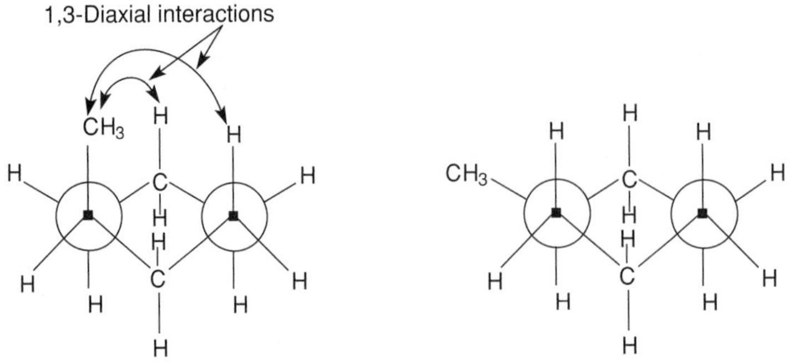

1,3-Diaxial interactions

4-30a

Methyl group is axial,
two 1,3-diaxial interactions

4-30b

Methyl group is equatorial,
no 1,3-diaxial interactions

Fig. 4-30. Newman projections of methylcyclohexane.

ANSWER 4-14
No, there is no other group for the methyl group to be cis or trans to.

In chair conformation Structure 4-29b the methyl group is in the equatorial position. There are no 1,3-diaxial steric strain interactions since all diaxial positions are occupied by hydrogen atoms. Also, there are no torsional strain interactions since all adjacent bonds are staggered.

Newman projections

Newman projections can also be used to calculate ring strain. Newman projections for both chair conformations of methycyclohexane are shown in Fig. 4-30. Projections are shown for the methyl group in the axial (Structure 4-30a) and equatorial (Structure 4-30b) positions. When the methyl group is axial, there are two 1,3-diaxial interactions between the methyl group and the hydrogen atoms as shown in Structure 4-30a. When the methyl group is equatorial (4-30b), all 1,3-diaxial positions contain hydrogen atoms and there is no steric strain.

An energy analysis shows that the conformation with the methyl group in an equatorial position is more stable than the conformation with the methyl group in an axial position. That would make sense since there can be no methyl-hydrogen diaxial interactions if the methyl group is not in an axial position. *A key point: a bulky group, something larger than hydrogen, is more stable in an equatorial position than in an axial position.*

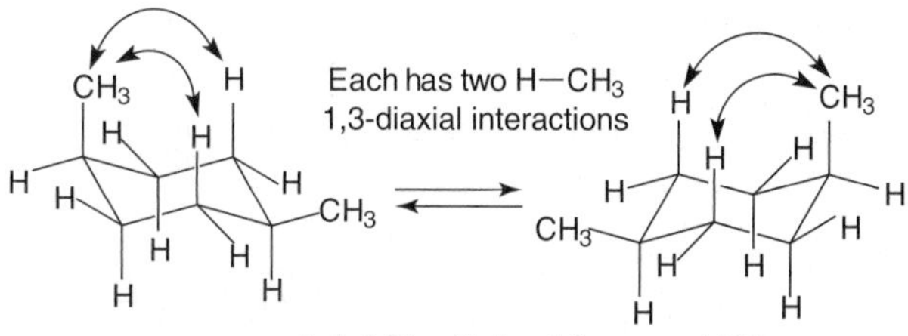

4-31a *cis*-1,4-Dimethylcyclohexane 4-31b

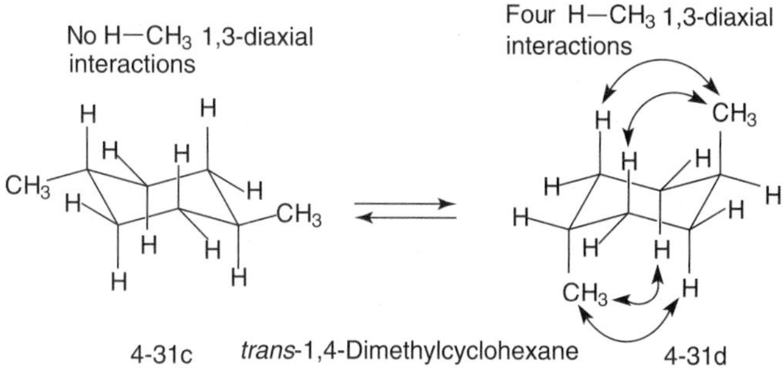

4-31c *trans*-1,4-Dimethylcyclohexane 4-31d

Fig. 4-31. Conformations of *cis*- and *trans*-dimethylcyclohexane.

CIS- AND TRANS-DIMETHYLCYCLOHEXANES

The four chair conformations for cis and trans isomers of 1,4-dimethylcyclohexane are shown in Fig. 4-31. Each chair conformation (Structures 4-31a and 4-31b) of the cis isomer has one methyl group axial and one methyl group equatorial, resulting in two H–CH_3 1,3-diaxial interactions in each structure. Each conformation has the same number of steric and torsional strain interactions and therefore both are of equal strain energy.

One chair conformation (Structure 4-31c) of the trans isomer has both methyl groups in equatorial positions. The other conformation (Structure 4-31d) has both methyl groups in axial positions. Conformer 4-31c has no H–CH_3 1,3-diaxial interactions while conformer 4-31d has four H–CH_3 1,3-diaxial interactions. All these interactions are shown in Structure 4-31d. One might then predict that the trans isomer, with no 1,3-diaxial interactions, is the most stable of the four conformations and the trans conformation with four 1,3-diaxial interactions is the least stable of the four conformations. The cis compound has two 1,3-diaxial interactions in both the chair conformations. Values for these strain energies are given in Table 4-3.

We do not know what the total strain energies of each conformer are without doing an energy analysis by adding strain energy values of all the torsional and steric interactions. But we can predict which conformers are the most and least stable by identifying all the strain interactions. A more complete strain energy analysis will not be done here, but such a discussion can be found in most organic chemistry textbooks.

QUESTION 4-15
Which of the four conformations of *cis*- and *trans*-1,3-dimethylcyclohexane is the most stable?

ANSWER 4-15
cis-1,3-Dimethylcyclohexane, with both groups equatorial, is the most stable.

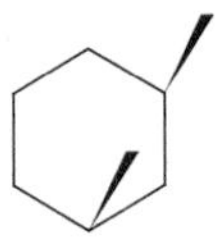

GIBBS ENERGY

As shown above, some conformations are more stable than others. If two conformations are in equilibrium (see the equilibrium equations shown in Fig. 4-31), we can calculate the ratios of the two chair conformations at equilibration conditions. The *differences* of the Gibbs strain energy, ΔG, for both conformations can be calculated with the data shown in Table 4-3. The value of ΔG can then be used in the following equation to calculate the ratio of the two conformations where one conformation represents the products and the other conformation represents the reactants.

$$\Delta G = -2.3RT \log K_{eq}$$

Here R is a constant (= 2.31 J/mol-K), T is the temperature in Kelvin, and K_{eq} is the ratio of the two conformations, the conformation on the right in the equilibrium equation is the numerator and the conformation on the left is the denominator. An example of K_{eq} for *trans*-dimethylcyclohexane would be a ratio of the concentrations of Structure 4-31d/4-31c in Fig. 4-31.

Quiz

1.   Alkanes
     (a) are unsaturated
     (b) are saturated

(c) contain only carbon atoms
(d) are named for Al Kane

2. Alkyl groups are
    (a) halogens
    (b) alkanes without a hydrogen atom
    (c) alkanes without a carbon atom
    (d) a member of Al Kane's group

3. Substituents are
    (a) conformations
    (b) alkyl groups
    (c) never present in alkanes
    (d) players sitting on the bench

4. The parent name for an alkane ends with
    (a) -ide
    (b) -ane
    (c) -ene
    (d) -ate

5. Lewis structures
    (a) show how atoms are connected
    (b) do not show carbon atoms
    (c) do not show hydrogen atoms
    (d) indicate a carbon atom where two lines meet

6. When atomic orbitals mix (hybridize) one gets
    (a) no change
    (b) sp^4 orbitals
    (c) molecular orbitals
    (d) hybrid orbitals
    (e) a mess

7. Constitutional isomers
    (a) are identical compounds
    (b) vary in how atoms are connected
    (c) are mirror images
    (d) are a formal document about isomers

8. Monocycloalkanes have the formula
    (a) C_nH_{2n}
    (b) C_nH_n

(c) C_nH_{2n+2}
(d) C_nH_{2n-2}

9. Removing any one hydrogen atom from *n*-butane gives ____ different alkyl group(s).
    (a) zero
    (b) one
    (c) two
    (d) three

    C - C - C - C

10. A secondary carbon atom
    (a) has one carbon atom attached
    (b) has two carbon atoms attached
    (c) has three carbon atoms attached
    (d) is second best to a primary carbon atom

11. Nomenclature means
    (a) naming compounds
    (b) determining 3-D structures
    (c) forming alkanes
    (d) no men's clature

12. Alkanes ____ water.
    (a) float on
    (b) sink in
    (c) readily mix with
    (d) none of the above

13. Conformational structures
    (a) do not exist
    (b) differ as a result of bond rotation
    (c) differ in how atoms are connected
    (d) contain information about con(vict)s

14. Newman structures
    (a) show correct bond angles
    (b) are drawn in 3-D
    (c) show different conformations
    (d) are a type of salad dressing

15. Torsional strain is a result of
    (a) bond angles less than ideal
    (b) electron clouds on two groups repelling each other

(c) repulsion of electrons in bonds
(d) straining one's back

16.   Steric strain is a result of
       (a) bond angles less than ideal
       (b) electron clouds on two groups repelling each other
       (c) repulsion of electrons in bonds
       (d) straining one's back

17.   Angle strain is a result of
       (a) bond angles less than ideal
       (b) repulsion of electrons surrounding the nucleus
       (c) repulsion of electrons in bonds
       (d) straining one's back

18.   The cyclohexane molecule is
       (a) flat
       (b) chair shaped
       (c) a five-membered ring
       (d) acyclic

19.   Ring flipping means
       (a) converting into different conformations
       (b) exchanging atoms in a ring
       (c) a nonequilibrium process
       (d) throwing rings in the air and catching them

20.   When a cyclohexane ring flips, cis groups
       (a) become trans
       (b) remain cis
       (c) are removed from the ring
       (d) move to an adjacent carbon atom

21.   When a cyclohexane ring flips, equatorial groups
       (a) become axial
       (b) remain equatorial
       (c) are removed from the ring
       (d) move to an adjacent carbon atom

22.   1,3-Diaxial strain interactions refer to the interaction of
       (a) groups in axial and equatorial positions
       (b) axial groups on adjacent carbon atoms

(c) axial groups on alternate carbon atoms
(d) equatorial groups on alternate carbon atoms

23. The difference in strain energy of two conformations in equilibrium
    (a) is a measure of the relative amounts of each
    (b) can be estimated from a Newman projection
    (c) can be calculated from the individual strain energies
    (d) all of the above

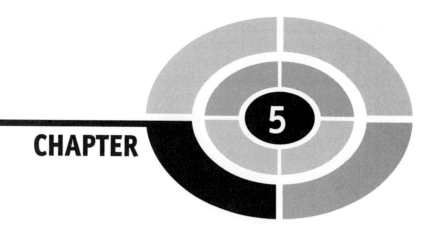

CHAPTER 5

Stereochemistry

To have a better understanding of the properties of organic molecules, one has to study their three-dimensional (3-D) structure. Why is this important? Our perception of smell and taste depends, in many instances, on the 3-D structure of molecules. Enzymes are very selective in the 3-D structure of the molecules they interact with. The effectiveness of drugs is highly dependent on their 3-D structure. Organic chemists need to be able to determine the 3-D structures (stereochemistry) of new and existing molecules to relate 3-D structure to reactivity.

Isomers

Isomers are compounds that have the same formula but different structures. Types of isomers are summarized in Fig. 5-1. The three main types of isomers are constitutional isomers (discussed in the section entitled *Constitutional Isomers* in Chapter 4), conformational isomers (discussed in the section entitled *Conformations of Alkanes* in Chapter 4), and stereoisomers. *Stereoisomers* have

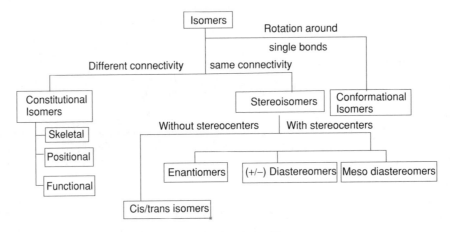

Fig. 5-1. Classification of isomers.

the *same connectivity* but a different spatial orientation. Stereoisomers do not differ from each other as a result of rotation around single (sigma) bonds. They are different compounds.

CONNECTIVITY

The term *same connectivity* means all the atoms in a molecule are connected in the same sequence. For a molecule containing atoms A, B, and C, one connectivity is A—B—C while A—C—B is a different connectivity. Figure 5-2 shows an example of two stereoisomers. Both have the same connectivity but the atoms have a different orientation in space. One would have to break the H—C and HO—C bonds and reverse their connectivity to convert stereoisomer 5-2a into stereoisomer 5-2b. Rotating the entire molecule or rotation around single bonds (for a review of bond rotation see the section entitled *Conformations of Alkanes* in Chapter 4) does not convert one stereoisomer into another stereoisomer.

Chiral Compounds

All objects have a mirror image. Some mirror images are *superimposable* (or superposable if you prefer) and some are not. Hold a ping-pong ball in front of a mirror. Now take its mirror image "out of the mirror" and you can put it right on top of (into) the "real" ping-pong ball. The ping-pong ball and its mirror

Fig. 5-2. Stereoisomers. The same atoms or groups are connected to the central carbon atom. The connectivity is the same, but OH and H are oriented differently in space. To change the positions of H and OH, one would have to break the H—C and HO—C bonds and connect them in the reverse order.

image are identical and are superimposable. How about a coffee cup without any writing or design on it? The cup and its mirror image are superimposable.

What about a coffee cup with your name on it? Its mirror image is not superimposable. Get a coffee cup with a design on it and put it in front of a mirror. Does the mirror image of the design look the same as the original or is the design backwards? Objects that *are not* superimposable on their mirror image are *chiral* (ki ral, rhymes with spiral). Objects that *are* superimposable on their mirror image are *achiral* (not chiral). Chiral comes from the Greek word *for hand*, so we talk about the *handedness* of some molecules.

QUESTION 5-1
Are the following chiral or achiral? Screw driver, wood screw, tube sock, right shoe, fork, pencil with your name on it, and this page.

ANSWER 5-1
Wood screw, right shoe, pencil, and this page are chiral. The others are achiral.

Are your hands chiral? Hold your right hand before a mirror. You see reflected in the mirror the structure of your left hand. Your hands are chiral. Try superimposing your hands. Have the palms facing the same direction. Your thumbs are on opposite sides. OK, I'm sure you get the picture. Now what does this have to do with real molecules?

Stereocenters

A *stereocenter* exists in a compound if the interchange of two groups attached to the same atom gives a different stereoisomer. Three examples of these compounds are cis and trans cyclic compounds (discussed in Chapter 4), cis and trans alkenes (discussed in Chapter 6), and compounds containing asymmetric carbon atoms (discussed in this chapter). Each of these three types of compounds has one or more stereocenters. Stereocenters can be atoms other than carbon but only carbon stereocenters will be discussed.

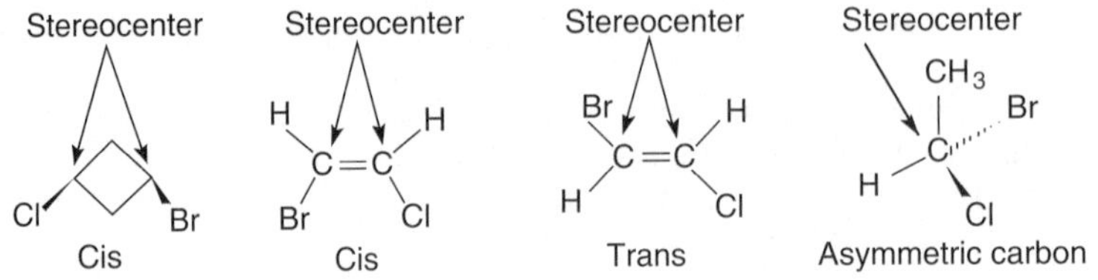

Chirality Centers

An asymmetric carbon atom is bonded to four *different* groups: e.g., A, B, D, and E. A compound containing one asymmetric carbon atom is necessarily chiral. The molecules shown in Figs. 5-2 and 5-3 are chiral. The carbon atom bonded to four different groups is called by various names: *a stereocenter, a stereogenic center, a chiral center, a chiral carbon, a chirality center, an asymmetric carbon, and an asymmetric center*. All the terms refer to the same species. (Organic chemists seem to have a difficult time using one name for a given species.) Not all compounds with stereocenters contain chiral atoms. Cis and trans alkenes contain stereocenters but are not necessarily chiral.

A carbon atom attached with single, sigma bonds to four atoms or groups is sp^3 hybridized (for a review see the section entitled *sp^3 Orbitals* in Chapter 1); all bond angles are ideally 109.5°; and the molecule has a tetrahedral structure. Practice drawing 3-D structures containing sp^3 hybridized carbon atoms. Some authors call these *perspective structures*. Use A, B, D, and E as four different

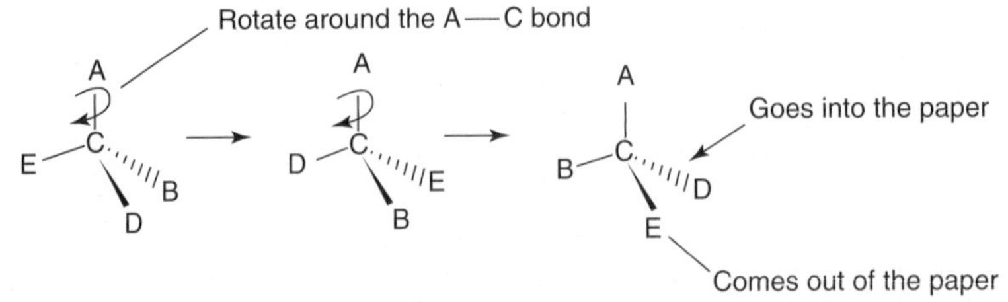

Fig. 5-3. Rotation around single bonds. Looking down the A—C bond, the D—E—B groups rotate clockwise, but still keep the same order. All molecules are identical.

groups attached to a carbon atom. See Fig. 5-2 as an example. *Solid lines* (–)
represent bonds in the plane of this page. A *solid wedge* (▬) represents a bond
coming out of the page toward you. A *hatched* or *broken wedge* (⸴⸴⸴⸴) represents
a bond going into the page. If the entire molecule is rotated, it will, of course,
remain the same molecule. One can rotate around a single, sigma bond and the
molecule still remains the same molecule as no bonds are broken and remade
with a different connectivity.

QUESTION 5-2
Is bromochloromethane, CH_2BrCl, a chiral compound?

ANSWER 5-2
No, it does not have four different groups/atoms attached to the central carbon
atom.

   The example in Fig. 5-3 shows rotation around the A–C bond. Rotating around
the A–C bond changes the spatial positions of B, D, and E, but all orientations
still represent the same stereoisomer.
   Use any of the structures in Fig. 5-3 and draw equivalent structures by rotating
around any of the other σ bonds. You need to become familiar with these types
of rotations. Figure 5-4 shows the same molecule when the entire molecule is
rotated. All bond angles are ideally 109.5°, the overall structure is still tetrahedral, and the connectivity is still the same. All structures represent the same
molecule.

GUMDROP MODELS

Models are extremely helpful for visualizing 3-D structures. If you cannot get
to a college bookstore to buy a molecular model set, try the following: Buy

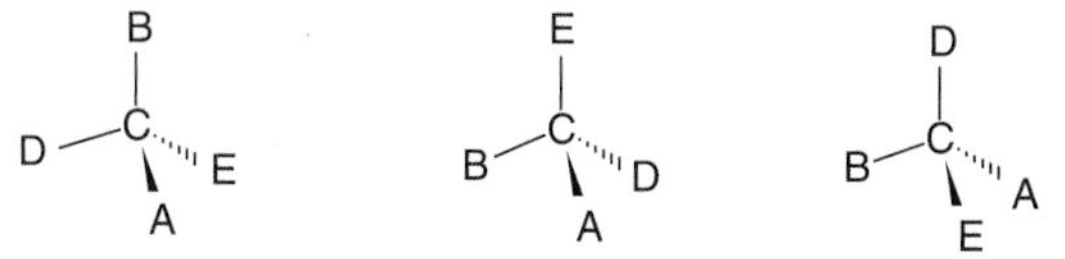

Fig. 5-4.   Rotation of the entire molecule. One can use the method shown in Fig. 5-3
to show different orientations, by holding one bond steady and rotating the remaining
groups clockwise or counterclockwise around that bond. Looking down the A–C bond in
5-4a, rotate counterclockwise to go from 5-4a to 5-4b. Now looking down the B–C bond
in 5-4b, rotate clockwise to go from 5-4b to 5-4c.

some gumdrops, large and small ones. Use a large one for the central atom, the stereocenter, and small ones of different colors for the various atoms or groups attached to the center gumdrop. Use tooth picks as bonds to attach the gumdrops. Other candies may work as well. Check out the candy section at a food mart. Use your imagination. Experiment, that's what science is all about. (What you don't use you can always eat during a break.)

Tetrahedral template

When you make a gumdrop model, you need to make all the angles as close to 109.5° as possible. Use a different color for each gumdrop. Use the templates shown in Fig. 5-5. It may be helpful to draw larger templates on another sheet of paper. Use a large gumdrop for the center "carbon atom" and two smaller gumdrops as atoms #1 and #2. Rotate this structure 90° in the plane of the paper about the large gumdrop. Now use the template to insert the last two smaller gumdrops. You will have to keep the center "carbon" above the plane of the paper for this step. With a little practice, you should have a tetrahedral structure. Make a second molecule and reverse any two atoms (different colored gumdrops). Compare the two structures. They should be mirror images. If not, do the exercise again. Now that you have mirror images, try to superimpose the two molecules. It can't be done.

Enantiomers

Compounds with stereocenters can be divided into two groups: enantiomers and diastereomers. Molecules that are *nonsuperimposable* mirror images are called *enantiomers*. Each molecule is an enantiomer. Enantiomers may contain one

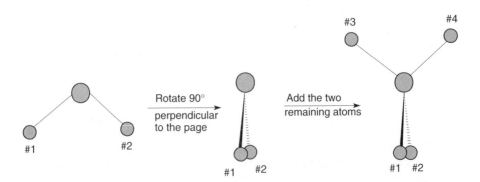

Fig. 5-5. Tetrahedral template.

or more stereocenters. Where do chemists get such unfamiliar names? Many words, as you probably have already found out, are based on Latin or Greek words. Enantiomer comes from the Greek word en*anti*on meaning opposite. If it helps, think of anti as meaning opposite.

SYMMETRY PLANES

There are many examples of carbon atoms bonded to four groups that have a tetrahedral structure. Are all of these molecules enantiomers? Only if all four groups are different. (There are a few exceptions but an explanation is beyond the scope of this text.) If the molecule has a plane of symmetry, the molecule will *not* be chiral. A plane of symmetry exists if you can put a plane (or mirror) anywhere through a molecule and an identical image appears on both sides of the plane. Look at the example of C bonded to A, B, and two Ds in Fig. 5-6. If the plane/mirror bisects the A, B, and C atoms, half of each atom will be on each side of the mirror plane. The D atoms appear opposite to each other on each side of the mirror plane. This molecule contains a plane of symmetry.

The mirror image of this molecule is superimposable on itself. The mirror image 5-6b is identical to Structure 5-6a. This molecule is achiral and the C atom is not an asymmetric center. Also, the compound is not chiral because it does not have four different groups bonded to the central C atom. Get out the gumdrops and make a model so you can visualize the mirror plane bisecting the molecule.

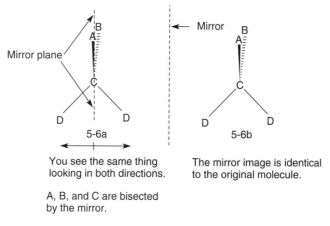

Fig. 5-6.   Mirror plane/plane of symmetry.

QUESTION 5-3
Draw an isomer of 2,3-dibromobutane that has a mirror plane.

ANSWER 5-3

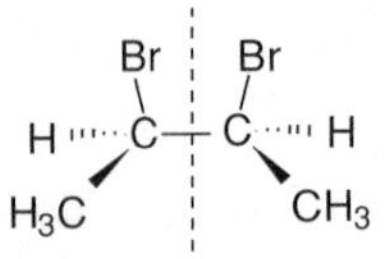

The mirror plane perpendicular
to the page bisects the molelcule

Our bodies are very good at selectively reacting with different enantiomers. Identify the chiral carbon in limonene, Structure 5-7a in Fig. 5-7. We perceive one enantiomer as orange flavor and the other as lemon flavor. Amazing! Structure 5-7b identifies the stereocenter with an asterisk. Identifying different groups when a ring is involved is explained in a following section entitled *Cyclic Structures*.

QUESTION 5-4
Consider 2-chlorobutane, $CH_3CHClCH_2CH_3$. How many stereocenters are in this molecule? Remember, a carbon stereocenter has four different atoms or groups attached to it. It will be easier to recognize stereocenters if you draw a Lewis structure.

ANSWER 5-4
There is one stereocenter.

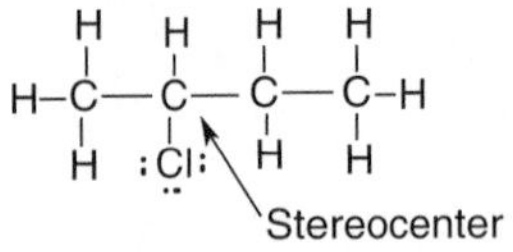

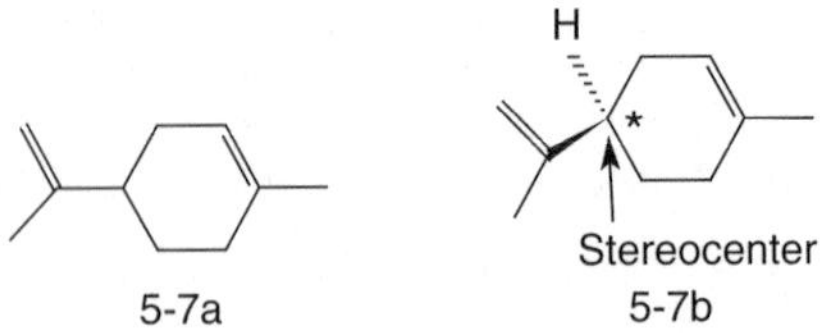

5-7a   5-7b

Fig. 5-7.   Limonene.

PHYSICAL AND CHEMICAL PROPERTIES OF ENANTIOMERS

How do the physical and chemical properties of two enantiomers differ? Mirror image enantiomers have identical melting points, boiling points, refractive indexes, heats of combustion, solubilities in achiral solvents, and chemical reactivities with achiral reagents. If all the physical and chemical properties are identical, how can we prove they are different molecules? There are instances when the chemical and physical properties are *not* the same.

Stereoselectivity

Enantiomers have different reactivities under special conditions. One enantiomer can react preferentially with a specific stereocenter of another reagent. As an analogy, consider your two (enantiomeric) hands. They are going to react with a reagent, a left-handed (chiral) glove. Your left hand (one enantiomer) easily fits into the glove. Your right hand (the mirror image enantiomer) does not easily fit into a left-handed glove. Like the hand-glove analogy, chiral reagents react selectively with other chiral reagents.

OPTICAL PROPERTIES OF ENANTIOMERS

There is one very important *physical property* that is used to differentiate enantiomers. Enantiomers differ in the way they interact with *plane polarized light*. What is plane polarized light? First, we need to review the properties of light. Ordinary light can be described in terms of an oscillating wave, like a guitar string that is plucked. (It may be helpful to review the section entitled *Nature of Waves* in Chapter 1.) A plucked guitar string just vibrates up and down; that is, it oscillates in one direction. Ordinary light oscillates in all directions, up and down (vertical), sideways (horizontal), and at all angles in between. If ordinary light is directed through a filter, like a polaroid lens, the oscillations can be limited to one plane, as depicted in Fig. 5-8. That is where the *plane polarized light* term comes from.

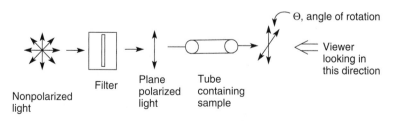

Fig. 5-8. Schematic of a polarimeter.

Polarized light is directed through a tube containing a solution of the sample and you look through the tube from the end opposite the light source. This would be like holding a flashlight in front of you and shining it toward your eyes with a glass of water (the sample) between the flashlight and your eyes. The sample rotates the plane of the polarized light. The instrument used to make these measurements is called a polarimeter. A schematic of a polarimeter is shown in Fig. 5-8.

USE OF A POLARIMETER

A polarimeter is very easy to use. To test a sample, dissolve the sample in a solvent and fill the polarimeter sample tube with this solution. For this discussion, the sample will consist of one enantiomer. As the polarized light passes through the tube, the plane of the polarized light is rotated. The amount of rotation, Θ, measured in degrees, depends upon how many molecules of enantiomer interact with the light beam. Therefore one needs to consider enantiomer concentration and length of the sample tube.

The amount of rotation also depends on the wavelength of light used and the temperature. The emerging light will be rotated either clockwise (right) or counterclockwise (left) relative to the original beam. You *cannot* predict the direction of rotation from the structure of the enantiomer. If the light is rotated clockwise, it has a (+) designation and if rotated counterclockwise, a (−) designation. Each enantiomer has a value called the *specific rotation* [α]. This is a characteristic of a material just as a melting point is. Since two enantiomers interact with light differently they are called *optical isomers*.

SPECIFIC ROTATION

The equation for *specific rotation* is:

$$[\alpha]_{\lambda}^{t} = \frac{\text{observed rotation}}{(\text{concentration})(\text{path length})}$$

The concentration is in grams per milliliter and the path length (length of the tube) is in decimeters. The wavelength λ, often a specific wavelength (the D line) from a sodium light source, and temperature t, in degrees centigrade, need to be given since the value of α depends on all of these variables.

ENANTIOMER RELATIONSHIPS

What is the relationship between the specific rotations of a pair of enantiomers? The mirror image enantiomer will have a specific rotation equal in magnitude and opposite in sign from the original enantiomer. If you know the specific rotation of one enantiomer, you know the specific rotation of its mirror image. It is stated above that you could not predict the direction of rotation by looking at the 3-D structure of a molecule. This is still true, but once you determine the specific rotation of one enantiomer, you also know the direction and magnitude of specific rotation of the other enantiomer.

QUESTION 5-5
The specific rotation of (−)-2-butanol is −13.52°. What is the specific rotation of (+)-2-butanol under the same experimental conditions?

ANSWER 5-5
+13.52°.

NAMING CRITERIA

The name of an enantiomer includes the (+) or (−) sign. An older naming system uses d instead of (+) and l instead of (−). This is a lowercase d or l. (An uppercase D and L have different meanings.) The d stands for dextrorotatory and l for levorotatory, terms derived from Latin words for right and left respectively. (Now don't you wish you would have studied Latin and/or Greek?) For example, (+)-tartaric acid (*d*-tartaric acid) is the mirror image of (−)-tartaric acid (*l*-tartaric acid). The *R/S* system (explained below) is also used to name enantiomers.

Racemic Mixtures

What would you predict the specific rotation to be if you put equal numbers (i.e., equal concentrations) of mirror-image enantiomers in the *same* sample tube and measured the optical activity of the mixture? The optical rotation is zero since the rotation of one enantiomer is "cancelled" by the rotation of its mirror image enantiomer. A mixture of equal concentrations of mirror-image enantiomers is called *a racemic mixture, a racemate, a d/l pair,* or *a +/− pair*. Sometimes both enantiomers are present, but not in equal amounts. Unequal amounts are described by the term *enantiomeric excess*, or *ee*. There is an excess of one isomer.

Why spend so much time talking about stereocenters? As you will see in Chapter 7 and in various reactions discussed throughout this text, stereocenters are made and destroyed in many reactions.

The R/S System

There are methods of determining the absolute 3-D structure of stereoisomers. *Absolute structure* means the actual orientation in space of the four atoms or groups bonded to the stereocenter(s). Structures drawn in this text use solid lines, solid wedges, and hatched wedges to show atoms in the plane of the page and those that go above or below the plane of the page. When the absolute structure is known, we need to be able to describe it in words. The *R/S* system was developed by Cahn, Ingold, and Prelog for naming the absolute structure of stereoisomers. *R* and *S* are based on the Latin terms for right and left respectively. *R* and *S* designations are made using the following rules:

(a) Assign a priority (#1 to #4) to each of the four groups bonded to a carbon stereocenter: #1 has the highest and #4 the lowest priority.

(b) Priority is based on *atomic number*: the higher the atomic number, the higher the priority, e.g., Cl (17) > F (9) > O (8) > N (7) > C (6) > H (1), etc. In the case of isotopes, higher atomic weight has higher priority, e.g., ^{2}H > ^{1}H.

(c) If the priority of two or more atoms bonded to the stereocenter is the same (e.g., three carbon atoms), consider the atoms bonded to these carbon atoms. Continue to work outward (away from the stereocenter) until there is a difference in priority.

AN R/S EXAMPLE

The structures in Fig. 5-9 show an example of these rules. Consider the four *atoms* or *groups* —H (5-9a), —$CH_2CH_2CH_3$ (5-9b), —$CH(OH)CH_2CH_3$ (5-9c), and —$CH(CH_3)_2$ (5-9d) attached to the stereocenter. The atoms directly bonded to the stereocenter are H, C, C, and C. H is lower in priority than C (atomic #1 vs. atomic # 6), so H is priority #4. The other three carbon atoms bonded directly to the stereocenter have the same priority, and therefore one needs to consider what is bonded to each of these three carbon atoms.

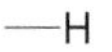

C₁ bonded to C, H, and H

C₁ bonded to O, C, and H

—H
(#4)

5-9a

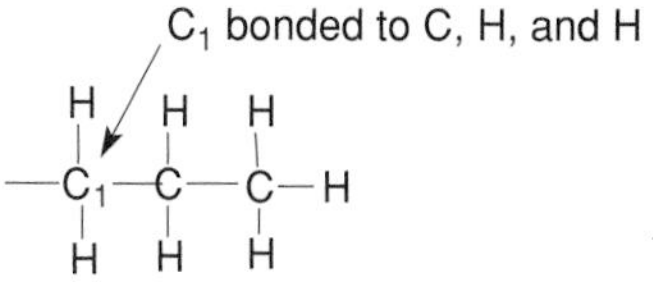

5-9b (#3)

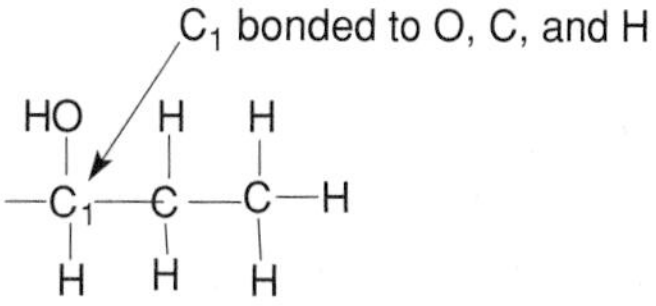

5-9c (#1)

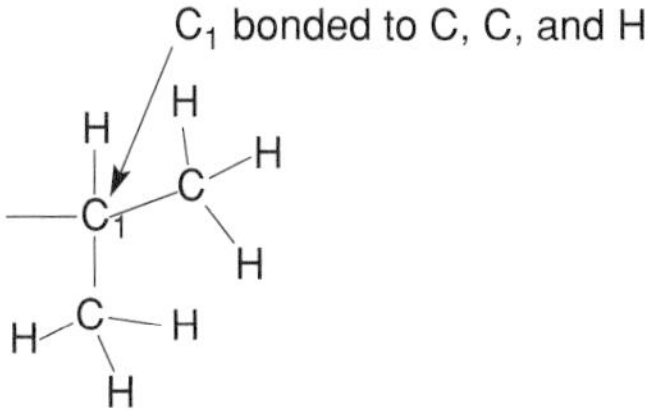

C₁ bonded to C, C, and H

	bonded to C₁
5-9b	-C, -H, -H
5-9c	-O, -C, -H
5-9d	-C, -C, -H

5-9d (#2)

Fig. 5-9.   Assigning *R/S* priority.

The C in group 5-9b is bonded to one C and two Hs (C, H, H). In group 5-9c the C is bonded to one O, one C, and one H (O, C, H). In group 5-9d, C is bonded to two Cs and one H (C, C, H). One now compares the atoms bonded to each of these three carbon atoms: (C, H, H), (O, C, H), and (C, C, H) arranged from highest to lowest priority in each individual group. The first atom in each of the three groups is a C, an O, and a C. Since O has the highest priority, group 5-9c is the highest priority (#1) group. In the two remaining groups, each contains a C (of equal priority), so one has to consider the next highest priority atom for each group. One then compares an H and a C. The C has a higher priority than H, so 5-9d is the second highest priority (#2) group, and 5-9b is therefore the third highest priority (#3) group.

IMAGINARY STRUCTURES

In the case of multiple (double and triple) bonds an imaginary structure is drawn to determine group priority, using the same rules already stated. This is most easily understood by following along in Fig. 5-10 as you read the following: First, break the π multiple bonds and remake the bonds as single bonds to a new (imaginary) atom of the same type. In Structure 5-10a, the π bond between C and O is broken. Since C had two bonds to O, an O (imaginary atom) is added and C will still have two (now single) bonds to two O atoms. A (imaginary) C is also added to one of the O atoms so there are still two (now single) bonds

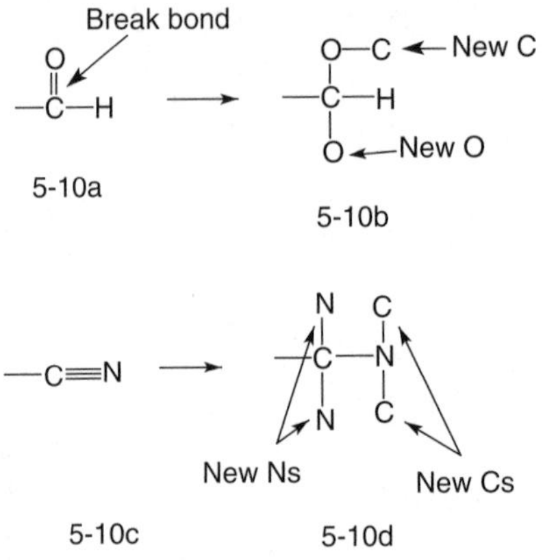

Fig. 5-10. Substitutents with multiple bonds.

from O to C. The bond between C and H is unchanged since it was originally a single bond.

An imaginary structure is drawn for Structure 5-10c. Since C has three bonds to N, two additional Ns are added in imaginary Structure 5-10d and C still has three (now single) bonds to N. N has three bonds to C in Structure 5-10c, and two additional Cs are added in 5-10d and hence N still has three (now single) bonds to C.

QUESTION 5-6
Assign priorities to the following:

(a) $-CH_2CH_3$, $-H$, $-CH_3$, $-CH(CH_3)_2$

(b) $-OH$, $-CH_3$, $-Br$, $-CH_2OH$

(c) $-OH$, $-\overset{O}{\underset{\|}{C}}OH$, $-\overset{O}{\underset{\|}{C}}OCH_3$, $-C\equiv N$

ANSWER 5-6
(a) #2, #4, #3, #1; (b) #2, #4, #1, #3; (c) #1, #3, #2, #4.

(d)   Once priorities are established, draw the actual 3-D structure of the molecule placing the lowest priority group or atom (#4) in the back (going behind the paper). Then assign priorities to the other three groups. Follow the priority ratings from #1 to #2 to #3. If this order is clockwise, the configuration is R. If the order is counterclockwise, the configuration

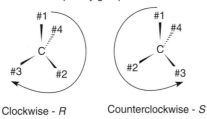

Lowest priority group is in the back

Clockwise - *R*   Counterclockwise - *S*

Fig. 5-11.   *R/S* assignments.

is *S*. Think of turning the steering wheel in a car. If you turn clockwise, you turn to the right (*R*) and if you turn counterclockwise, you turn to the left (*S*). Figure 5-11 shows an example of *R/S* assignments.

QUESTION 5-7

Are the following identical molecules or enantiomers: use your gumdrop models.

(a)

(b)

(c)

ANSWER 5-7

(a) Identical; (b) enantiomers; (c) identical.

When the Lowest Priority Group
Is Not in the Back

The *R/S* assignment described above (Rule d) places the lowest priority group in the back (behind the plane of the page). Often molecules are drawn where

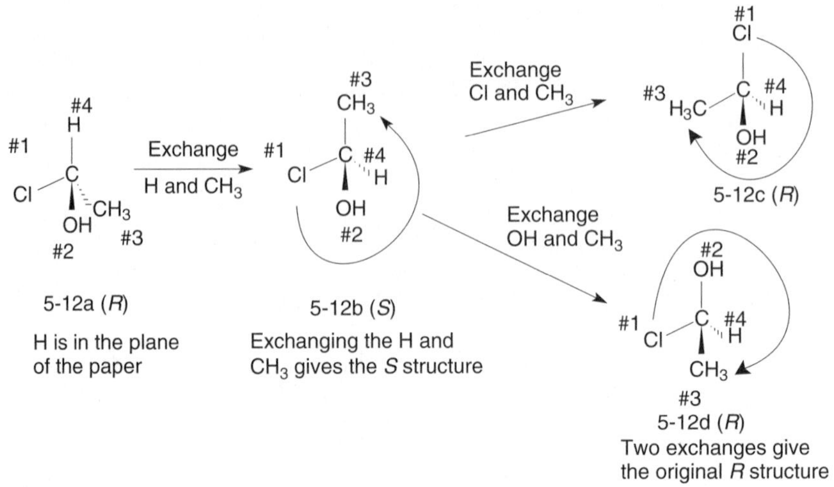

Fig. 5-12. Exchanging groups.

the lowest priority group is *not* in the back. This is probably more common in "real life." You can rotate the molecule on paper, but only in certain ways. Here is a key point: *if you interchange any two of the four groups on a stereocenter, you will make its mirror image (its enantiomer)* and you no longer have the same molecule. Then, if you again interchange *any two* of the four groups you will be back to the original structure. By using this method, you can make two interchanges and put the lowest priority atom or group (which may not necessarily be H) in the back. Follow atom or group exchanges on the molecule in Fig. 5-12. First H and CH_3 in Structure 5-12a are reversed, putting the lowest priority atom in the back, giving Structure 5-12b. This gives the mirror image. Now exchange the Cl and CH_3 groups, giving Structure 5-12c or exchange the OH and CH_3 groups to give Structure 5-12d. Both Structures 5-12c and 5-12d are identical to the first Structure 5-12a, but have the lowest priority group in the back.

ASSIGNING *R/S* CONFIGURATIONS

Assigning the R/S configuration to Structure 5-12a may be a challenge, as the lowest priority group is in the plane of the paper. The previous paragraph described assigning priorities to the four groups: Cl > OH > CH_3 > H. Looking up from the directly *beneath* the molecule (which places H in the back), one sees a clockwise rotation for decreasing priority of the groups, or one can look at

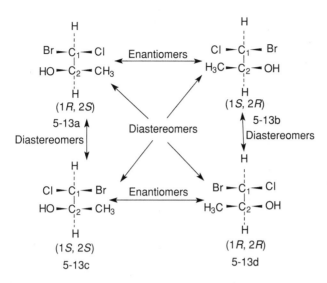

Fig. 5-13.   A compound with two stereocenters.

identical Structures 5-12c and 5-12d. Each of these also has a clockwise rotation of decreasing priority. All have *R* configurations, as they should, since they are all representations of the same molecule.

QUESTION 5-8
Assign *R/S* configurations to the structures in Question 5-7.

ANSWER 5-8
(a) *S*, *S*; (b) *R*, *S*; (c) *R*, *R*.

Molecules with Multiple Asymmetric Centers

Many molecules have more than one asymmetric center. The *maximum number of stereoisomers* is 2^n, where *n* is the number of stereocenters in a molecule. Some molecules have fewer than the maximum number of stereoisomers.

Consider the compound shown in Fig. 5-13. Four different groups are bonded to C-1 and four different groups are bonded to C-2. Four stereoisomers are shown: Structures 5-13a, 5-13b, 5-13c, and 5-13d. Can you draw more? If you can, they will be duplicates of existing structures. Remember the formula 2^n.

Enantiomers

Structures 5-13a and 5-13b are nonsuperimposable mirror images. They have the opposite configuration at each stereocenter and are a pair of enantiomers. Structures 5-13c and 5-13d are nonsuperimposable mirror images. They are also a pair of enantiomers.

Diastereomers

What is the relationship between diagonal pairs (5-13a and 5-13d) and (5-13b and 5-13c)? They have the same connectivity but are not mirror images and not identical. They are *diastereomers*. Diastereomers are stereoisomers that are not identical and are not enantiomers. This may seem like a vague definition. Consider the structures in Fig. 5-13. A pair of diastereomers has the *same* configuration around *at least* one stereocenter and the opposite configuration around the other stereocenters. This can also be stated by saying diastereomers have the *opposite* configuration around *at least* one stereocenter and the same configuration at the other stereocenters. Since there are only two stereocenters in this example, diastereomeric pairs will have the same configuration at one stereocenter and the opposite (mirror image) configuration at the other stereocenter. For example $(1R, 2S)$ and $(1R, 2R)$ are diastereomers as are $(1S, 2R)$ and $(1S, 2S)$.

What is the relationship between vertical pairs in Fig. 5-13? They are also diastereomers. They have the same configuration at one stereocenter and a different configuration at the other stereocenter.

PHYSICAL PROPERTIES OF DIASTEREOMERS

Diastereomers are different compounds. They have different melting points, boiling points, solubilities, etc. Unlike enantiomers, mixtures of diastereomers can be separated much more easily by physical means.

QUESTION 5-9
What is the relationship between two isomers that have the same configuration at each stereocenter, $2S, 3S$ for one isomer and $2R, 3R$ for the other isomer (2 and 3 refer to the numbered carbon atoms)?

ANSWER 5-9
They are enantiomers since they have the opposite configuration at each corresponding stereocenter.

SUMMARY OF *R/S* RELATIONSHIPS

The relationships between all stereoisomers can be determined if we know the *R/S* configurations of just one stereoisomer. In this example, the second and third carbon atoms (C-2 and C-3) are the only stereocenters in this molecule.

Known stereoisomer	Enantiomer	Diastereomers
$2R, 3S$	$2S, 3R$	$2S, 3S$ or $2R, 3R$
$2S, 3S$	$2R, 3R$	$2R, 3S$ or $2S, 3R$
$2R, 3R$	$2S, 3S$	$2R, 3S$ or $2S, 3R$
$2S, 3R$	$2R, 3S$	$2R, 3R$ or $2S, 3S$

An enantiomer is the mirror image of the known isomer, and it will have the opposite configuration at *each* stereocenter. In the special case where *the same four groups* are bonded to both C-2 and C-3, stereoisomers $2R, 3S$ and $2S, 3R$ are identical compounds. These compounds are called meso compounds (discussed below).

QUESTION 5-10
Draw all stereoisomers (with no duplicates) for $CH_3CHClCHBrCH_3$.

ANSWER 5-10
With two stereocenters there are a maximum of 2^n or 4 stereoisomers.

Meso Compounds

Louis Pasteur identified the stereoisomers of tartaric acid that he isolated from wine. (Perhaps a little sip stimulated his research.) As an exercise, draw all the stereoisomers of tartaric acid, $HO_2CCH(OH)CH(OH)CO_2H$. Four (2^n)

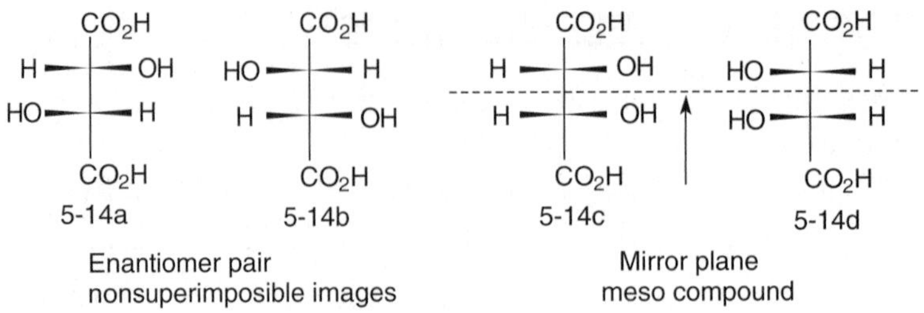

	CO_2H		CO_2H		CO_2H		CO_2H
H —	— OH	HO —	— H	H —	— OH	HO —	— H
HO —	— H	H —	— OH	H —	— OH	HO —	— H
	CO_2H		CO_2H		CO_2H		CO_2H
	5-14a		5-14b		5-14c		5-14d

Enantiomer pair
nonsuperimposible images

Mirror plane
meso compound

Fig. 5-14.   Tartaric acid.

stereoisomer structures can be drawn. Compare your structures to those in Fig. 5-14.

PLANE OF SYMMETRY

Note the last two structures, 5-14c and 5-14d, have a plane of symmetry right through the center of the molecule. A plane of symmetry means that the *molecule is achiral*—even though it has two chiral stereocenters. Such molecules are called *meso* (me zo) compounds. Structures 5-14c and 5-14d are identical. Rotate Structure 5-14c 180° in the plane of the paper. It is superimposable on Structure 5-14d. Structures 5-14c and 5-14d represent the *same molecule*.

OPTICAL INACTIVITY

Meso compounds with two stereocenters will have the *same four groups* attached to each of the two stereocenters and each stereocenter has the opposite stereochemistry. In Fig. 5-14, each stereocenter is bonded to the same four groups: $-H$, $-OH$, $-CO_2H$, and $-CH(OH)CO_2H$. If one stereocenter is R, the other stereocenter in the same molecule is S, and vice versa. What about optical activity? If each stereocenter is connected to the same four groups and one center rotates plane-polarized light $x°$ in one direction then the other stereocenter will rotate light by the same amount ($x°$) in the opposite direction. The total rotation of plane polarized light for a meso compound is zero. Isn't a meso compound somewhat like a racemic mixture? A single meso compound is *not* called a racemic mixture even though the end result (zero light rotation) is the same. The above example described a molecule with two stereocenters. Meso compounds also exist for compounds with more than two stereocenters as long as a plane of symmetry exists.

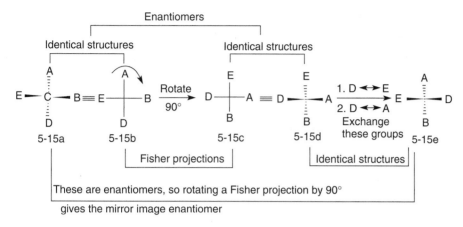

Fig. 5-15.   Fisher projections.

Fisher Projections

The 3-D structures of large molecules are sometimes drawn as *Fisher projections*. The bonds in a Fisher projection are all solid lines. By convention, all horizontal lines (bonds) are coming out of the plane of the paper toward you and all vertical lines (bonds) are going into (behind) the plane of the paper. Figure 5-15 shows the relationship of a 3-D structure to a Fisher projection. Structures 5-15a (3-D) and 5-15b (Fisher) are identical ($\equiv$).

Structures 5-15c (Fisher) and 5-15d (3-D) result if 5-15b is rotated clockwise by 90°. How can you determine if 5-15a/5-15b and 5-15c/5-15d are the same stereoisomer? Switch any two groups in either pair two times. This rotates the molecule but does not change its 3-D structure. Structure 5-15e results by first switching D and E and then switching D and A in Structure 5-15d. Structures 5-15d and 5-15e are identical structures. Is 5-15e identical to 5-15a? No, 5-15e differs from 5-15a by the positions of B and D. They are enantiomers. *A key point*: *rotating a Fisher projection by 90° about a stereocenter converts one enantiomer into its mirror image enantiomer*. You may need to make a gumdrop model to convince yourself. A Fisher projection can be rotated 180° in the plane of the paper and it will remain the same stereoisomer.

A challenge in stereochemistry is to determine if two structures are identical or nonidentical. Various ways have been discussed to make these comparisons: molecular rotations, exchanging pairs of atoms or groups two times, and assigning *R/S* configurations. If all the stereocenters have the same (*R* or *S*) configuration, the molecules are identical.

QUESTION 5-11

Are the following pairs identical or enantiomers:

(a)

(b)

ANSWER 5-11

(a) Enantiomers; (b) enantiomers.

Rotating Fisher Projections/Structures

Fisher projections are generally drawn so that the carbon chain (backbone) is vertical. Thus, if a hydrogen atom is attached to a stereocenter, it is usually in a horizontal position, projecting out of the paper toward you. In assigning *R/S* configurations, the lowest priority group (often hydrogen) is usually put in the back, or behind the plane of the paper. There are two ways to get the lowest priority group in the back. Exchange two groups two times. Or, select a bond to one of the four groups or atoms attached to the stereocenter and rotate about that bond until the lowest priority atom or group is in the back. Rotation can be clockwise or counterclockwise. (It may help to get out the gumdrops and make some models to follow this discussion.)

Figure 5-16 demonstrates this rotation process. In Line (a), rotation is around the A–C-1 bond. In the first rotation, B replaces D, D replaces E, and E replaces B. Rotate in this manner two more times and you return to the original structure. All the structures in Line (a) represent the same molecule. In Line (b), rotation is around the B–C-1 bond and all structures in this line represent the same molecule. Mentally switch two groups two times in any of the structures if you need to convince yourself the structures in each line are identical.

REVERSED *R/S* ASSIGNMENTS

R/S configurations can be assigned if the lowest priority group is projecting outward toward you. The atoms or groups are assigned priorities, as discussed previously. Decreasing priority is again assigned as clockwise or counterclockwise. The difference is that the *R/S assignment is now reversed when the lowest*

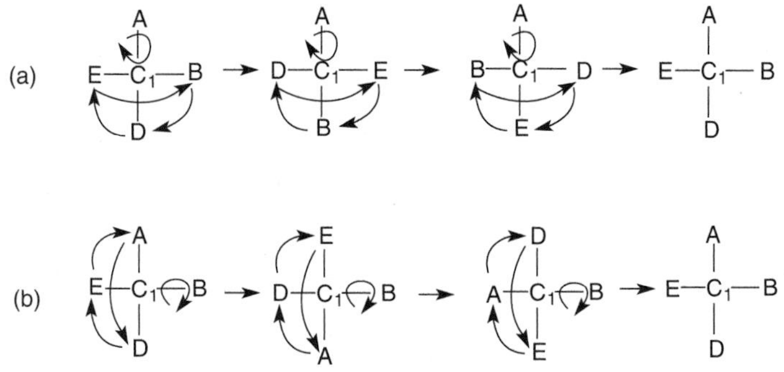

(a)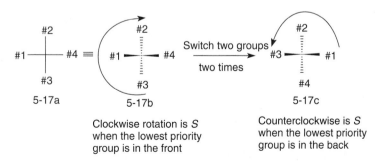

(b)

Fig. 5-16. Rotating Fisher projections. (a) Three rotations around the A–C_1 bond give the starting structure. (b) Three rotations around the B–C_1 bond give the starting structure.

priority group is coming out of the page toward you. If the direction of rotation is to the right (clockwise), the stereocenter is *S* and if rotation is to the left (counterclockwise), the stereocenter is *R*. They are just opposite the assignments made when the lowest priority group is in the back.

Figure 5-17 shows an example where the groups are listed with their priority designations. Structures 5-17a and 5-17b show the Fisher and 3-D structures. Since priority group #4 comes out toward you, the clockwise rotation for decreasing priority is assigned an *S* conformation. If you need to confirm this assignment, exchange two groups two times (first exchange #4 and #3 then switch #1 and #3) to get group #4 in the rear and assign *R/S* configuration. This is shown in Structure 5-17c.

Your decision

Which of the various methods discussed should you use to determine *R/S* assignments? It's up to you. Use the method you are most comfortable with. Each method will give the same result.

Fig. 5-17. *R/S* assignment with lowest priority group in front.

Rotate 90°

Mirror plane

5-18a   5-18b

Go around the ring to the left

Go around the ring to the right

Fig. 5-18. Chlorocyclohexane.

Cyclic Stereoisomers

Substituted cyclic compounds with rings greater than three atoms are puckered to lessen angle strain. (For a review of angle strain, see Chapter 4.) It is usually easier to identify stereocenters if the ring is planar and therefore the following examples will use planar structures. First consider chlorocyclohexane, Structure 5-18a, shown in Fig. 5-18. Is C-1 a stereocenter? It has one bond to a hydrogen atom and one bond to a chlorine atom. To determine if the ring represents two different groups, identify the ring atoms bonded to each side of the potential stereocenter. As shown in Structure 5-18b, starting on the left of C-1 and going around the ring to the left gives the sequence $-CH_2-CH_2-CH_2-CH_2-CH_2-$. Now starting again at C-1 and going around the ring to the right gives the same $-CH_2-CH_2-CH_2-CH_2-CH_2-$ sequence. When starting at C-1, one obtains the same sequence of attached groups in going around the ring in either direction. These two groups are *identical*. Thus C-1 is achiral. It does not have four different groups or atoms bonded to it. Also, this molecule has a mirror plane perpendicular to the plane of the ring and bisects C-1 and C-4, as shown in Structure 5-18b. Molecules with a plane of symmetry are achiral.

QUESTION 5-12
Is C-1 a stereocenter in hydroxycyclopentane?

ANSWER 5-12
No, there is a plane of symmetry and there are not four different groups attached to C-1.

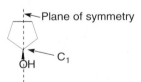

Plane of symmetry

C_1

OH

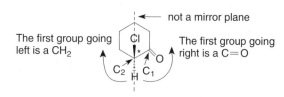

The four different atoms or groups are H, Cl, CH_2, and C=O.

Fig. 5-19.   2-Chlorocyclohexanone.

DISUBSTITUTED RINGS

Figure 5-19 shows the structure of 2-chlorocyclohexanone. It is a disubstituted molecule. How many stereocenters are there? Just one. The carbon atom attached to the oxygen atom has *only three* atoms attached to that carbon atom and it is *not* a stereocenter. In Fig. 5-19, start at the carbon attached to the chlorine atom and go around the ring in each direction. Going to the left is a CH_2 group and going to the right is a C=O group. CH_2 and C=O are two different groups attached to that carbon atom. The H and the Cl make a total of four different atoms or groups attached to carbon atom C-2 and therefore C-2 is a chiral center. Also note there is *no* plane of symmetry in this molecule. (An asterisk (*) is often used to indicate a stereocenter.)

DISUBSTITUTED CYCLIC COMPOUNDS WITH TWO STEREOCENTERS

1,4-Disubstituted Cyclohexanes

Does 1,4-dimethylcyclohexane (Fig. 5-20) contain stereocenters? Does it contain a plane of symmetry? Another point to consider is whether the methyl groups are cis or trans to each other. Consider the cis compound first. Look at the planar structure shown in Fig. 5-20. A plane of symmetry (a mirror plane)

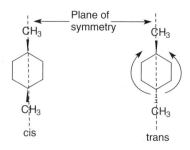

Fig. 5-20.   *cis-* and *trans-*1,4-Dimethylcyclohexane.

can be drawn perpendicular to the ring, bisecting C-1 and C-4 and the attached methyl groups. The molecule has a plane of symmetry and is therefore achiral.

Are C-1 and C-4 stereocenters? Certainly methyl and hydrogen are different groups. Going around the ring to the left or right from C-1 *or* C-4 gives $-CH_2-CH_2-CH(CH_3)-CH_2-CH_2-$, and so C-1 and C-4 are not stereocenters. Since the molecule has no stereocenters, it is achiral, as the mirror plane confirmed.

QUESTION 5-13
Consider *trans*-1,4-dimethylcyclohexane as shown in Fig. 5-20. Is there a plane of symmetry? Are C-1 and C-4 stereocenters? Is the molecule chiral?

ANSWER 5-13
There is a plane of symmetry, as shown in Fig. 5-20. C-1 and C-4 are not stereocenters. The molecule is achiral.

1,2-Disubstituted Cyclohexanes

Now consider 1,2-dimethylcyclohexane. The cis and trans structures are shown in Fig. 5-21. C-1 and C-2 are both stereocenters in the cis and trans molecules. Going around the ring in either direction gives a different sequence of groups. Start at either of the carbon atoms bonded to a methyl ($-CH_3$) group and a hydrogen atom. Going around the ring in one direction gives, initially, a $-CH_2-$ group and going around the ring in the opposite direction gives, initially, a $-CH(CH_3)-$ group. Each carbon atom (C-1 and C-2) is bonded to four different groups and each is a stereocenter.

Does the cis or trans isomer have a plane of symmetry? The cis compound has a plane of symmetry, as shown in Fig. 5-21. A plane perpendicular to the ring bisecting the bonds between C-1 and C-2 and C-4 and C-5 is a mirror plane. So the cis compound is achiral even though it contains two stereocenters. It is a meso compound. There is *no* mirror plane in the trans compound. One side of the mirror plane has the methyl group "above" the ring and the other side of the

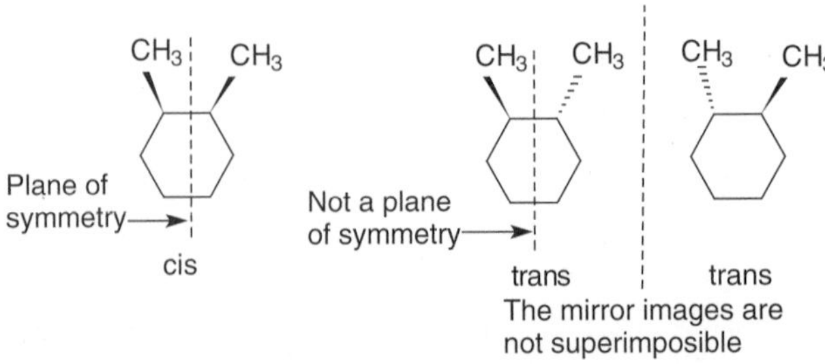

Fig. 5-21.   *cis*- and *trans*-1,2-Dimethylcyclohexane.

mirror has the methyl group "below" the ring. The trans compound is a chiral compound.

How many stereoisomers are possible for 1,2-dimethylcyclohexane? Using the 2^n formula, there is a maximum of four. There are two stereoisomers (enantiomers) for the trans isomer. Only one stereoisomer exists for the meso cis isomer, since its mirror image is superimposable upon itself. There is a total of only three stereoisomers for 1,2-dimethylcyclohexane.

QUESTION 5-14

Draw the cis and trans structures for 1,3-dimethylcyclohexane. How many stereocenters are in the cis isomer and in the trans isomer? Does either have a plane of symmetry? Is either a chiral molecule? How many stereoisomers exist for 1,3-dimethylcyclohexane?

ANSWER 5-14

There are two stereocenters in the cis and two in the trans isomer. The cis isomer has a plane of symmetry and the trans isomer does not. The trans isomer is a chiral molecule. There is a total of three stereoisomers. If this problem is still unclear, reread the discussion for 1,2-dimethylcyclohexane. The explanation is the same.

Naming Cyclic Stereoisomers

The two stereoisomers for *trans*-1,2-dimethylcyclohexane are different compounds and require different names. *R/S* designations are assigned to stereocenters C-1 and C-2 in each isomer and included in their names. Both stereoisomers are shown in Fig. 5-22. In the first structure, the ring is drawn in the plane of the paper. The bonds from carbons C-1 and C-2 to H and CH_3 that are shown "up" (solid wedge) are coming out toward you and those shown "down" (hatched wedge) are going behind the plane of the paper. The two names showing *R/S* designations for C-1 and C-2 are shown in Fig. 5-22.

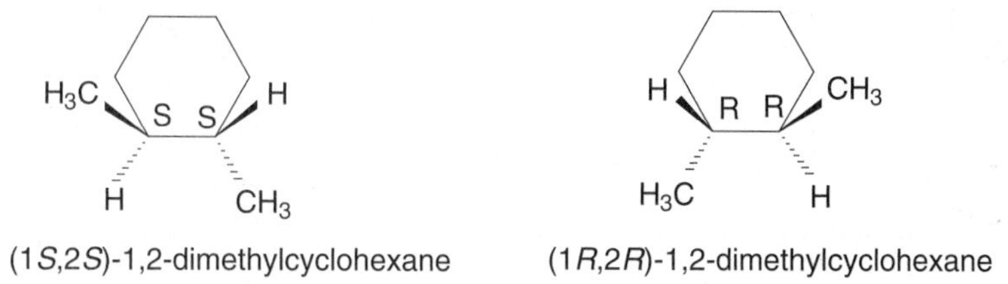

(1*S*,2*S*)-1,2-dimethylcyclohexane   (1*R*,2*R*)-1,2-dimethylcyclohexane

Fig. 5-22. Assigning *R/S* designations to *trans*-1,2-dimethylcyclohexane.

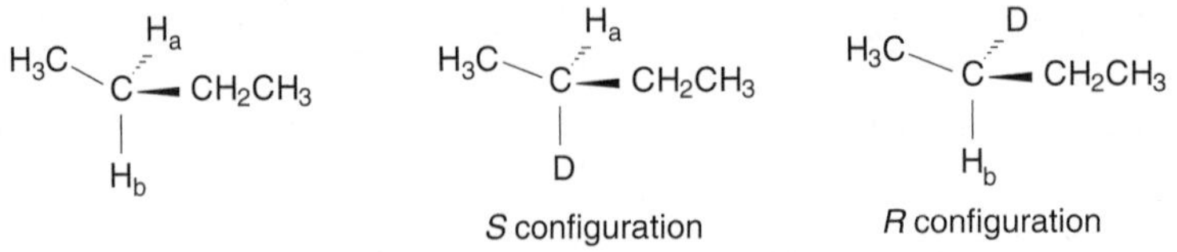

Fig. 5-23. Prochiral atoms.

Prochiral Carbons (Wanabees)

Prochiral carbon atoms are not chiral but could become chiral carbon atoms with appropriate substitution. Consider butane shown in Fig. 5-23. C-2 and C-3 are prochiral. Just consider C-2. It is bonded to two hydrogen atoms, labeled H_a and H_b. These two hydrogen atoms are called *enantiotopic hydrogens*. Replace H_b with a deuterium atom. A deuterium atom (D) is an isotope of the hydrogen atom. Its nucleus contains one proton and one neutron and it has a slightly higher priority than a hydrogen atom in the *R/S* priority system. With this replacement, *C-2 is now chiral*. It was prochiral before the substitution. Compounds that have prochiral atoms (they do not necessarily have to be carbon atoms, although that is all that is discussed here) are called *prochiral compounds*. If D is substituted for H_b, an *S* configuration results and H_b is called pro-*S*. Replacement of H_a with D gives an *R* configuration and H_a is called pro-*R*.

Quiz

1. Isomers are
    (a) identical compounds
    (b) made of many mer units
    (c) molecules that freeze at 0°
    (d) compounds with the same formula and different structures

2. Stereoisomers are
    (a) constitutional isomers
    (b) different compounds with the same connectivity
    (c) conformational isomers
    (d) identical isomers

3. Chiral compounds are
   (a) not superimposable
   (b) superimposable
   (c) have a plane of symmetry
   (d) both (a) and (c)

4. Which one of the following is chiral?
   (a) A hammer
   (b) A candle
   (c) A catcher's mitt
   (d) A screwdriver

5. Stereocenters are also called
   (a) chiral centers
   (b) asymmetric centers
   (c) stereogenic centers
   (d) all of the above

6. The maximum number of isomers of a compound with two stereocenters is
   (a) one
   (b) two
   (c) four
   (d) eight

7. Enantiomers are
   (a) identical compounds
   (b) superimposable mirror images
   (c) nonsuperimposable mirror images
   (d) diastereomers

8. Enantiomers have the same
   (a) melting points
   (b) boiling points
   (c) heats of combustion
   (d) all of the above

9. An example of plane polarized light is
   (a) light from the sun
   (b) light that oscillates in one plane
   (c) light that oscillates in two planes
   (d) light that oscillates in all directions

10.   The optical rotation depends upon
       (a) the concentration
       (b) the temperature
       (c) the wavelength of the light source
       (d) all of the above

11.   The specific rotation of (+)-2-butanol is
       (a) clockwise
       (b) counterclockwise
       (c) neither (a) nor (b)
       (d) both (a) and (b)

12.   A racemic mixture rotates plane-polarized light
       (a) clockwise
       (b) counterclockwise
       (c) in neither direction
       (d) all of the above

13.   Which has the highest priority in the R/S system?
       (a) $-CN$
       (b) $-CH_3$
       (c) $-CH_2Br$
       (d) $-CH_2Cl$

14.   Which combination is a diastereometric pair?
       (a) $2S, 3S$ and $2S, 3S$
       (b) $2S, 3S$ and $2R, 3R$
       (c) $2S, 3S$ and $2S, 3R$
       (d) $2S, 3R$ and $2R, 3S$

15.   In a Fisher projection, the horizontal positions
       (a) are in the plane of the page
       (b) go into the plane of the page
       (c) come out of the plane of the page
       (d) have no special arrangement

16.   A monosubstituted cycloalkane is
       (a) achiral
       (b) chiral
       (c) either chiral or achiral
       (d) both chiral and achiral

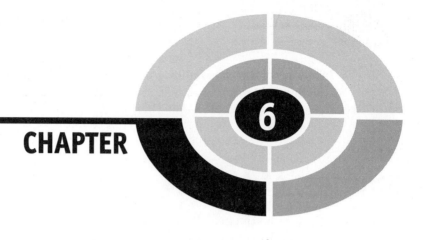

6

Structure and Properties of Alkenes

Introduction to Alkenes

The *alkene family* of hydrocarbons contains one or more carbon-carbon double bonds (C=C). Alkenes are also called olefins. A generic structure would be $R_2C=CR_2$, where R can be a hydrogen atom or any alkyl or aryl (aromatic) group. Many molecules contain more than one functional group but in this chapter only compounds that contain alkene groups will be discussed. The simplest alkene is ethylene ($H_2C=CH_2$). Some ethylene occurs naturally in oil wells but most is produced by thermal degradation of crude oil. Ethylene is converted into a variety of other compounds including polyethylene, polyester, and automobile radiator coolant. Ethylene is a common starting material for many other products. About 5 million pounds are produced annually.

Alkenes are called *unsaturated compounds* since they contain fewer bonded hydrogen atoms than do alkanes, which are called saturated compounds. For example ethylene ($H_2C=CH_2$) contains two less hydrogen atoms than does ethane (H_3C-CH_3).

Structure of Alkenes

Each carbon atom in the double bond of an alkene is sp^2 hybridized. (For a review of hybridization see the section entitled *Hybrid Atomic Orbitals* in Chapter 1.) The three sp^2 hybrid orbitals on a carbon atom are in a common plane and the angle between orbitals is about 120° as shown in Fig. 6-1, Structure 6-1a. One p orbital on each carbon atom remains unhybridized and is perpendicular to the plane of the sp^2 orbitals. Two sp^2 orbitals, one from each carbon atom in the alkene double bond, overlap in a head-to-head manner to form a sigma bond, as shown in Structure 6-1a. The unhybridized p orbitals can overlap in a side-to-side manner (Structure 6-1b) if they are in the same plane. This overlap occurs above and below a line drawn between the nuclei of the two carbon atoms. When

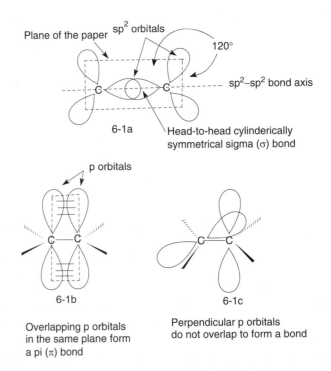

Fig. 6-1.   Bond formation between two sp^2 hybridized carbon atoms.

the p orbitals overlap, they can share a pair of electrons to form a *pi (π) bond*. (Most of the structures in books showing adjacent p orbitals in alkenes imply the p orbitals are too far apart to overlap forming π bonds, but this is just a limitation in the drawings.)

BOND STRENGTH

The carbon-carbon double bond is usually drawn such that the two bonds appear equivalent, e.g., C=C. They are not equivalent. One bond is a σ bond and the other bond is a π bond. The σ bond is stronger than the π bond. In the case of ethylene 268 kJ/mol (64 kcal/mol) of energy is required to break the π bond and 368 kJ/mol (88 kcal/mol) of energy is required to break the sigma bond. The π bond is still quite strong as an ethylene molecule would have to be heated to about 300°C to break a π bond. The sp^2 orbitals overlap to a greater extent forming the σ bond than do the p orbitals forming the π bond and consequently the σ bond is stronger than the π bond. Bond strength is generally directly proportional to the extent of orbital overlap.

BOND ROTATION

Rotation around the C—C σ bond in ethane occurs readily at room temperature. About 12 kJ/mol (3 kcal/mol) is required for this rotation and there is sufficient thermal energy at room temperature for this rotation to occur. The σ bond is cylindrically symmetrical along an axis connecting the two carbon atoms (see the dashed line in Structure 6-1a). Rotation around this axis does not change total orbital overlap or bond strength. The situation is very different for the overlap of p orbitals on adjacent carbon atoms.

   The p orbitals have maximum overlap if they are in the same plane, as shown in Structure 6-1b. If p orbitals on adjacent carbon atoms are perpendicular to each other, as shown in Structure 6-1c, there is no orbital overlap and a bond is not formed. Orbital overlap and subsequent bond formation is energetically favorable. Molecules are more stable (have a lower energy content) when orbital overlap and bond formation occurs. Although the entire ethylene molecule can rotate, rotation does not readily occur around the double bond axis as is the case for single, σ bonds.

QUESTION 6-1
Which molecule undergoes C_1—C_2 bond rotation more easily?

ANSWER 6-1

The energy required for bond rotation, due to steric strain interactions, in the first molecule requires much less than the energy required to break a pi bond, allowing bond rotation around the remaining sigma bond. Therefore the first molecule undergoes C_1–C_2 bond rotation more readily.

Naming Alkenes

Naming alkenes is very similar to naming alkanes. IUPAC rules are used for assigning formal names. Follow along with the structures in Fig. 6-2 as you read each of the following rules for naming compounds.

1. Select the longest *continuous* (parent) carbon atom chain that contains the largest number of double bonds. The longest continuous chain containing the double bond in Structure 6-2a has 10 carbon atoms. The longest continuous carbon atom chain in Structure 6-2b that contains both double bonds has four carbon atoms. Note this is not the longest continuous carbon atom chain in this molecule, but it does contain all the double bonds.

2. Name the longest chain (that contains the double bond(s) as you would an alkane, but change the ending from –ane to –ene. Structure 6-2a is a decene and Structure 6-2b is a butene. If there is more than one double bond, you need to state how many double bonds are present. Structure

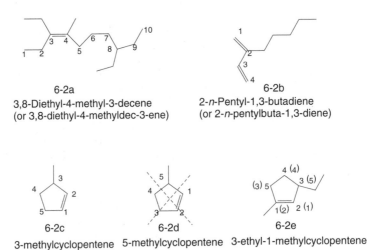

6-2a
3,8-Diethyl-4-methyl-3-decene
(or 3,8-diethyl-4-methyldec-3-ene)

6-2b
2-*n*-Pentyl-1,3-butadiene
(or 2-*n*-pentylbuta-1,3-diene)

6-2c
3-methylcyclopentene

6-2d
5-methylcyclopentene

6-2e
3-ethyl-1-methylcyclopentene

Fig. 6-2. Naming alkenes.

6-2b has two double bonds and is called a *diene*. An "a" prefix is added to diene (adiene) to help pronunciation (butadiene vs. butdiene). Structure 6-2b is a butadiene.

3. Number the carbon atoms in the parent chain starting at the end that gives the first carbon atom in the double bond the lowest number. The carbon atom chains are numbered in the examples given in Fig. 6-2.

4. State the position of the double bond. You need only number the first carbon atom of a double bond since the second carbon atom will necessarily be the next higher number. The first carbon atom in the double bond in Structure 6-2a is carbon atom number 3 (C-3). The first carbon atoms in each of the double bonds in Structure 6-2b are C-1 and C-3.

5. Name substituents alphabetically. Precede the substituent with the number of the carbon atom in the parent chain to which the substituent is bonded. Structure 6-2a has an ethyl substituent on carbon atoms C-3 and C-8 and a methyl group on C-4. Structure 6-2b has an *n*-pentyl (or just pentyl) substituent on carbon atom C-2. The parent chain is numbered so the substituents have the lowest possible numbers. The names for Structures 6-2a and 6-2b are given in Fig. 6-2.

6. Cyclic compounds are named using similar rules. Count the number of carbon atoms in the ring and name it as a cycloalkane but change the suffix to -ene. Structures 6-2c and 6-2e, five-membered rings with one double bond, are cyclopentenes.

7. The location of the double bond in a ring does not have to be specified if there is only one double bond, since one of the carbon atoms in the double bond will necessarily be C-1.

8. For rings with one double bond, number the carbon atoms in the ring in the direction that gives the lowest combination of numbers to the substituents. Then alphabetize the substituents in the name. Numbering the ring in Structure 6-2c places the substituent on C-3 while the numbering sequence in Structure 6-2d places the substituent on C-5. Structure 6-2c has the correct numbering scheme since the substituent has a lower number. The substituents in Structure 2-6e have the lowest combination of numbers when bonded to atoms C-1 and C-3. Numbering the ring in the reverse order would put the substituents on C-2 and C-5. (1 and 3 is a lower combination than 2 and 5.) The names for the cyclic compounds are given in Fig. 6-2.

9. A newer naming system for alkenes proposed by IUPAC places the position number for the double bond before the -ene ending rather than as a prefix. 2-Butene would become but-2-ene. Most texts and articles use the older system but you should be familiar with both naming systems.

Table 6-1. Common and formal (IUPAC) names.

Common	IUPAC	Structure
Ethylene	Ethene	$CH_2{=}CH_2$
Propylene	Propene	$CH_3CH{=}CH_2$
Isobutylene	2-Methylpropene	$(CH_3)_2C{=}CH_2$
Vinyl	Ethenyl	$-CH{=}CH_2$
Allyl	2-Propenyl	$-CH_2CH{=}CH_2$

QUESTION 6-2

Why isn't Structure 6-2e in Fig. 6-2 named 1-methyl-3-ethylcyclopentene? Another possible name is 5-ethyl-2-methylcyclopentene. Why are these names incorrect?

ANSWER 6-2

In the first name the substituents are not listed alphabetically. In the second name the substituents do not have the lowest possible position numbers.

Common Names

Common names are used for many of the lower molecular weight alkenes. A few examples of common names are given in Table 6-1. Using IUPAC rules, ethylene would be called ethene. Common names are accepted by the IUPAC system and both naming systems can be used. This text will use the common names that are most often found in the literature.

Cis and Trans Isomers

Alkenes with two different groups on *each carbon* atom of a double bond can exist as diastereomers, also called *geometric isomers*. In Fig. 6-3, each carbon

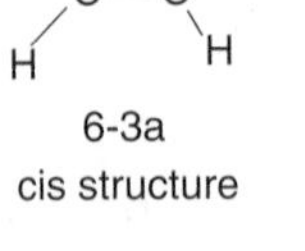

6-3a   6-3b

cis structure   trans structure

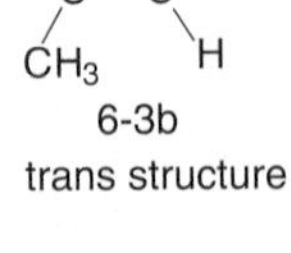

Fig. 6-3. Cis and trans structures.

atom is bonded to a methyl group and a hydrogen atom. Structure 6-3a has both methyl groups (and both hydrogen atoms) on the "same side" of the molecule and Structure 6-3b has the methyl groups on "opposite sides" of the molecule. Since rotation around double bonds does not readily occur, molecules 6-3a and 6-3b are two different compounds. A *cis* structure results when two identical (or similar) groups are on the "same side" of a double bond. A *trans* structure results when the identical groups are on "opposite sides" of a double bond. The cis/trans nomenclature works quite well when an alkene is *disubstituted*, that is when *each* carbon atom in a double bond has one hydrogen atom and one group other than a hydrogen atom bonded to it.

The *E/Z* (Easy) System

Naming alkenes using the cis and trans nomenclature can be ambiguous if an alkene is *tri-* or *tetrasubstituted*. One has to determine which two atoms or groups to compare as being cis or trans to each other. To avoid confusion, the *E/Z system* is used. *E* is derived from the German word *entgegen*, meaning opposite, and *Z* is derived from the German word *zusammen*, meaning together or on the same side. Another way to remember the definition of *Z* is to say "*za zame zide*" (the same side). Disubstituted cis isomers are *Z* isomers and disubstituted trans isomers are *E* isomers. The *E/Z* system can also be used for disubstituted alkenes.

ASSIGNING PRIORITIES

The key is to determine which atoms or groups to compare as being on the same or opposite side for assigning E or Z designations. First prioritize the atoms or groups bonded to each alkene carbon atom. The Cahn-Ingold-Prelog

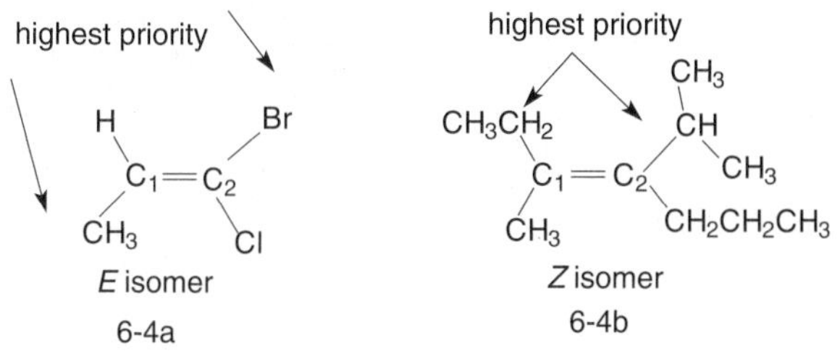

Fig. 6-4. *E/Z* isomers.

(C-I-P) system is used for prioritizing. (For a review of the C-I-P system see the section entitled *The R/S System* in Chapter 5.) Consider the two alkenes in Fig. 6-4. In Structure 6-4a, alkene carbon atom C-1 is bonded to a methyl group and a hydrogen atom. The methyl group has a higher priority than a hydrogen atom (carbon has a higher atomic number than does hydrogen). Alkene carbon atom C-2 is bonded to a chlorine atom and a bromine atom. Bromine has a higher priority since it has a higher atomic number than does chlorine. Use the highest priority atom/group on *each* carbon atom, methyl and bromine in this example, to determine *E* or *Z*. Since the methyl group and the bromine atom are on opposite sides of the double bond, the molecule has an *E* (opposite side) configuration.

In Structure 6-4b, the ethyl group on C-1 has a higher priority than the methyl group and the isopropyl group on C-2 has a higher priority than the *n*-propyl group. Since the ethyl and isopropyl groups have the highest priorities and are on the same side of the molecule, Structure 6-4b has a *Z* (same side) configuration.

QUESTION 6-3

Does the following have an *E* or *Z* configuration?

ANSWER 6-3

Z, as $CH_3CH_2 > CH_3$ and $Br > OH$.

Degrees of Unsaturation

Acyclic *monoalkenes* are unsaturated compounds that have the generic formula C_nH_{2n}. *Monocycloalkanes* have the same generic formula, C_nH_{2n}. Compounds to which pairs of hydrogen atoms can be added to give a saturated *acyclic* (*noncyclic*) alkane are said to have a *degree(s) of unsaturation*. The term degree of unsaturation has a different meaning than the term *unsaturated compound*. A monocycloalkane has one degree of unsaturation but is considered a saturated cyclic molecule. Synonymous terms for degrees of unsaturation are *elements of unsaturation* and *index of hydrogen deficiency*.

One degree of unsaturation is defined as the absence of two hydrogen atoms relative to an acyclic saturated alkane with the same number of carbon atoms. The compound C_2H_4 ($H_2C{=}CH_2$) has one degree of unsaturation since adding two hydrogen atoms would give C_2H_6 (H_3C-CH_3), an acyclic saturated alkane. Acyclic alkanes have the generic formula C_nH_{2n+2}.

CALCULATING DEGREES OF UNSATURATION

To determine the degrees of unsaturation, one determines how many *pairs* of hydrogen atoms need to be added to the compound under study to give the formula of a saturated acyclic alkane. Consider the formula C_3H_6. The formula for a saturated acyclic three-carbon atom molecule is C_3H_8. The compound under study is deficient of two hydrogen atoms; that is, it has one degree of unsaturation:

$$C_3H_8 - C_3H_6 = 2H, \text{ or one degree of unsaturation.}$$

QUESTION 6-4
How many degrees of unsaturation does C_4H_8 have?

ANSWER 6-4
One.

MULTIPLE BONDS AND RINGS

Consider the compound C_4H_6. The saturated acyclic compound would have the formula C_4H_{10} (C_nH_{2n+2}). Four hydrogen atoms would need to be added to C_4H_6 to give C_4H_{10}, so there are two degrees of unsaturation. Degrees of unsaturation can result from double bonds (one degree of unsaturation for each double bond), triple bonds (two degrees of unsaturation for each triple bond, $-C{\equiv}C-$), and rings (one degree of unsaturation for each ring).
   The possible structures can be summarized as follows:

One degree of unsatuation = one double bond or one ring

Two degrees of unsaturation = two double bonds, two rings,
one triple bond, or one double
bond and one ring

The compound C_4H_6, with two degrees of unsaturation, could have two double bonds, one triple bond, two rings, or one double bond and one ring. There are nine possible structures, as discussed in Question 6.5.

QUESTION 6-5
Draw the possible structures for C_4H_6.

ANSWER 6-5

There are nine possible structures:

EMPIRICAL FORMULAS

Why is it important to determine the degrees of unsaturation? Assume a compound produced by a plant in a rain forest of South America is shown to be an excellent drug for treating cancer. The structure of the compound is initially unknown, but needs to be determined so that it can be made more economically in the laboratory. There are laboratory techniques that can be used to determine the molar ratio of the elements in the compound. Assume in this case the C:H:O molar ratio is 4:6:1, corresponding to C_4H_6O. This is called the *empirical formula*. The *molecular formula* may be C_4H_6O or some multiple of C_4H_6O, such as $C_8H_{12}O_2$, $C_{12}H_{18}O_3$, etc. An instrument, called a *mass spectrometer* (discussed in Chapter 10), can be used to determine the molecular weight and hence the molecular formula. The number of degrees of unsaturation can be determined from the molecular formula and subsequently the possible number of double bonds, triple bonds, and/or rings. Often several pieces of information are needed from various tests to determine the structure of some unknown compound.

HETEROATOMS

Additional steps are needed to determine the degrees of unsaturation if the compound contains halogen (F, Cl, Br, or I), oxygen, or nitrogen atoms.

Halogen atoms

If the compound contains halogen atoms, one substitutes one hydrogen atom for each halogen atom to get an *equivalent formula*. Consider the compound C_2H_3Cl (with the condensed structure $H_2C=CHCl$). The compound has one double bond and, therefore, one degree of unsaturation. Substituting an H for the Cl in C_2H_3Cl gives the equivalent formula C_2H_4 (with the condensed structure $H_2C=CH_2$). Making this substitution does not change the number of degrees of unsaturation. Now compare the equivalent formula to the formula of a saturated compound, C_2H_6 in this case, to determine the number of degrees of unsaturation. The compound is deficient in two hydrogen atoms and has one degree of unsaturation (as confirmed by the condensed structure $H_2C\!=\!CHCl$).

QUESTION 6-6
How many degrees of unsaturation are in C_5H_4BrCl?

ANSWER 6-6
Three. Replace the two halogen atoms with two hydrogen atoms to give the equivalent formula C_5H_6. Compare this with the saturated acyclic alkane C_5H_{12}. The unknown compound is deficient in six hydrogen atoms (three pairs) and has three degrees of unsaturation.

Oxygen atoms

If the unknown compound contains oxygen atoms, simply ignore the number of oxygen atoms to get the equivalent formula. The compound C_2H_6O (with the condensed structure CH_3CH_2OH) is a saturated compound. It has no rings or multiple (double or triple) bonds. Dropping (ignoring) the oxygen atom from the formula gives the equivalent formula C_2H_6. This is the formula of a saturated acyclic alkane with no degrees of unsaturation. This is consistent with the condensed structure (CH_3CH_2OH) which has no degrees of unsaturation.

Nitrogen atoms

If the compound contains nitrogen atoms, drop the nitrogen atoms and subtract one hydrogen atom for each nitrogen atom dropped. Consider the compound C_3H_7N with the condensed structure $H_2C\!=\!CHCH_2NH_2$. Dropping the nitrogen atom and subtracting one hydrogen atom gives the equivalent formula C_3H_6. A three-carbon atom saturated acyclic hydrocarbon has the formula C_3H_8. The unknown compound has one degree of unsaturation ($C_3H_8 - C_3H_6 = 2\,H$). This is consistent with the condensed structure ($H_2C\!=\!CHCH_2NH_2$) that contains one double bond and one degree of unsaturation.

In summary, to determine the equivalent structure of a compound that contains a halogen, oxygen, or nitrogen atom: substitute a hydrogen atom for each halogen atom, ignore all oxygen atoms, and subtract a hydrogen atom for each nitrogen atom deleted.

QUESTION 6-7
How many degrees of unsaturation are in each of the following molecules: C_4H_9Cl, C_3H_6O, C_5H_9N, and C_4H_8NOCl?

ANSWER 6-7
Zero, one, two, and one.

Stability of Alkenes

Compounds have stored energy in the form of potential energy. If a material is burned, heat (energy) is given off. This energy was inherent in the compound and released upon combustion. If a series of hydrocarbon alkenes containing the same number of carbon atoms is burned, the total amount of heat given off per mole of hydrocarbon burned can be determined and compared. The end products of combustion, CO_2 and H_2O, are the same for each compound. The amount of heat given off gives information about the potential energy stored in the alkene molecules and hence their relative stability. The lower the potential energy of the initial molecule, the more stable is that molecule. Figure 6-5 shows the relative stability of a series of alkenes determined by combustion studies. In general, *the most highly alkyl-substituted alkenes are the most stable*. The order of *increasing* stability is: monosubstituted alkenes < cis-disubstituted alkenes < trans-disubstituted alkenes < trisubstituted alkenes < tetrasubstituted alkenes (the most stable).

HYPERCONJUGATION

Why are more highly substituted alkenes more stable? One explanation involves *hyperconjugation*: hyper meaning above/beyond and conjugation meaning getting together. Figure 6-6 shows the overlapping of the sp^3-s orbitals of a C—H bond with an empty antibonding π orbital of an adjacent alkene carbon atom. (Antibonding orbitals are explained in Ch. 16.) This overlapping of orbitals and sharing of the C—H bonding electrons, called hyperconjugation, increases the stability of the molecule. Highly alkyl substituted alkenes have more opportunities to undergo hyperconjugation and therefore have increased stablility.

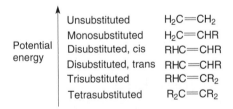

Potential energy	Unsubstituted	H_2C=CH_2
	Monosubstituted	H_2C=CHR
	Disubstituted, cis	RHC=CHR
	Disubstituted, trans	RHC=CHR
	Trisubstituted	RHC=CR_2
	Tetrasubstituted	R_2C=CR_2

Fig. 6-5. Relative stability of substituted alkenes.

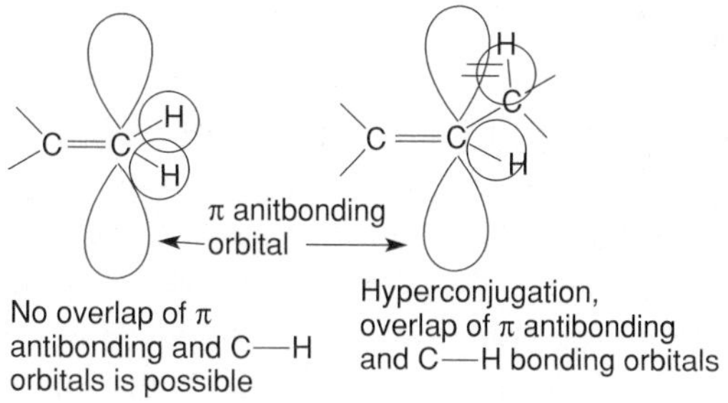

π anitbonding
◄— orbital —►

No overlap of π
antibonding and C—H
orbitals is possible

Hyperconjugation,
overlap of π antibonding
and C—H bonding orbitals

Fig. 6-6. Hyperconjugation stabilization.

Physical Properties

Physical properties of alkenes are similar to those of alkanes. Alkenes are less dense than water, their boiling points increase with increasing molecular weight, they are relatively nonpolar, and they are insoluble in water. Alkenes with less than five carbon atoms are gases at room temperature.

Chemical Properties

Alkenes are very different from alkanes in that alkenes readily undergo many chemical reactions. Alkanes are relatively inert. Alkenes have π electrons they are willing to share. Reagents that are electrophilic (electron loving), such as a proton (H^+), tend to react with the π electrons in a double bond. Figure 6-7 shows this reaction. This is also a Lewis acid-base reaction. Alkenes (nucleophiles) are electron-pair donors and protons (electrophiles) are electron-pair acceptors. These reactions are usually called electrophilic addition reactions rather than acid-base reactions.

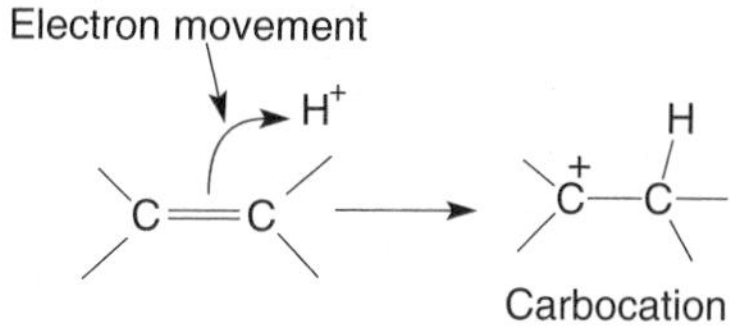

Electron movement

H^+

Carbocation

Fig. 6-7. Electrophilic addition to alkenes.

The Curved Arrow

Reactions take place when molecules randomly bump into each other. If the molecules have sufficient energy and the correct orientation when they collide, a reaction can occur. When the mechanism for these reactions is diagrammed, electrons will *always* be shown moving toward the electrophile. The electrophile is never shown attacking the electrons. Electron movement, the reaction mechanism, is shown in Fig. 6-7 by a *curved arrow*. Electrons move from the tail of the arrow toward the head of the arrow. The tail of the arrow starts at the electrons in the double bond, not at the atoms in the double bond.

A neutral alkene reacts with a positively charged electrophile to give a positively charged carbocation, as shown in Fig. 6-7. To which carbon atom in the double bond does the electrophile become bonded? How fast is the reaction? How much product is formed? These questions will be answered in the following chapters. Since reaction mechanisms are quite similar for the various electrophilic addition reactions, it is necessary to discuss reactions in general (Chapter 7) before discussing specific reactions.

Quiz

1. Alkene compounds
    (a) are saturated
    (b) are unsaturated
    (c) contain all single bonds
    (d) contain all triple bonds

2. Carbon atoms in a double bond are
    (a) sp hybridized
    (b) sp^2 hybridized
    (c) sp^3 hybridized
    (d) not hybridized

3. A double bond in an alkene consists of
    (a) two sigma bonds
    (b) two pi bonds
    (c) one sigma bond and one pi bond
    (d) one sigma bond and two pi bonds

4. Alkene names end in
    (a) -ane

(b) -ene

(c) -yne

5. Rotation readily occurs around C–C axis in double bonds.
   (a) True
   (b) False
   (c) Unfair question

6. 2-Pentene and pent-2-ene are
   (a) the same compound
   (b) different compounds
   (c) isomers
   (d) cyclic compounds

7. Cis geometric isomers have similar groups on
   (a) the same side
   (b) opposite sides
   (c) neither side

8. The *E/Z* system is used to
   (a) indicate the position of the double bond
   (b) indicate structure based on group priority
   (c) define the stability of an alkene
   (d) confuse students

9. The compound C_4H_6 has ____ degree(s) of unsaturation.
   (a) one
   (b) two
   (c) three
   (d) four

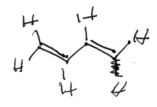

10. The compound C_3H_6NOCl has ____ degree(s) of unsaturation.
    (a) one
    (b) two
    (c) three
    (d) four

11. A trisubstituted alkene is ____ a monosubstituted alkene.
    (a) more stable than
    (b) less stable than
    (c) equal in stability to

12. Hyperconjugation ____ the stability of a compound.
    (a) increases

(b) decreases

(c) has no effect on

13. The curved arrow shows electron movement

(a) toward a nucleophile

(b) toward an electrophile

(c) away from an electrophile

(d) in either direction

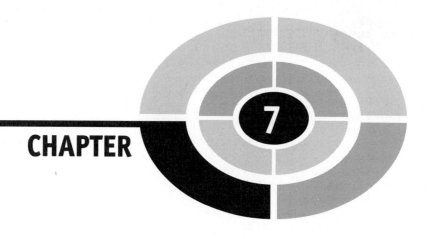

CHAPTER 7

Reaction Mechanisms

Introduction

Many different types of chemical reactions are discussed in this text. Which reactions go to completion? How fast are the reactions? How does one predict the stereochemistry of the intermediates and products? These are some questions that will be answered in this chapter.

Thermodynamics

Reactions can be discussed in terms of thermodynamics and kinetics. *Thermodynamics* is the study of energy and its transformation. *Kinetics* deals with reaction rates, that is, how fast reactions occur. Consider three reactions that proceed until *equilibrium* is achieved. Figure 7-1 shows the energy profiles of reactant A going to product B via some *reaction pathway*. The reaction pathway is the mechanism by which A is converted into B. In Graph 7-1a, product B has

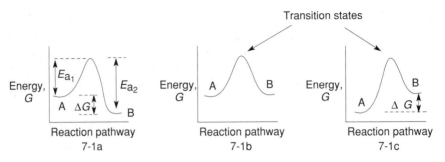

Fig. 7-1.   Reaction energy profiles.

a lower energy than starting material A. In Graph 7-1b, A and B are equal in energy. In Graph 7-1c, B has a higher energy than A.

If only A is present initially in the three examples in Fig. 7-1, will any reaction take place in each case? Assume the reaction continues until equilibrium is achieved. *At equilibrium*, the concentrations of A and B remain constant, but not necessarily equal. The system is dynamic; that is, A continues to be converted to B and B continues to be converted to A. These forward and reverse (backward) reactions take place at the same rate and therefore the concentrations of A and B remain constant. Thermodynamics tells *how much* reactant and product is obtained at equilibrium, but *not how fast* equilibrium is achieved.

EXERGONIC/ENDERGONIC REACTIONS

Graph 7-1a shows an *exergonic* (energy is given off) reaction for A being converted into B. At equilibrium, one would expect to have more B than A since B is more stable (it is lower in energy). In Graph 7-1b, A and B are of equal energy and one would expect equal amounts of both when equilibrium is reached. Graph 7-1c represents an *endergonic* (energy is absorbed) reaction in going from A to B. In this reaction one would expect more A (the lower energy product) than B at equilibrium. We can determine how much A and B are present when equilibrium is reached if we know the energy difference between the reactants and the products.

You are probably more familiar with the terms exothermic and endothermic that refer to heat energy H. Exergonic and endergonic refer to *Gibbs free energy* G. These quantities are related by the equation $\Delta G = \Delta H - T \Delta S$. H is enthalpy T is temperature in kelvins, and S is entropy.

QUESTION 7-1
Sketch the energy-reaction pathway diagram of reactants A + B going to product
C, where A and B are at a higher energy than is C.

ANSWER 7-1
The graph will look the same as Graph 7-1a in Fig. 7-1, where A and B are the
reactants and C is the product.

GIBBS FREE ENERGY

It is difficult to determine the absolute energy of a system, but it is relatively easy
to determine the *change* in Gibbs free energy, ΔG, in going from reactants to
products. The equation $\Delta G° = -2.3RT \log K_{eq}$ shows the relationship between
$\Delta G°$ and K_{eq}. The "°" in $G°$ refers to products and reactants in their standard
states: 1 atm for gases and 1 molar for solutions. R is a constant, 8.32×10^{-3} kJ/
mol-K (1.99×10^{-3} kcal/mol-K), T is the temperature in kelvins, and K_{eq} is the
equilibrium expression. The equilibrium expressions for A going to B for two
different reactions are shown in Fig. 7.2.

   If the differences in the Gibbs free energy between the products and the reactants can be measured, one can calculate K_{eq} and the ratio of the concentrations
of reactants and products present when equilibrium is achieved.

Optimizing yields

A reaction is usually run with the goal of obtaining as much product as possible.
How can the product yield be increased, especially in an endergonic reaction? If
the product is removed as it is formed, the equilibrium will shift to produce more
product. Two ways of removing the product are by precipitating or distilling the
product from the reaction mixture as it is formed.

$$A \rightleftharpoons B \qquad K_{eq} = \frac{[B]}{[A]}$$

$$A \rightleftharpoons 2B \qquad K_{eq} = \frac{[B]^2}{[A]}$$

Fig. 7-2.   Equilibrium expressions.

Kinetics

How rapidly is equilibrium obtained? This involves the study of kinetics. Look at Graph 7-1a in Fig. 7-1. Reactant A follows a reaction pathway where it climbs an "energy hill" as it is being converted to product B. The *activation energy*, E_a, is the minimum energy needed for starting material(s) to get over the "energy hill." Graph 7-1a shows the activation energies: E_{a_1} for A going to B and E_{a_2} for B going back to A. Since E_{a_1} is smaller than E_{a_2}, it takes less energy for A to go to B than for B to go to A and one would expect to have more B than A when equilibrium is reached.

THE TRANSITION STATE

The top of the activation energy hill in all three graphs in Fig. 7-1 represents the transition of A being converted into B, and is called the *transition state* (TS). The TS usually represents a species where bonds are being broken and/or bonds are being formed. These species are very unstable, have very short lifetimes, and cannot be isolated.

QUESTION 7-2
Is the activation energy E_a for a reaction the difference between the energy of the starting materials and the products?

ANSWER 7-2
No, it is the difference between the energy of the starting material(s) and the transition state leading to the products.

REACTIVE INTERMEDIATE

Figure 7-3 is an energy diagram that differs from Fig. 7-1 in that there is dip in the curve for A going to B. This dip represents an intermediate species C. This species is usually quite reactive (it is much higher in energy than is the starting material or the product) and is called a *reactive intermediate*. These species can be observed with special instruments but they are usually too unstable to be isolated. Since reactive intermediates are lower in energy than are the transition states leading to them, they are somewhat more stable than the two transition states TS-1 and TS-2.

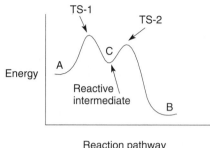

Fig. 7-3.   Reaction energy profile with an intermediate.

REACTION RATES

The rate of disappearance of starting materials or the rate of appearance of products can be measured to determine how fast a reaction proceeds. Rates are usually expressed as the change in concentration of the starting material per unit time. Equation 7-4a in Figure 7-4 shows the general rate law for a reaction. The *rate of reaction*, R, is usually expressed in moles/second. R is equal to a rate constant, k, times a *concentration* function, $[\ ]^x$. The concentration of the starting material(s) is measured at the beginning of the reaction, time 0, or t_0, and at the end of some time period, t_1. The change in concentration, $[\ ]$ at $t_1 - [\ ]$ at t_0, (or $\Delta[\ ]/\Delta t$) is a negative number, so a negative sign is used in Eq. 7-4a and the rate of reaction, R, is expressed as a positive number.

THE RATE CONSTANT

The *rate constant 'k'* is not always a constant. Equation 7-4b shows that the value of k depends on the temperature T and activation energy E_a. If the temperature increases, k increases. This is expected since heating a reaction usually speeds up the reaction. A reaction is also faster if the activation energy (the energy hill, E_a) is lowered. Decreasing E_a increases k. Catalysts are used to lower the activation energy. So "constant" k varies with temperature and activation energy.

(a) Rate $= -k\,[\text{ starting materials}]^x$

  Note negative sign

(b) $k = e^{-E_a/RT}$

Fig. 7-4.   Reaction rate expression.

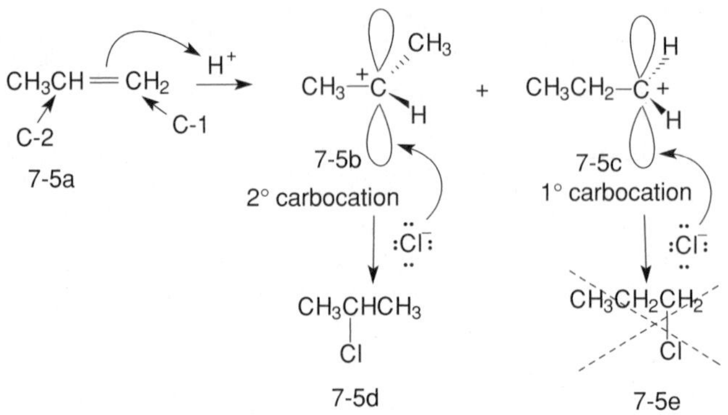

Fig. 7-5. Addition of HCl to propylene.

Carbocations

The addition of HCl to propylene is shown in Fig. 7-5. The electron-rich propylene molecule (Structure 7-5a) donates its π electrons to the electron pair-accepting electrophile H^+. The resulting species (Structure 7-5b) has a positive charge on a carbon atom and is called a *carbocation*. Carbocations tend to be very reactive and not very stable. Carbocations are often intermediates in chemical reactions and are called *reactive intermediates*. For this particular reaction, the dip in the energy curve in Fig. 7-3 represents the carbocation, 7-5b.

ELECTROPHILIC ADDITION

Carbocation 7-5b is electrophilic. It is electron deficient and seeks an electron pair. The chloride anion has nonbonding electron pairs it is willing to share with the carbocation to give Structure 7-5d. The overall reaction is the addition of one molecule of HCl to one molecule of propylene.

QUESTION 7-3
Do carbocations share their electrons readily with nucleophiles?

ANSWER 7-3
No, nucleophiles share their electrons with electrophilic carbocations.

Stereochemistry

In Fig. 7-5 the proton (H^+) can bond to either the terminal carbon atom (C-1) or the internal carbon atom (C-2) of the double bond. These two intermediate carbocations are shown as Structures 7-5b and 7-5c. Is there a preference for addition of H^+ to either carbon? If the proton bonds to C-1, the resulting carbocation is called a secondary (2°) carbocation (7-5b). A secondary carbocation has two carbon atoms bonded to the carbon atom bearing the positive charge. If the proton bonds to C-2, a primary (1°) carbocation (7-5c) results. A primary carbocation has only one carbon atom bonded to the carbon atom with the positive charge. The other two bonds in a 1° carbocation are to hydrogen atoms. There is a difference in stability of 1° and 2° carbocations.

CARBOCATION STABILITY

A series of studies have shown the order of stability of alkyl-substituted carbocations to be 3° > 2° > 1°. A tertiary (3°) carbocation has three different carbon atoms bonded to the carbon atom bearing the positive charge. Figure 7-6 gives the structures of a 3°(Structure 7-6a), 2°(Structure 7-6b), and 1°(Structure 7-6c) carbocation. One carbon atom of the carbocation is sp^2 hybridized and contains one unhybridized p orbital. This p orbital contains no electrons. The sp^2 hybridized carbon atom has a formal positive charge. (For a review see the section entitled *Formal Charges* in Chapter 1.)

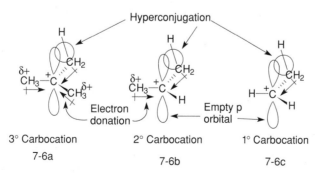

Fig. 7-6. Stabilization of carbocations.

INDUCTIVE EFFECT

There are two ways that a carbocation is stabilized. Ionic species are more stable if the *charge can be delocalized* (spread out) throughout the molecule. Alkyl groups (such as —CH_3) are electron donating. They donate electrons through a single, σ bond. The donation or withdrawal of electrons through a σ bond is called an *inductive effect*. This is shown in each structure in Fig. 7-6 by an arrow indicating the direction of electron movement. The tail of the arrow has a cross (+) indicating a partial positive charge. A delta plus (δ^+) is also used to indicate a partial positive charge.

By donating electrons to the electron-deficient sp^2 hybridized carbon atom, the positive charge is delocalized. The full positive charge on the sp^2 hybridized carbon atom is decreased somewhat and a small amount of positive charge is transferred to the alkyl groups, increasing the stability of the carbocation species. Structures 7-6a and 7-6b show the stabilizing effect of electron-donating alkyl groups. The 3° carbocation (7-6a) has three methyl groups that donate electrons. The 2° carbocation has two methyl groups that donate electrons. The 1° carbocation has only one methyl group that donates electrons. The greater the number of electron-donating groups, the greater the stability of the carbocation. This partially explains the order of stability of carbocations.

HYPERCONJUGATION

The carbocation is also stabilized by *hyperconjugation*. The sp^3-s orbitals containing the bonding electrons in the C—H bond adjacent to the sp^2 hybridized carbon atom can overlap with the unhybridized, empty p orbital on the sp^2 hybridized carbon atom and share the bonding electrons. Sharing electrons in this manner is called hyperconjugation. Each structure in Fig. 7-6 shows this orbital overlap. The 3° carbocation in Structure 7-6a has nine opportunities (there are nine adjacent C—H bonds) for hyperconjugation while a 1° carbocation (7-6c) has only three opportunities (there are three adjacent C—H bonds) for hyperconjugation. Each structure in Fig. 7-6 shows only one of the possible hyperconjugation interactions. The 1° carbocation is shown for comparison purposes although it is too unstable to be formed in typical solvent-based chemical reactions.

The Hammond Postulate

The *Hammond postulate states that the transition state resembles the structure (and energy) of the reactant or product that is closest to it in energy.* Look at transition state TS-1 in Fig. 7-7. Starting material A goes to "product" B, which

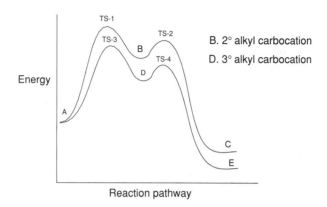

Fig. 7-7.   Stability of carbocations and transition states.

in this case is a reactive intermediate. Reactant B (the reactive intermediate) goes through transition state TS-2 to become product C. In the first part of the reaction, TS-1 is closer in energy to B than it is to A, and therefore TS-1 resembles (looks like) B more than it resembles A. Transition state TS-2 is closer in energy to B than C, and TS-2 looks more like B than C. Since TS-1 and TS-2 are similar in structure to B, *factors that stabilize B also stabilize TS-1 and TS-2.*

Figure 7-8 shows a reaction where the π bond in an alkene (7-8a) is broken and a C–H bond is formed, giving reactive intermediate 7-8c. In transition state 7-8b, the π bond is partially broken and the C–H σ bond is partially formed. Assume this reaction (7-8a going to 7-8c) is represented by the top curve in Fig. 7-7. Transition state TS-1 (7-8b) would look more like carbocation 7-8c than the starting alkene (7-8a) since 7-8b is closer in energy to 7-8c than it is to 7-8a. The π bond is completely broken and the new C–H bond is completely formed in 7-8c. If 7-8b looks more like 7-8c than 7-8a, 7-8b has a new C–H bond that is more than half formed and a π bond that is more than half broken.

Fig. 7-8.   Reaction pathway showing a transition state.

COMPETING REACTION PATHWAYS

If the order of stability of carbocations is $3° > 2°$ and the transition state is similar in structure to its corresponding carbocation, then the stability of the transition state TS-3 (going to the $3°$ carbocation) is greater than TS-1 (going to the $2°$ carbocation). Figure 7-7 shows the relative stability of transition states and their corresponding $2°$ and $3°$ carbocations. Reaction kinetics predict that a reaction will proceed via the lowest energy pathway. The reaction will prefer the pathway that goes over the lowest energy hill. If, as shown in Fig. 7-7, a reaction has the choice of two pathways to form a $2°$ or $3°$ carbocation, the $3°$ carbocation pathway will be preferred since it has the lowest energy hill to climb.

Regiochemical Reactions

The magnitude of the activation energies explains why the addition of HCl to propylene (shown in Fig. 7-5) leads primarily to one product, 2-chloropropane. Protonation of C-1 results in a $2°$ carbocation (7-5b) while protonation of C-2 leads to a $1°$ carbocation (7-5c). Reaction of the $2°$ carbocation with chloride anion gives 2-chloropropane (7-5d) while the reaction with the $1°$ carbocation would give 1-chloropropane (7-5e), which is not formed in this reaction. A reaction where two or more constitutional isomers can potentially be formed (as in the reactions in Figs. 7-5 and 7-9) is called a *regiochemical (ree gee oh chemical)* reaction. If one of the products predominates, the reaction is called *regioselective*. If one of the products is formed exclusively, the reaction is called *regiospecific*.

QUESTION 7-4
What is formed by the addition of electrophile E^+ to propene, $CH_3CH=CH_2$?

ANSWER 7-4
$CH_3\overset{+}{C}HCH_2E$, a reactive $2°$ carbocation.

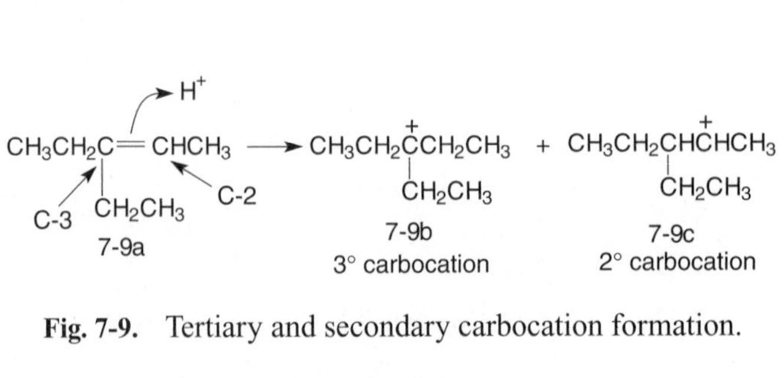

Fig. 7-9.   Tertiary and secondary carbocation formation.

The Markovnikov Rule

Figure 7-9 shows the reaction of HCl with 3-ethyl-2-pentene (7-9a). Addition of a proton to C-2 gives a 3° carbocation (7-9b) while addition of a proton to C-3 gives a 2° carbocation (7-8c). Since the energy hill for formation of a 3° carbocation is lower than the energy hill for formation of a 2° carbocation, the 3° carbocation is preferentially formed. The carbocations formed in the reactions in Figs. 7-5 and 7-9 result from the proton bonding to the carbon atom in the double bond that has the most hydrogen atoms bonded to it (before the reaction).

In Fig. 7-5, the proton preferentially bonds to the carbon atom that is already bonded to two hydrogen atoms rather than bonding to the carbon atom that is bonded to only one hydrogen atom. In Fig. 7-9, the proton becomes bonded primarily to the carbon atom that already is bonded to one hydrogen atom rather than bonding to the carbon atom bonded to no hydrogen atoms. This is known as the *Markovnikov rule* (or *Markownikoff rule*): *in electrophilic addition reactions a proton (the electrophile) will bond to the carbon atom in a double bond that already is bonded to the greater number of hydrogen atoms.*

Electrophiles other than protons can also react with alkenes and bond to the carbon atom containing the most hydrogen atoms. In thermodynamic terms, electrophiles react with alkenes to give the most stable carbocation intermediate (3° > 2° > 1°). This is a broader definition of the Markovnikov rule. Note that in Figs. 7-5 and 7-9 the most stable carbocation intermediate is formed in each case.

QUESTION 7-5

What would you expect to be the major product from the reaction of HCl with 2-methylpropene?

ANSWER 7-5

Protonation of C-1 gives a 3° carbocation and protonation of C-2 gives a 1° carbocation. Tertiary carbocation formation is a lower energy pathway. Chloride anion reacts with the 3° carbocation to give *tert*-butyl chloride (2-chloro-2-methylpropane).

```
┌──────────────────────────────────────────────┐
│  alkene + electrophile  ──▶  most stable carbocation │
│                               Markovnikov addition    │
└──────────────────────────────────────────────┘
```

Toolbox 7-1.

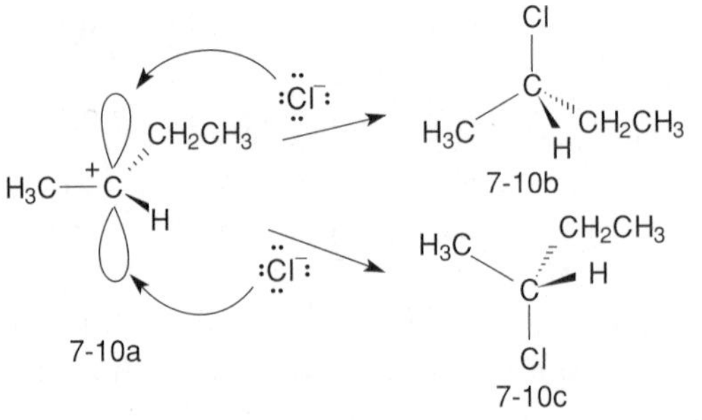

Equal probability of attack from top or bottom
Equal amounts of both enantiomers gives a racemic mixture

Fig. 7-10. Nucleophilic attack on a secondary carbocation.

Stereochemistry

The positively charged carbon atom of a carbocation is sp^2 hybridized and is bonded to three atoms or groups that are in a common plane. The unhybridized p orbital is perpendicular to the plane of the bonded groups. Figure 7-10 shows the attack of a chloride nucleophile on the carbocation. Since the carbocation (Structure 7-10a) is planar, there is equal probability of the chloride anion attacking from either side of the planar carbocation. Attack from the "top" side gives Structure 7-10b and attack from the "bottom" side gives Structure 7-10c. The stereochemistry of Structures 7-10b and 7-10c needs to be considered. Attack from both sides gives the possibility of forming enantiomers, nonsuperimposable mirror images. (For a review see the section entitled *Enantiomers* in Chapter 5.) Structures 7-10b and 7-10c are enantiomers formed in equal amounts. A solution containing equal amounts of two enantiomers is called a *racemic mixture*.

QUESTION 7-6
Does the addition of HCl to 1-methylcyclobutene give enantiomers or one product?

ANSWER 7-6
One product, 1-chloro-1-methylcyclobutane is obtained, since C-1 is not a stereocenter, only one isomer is obtained.

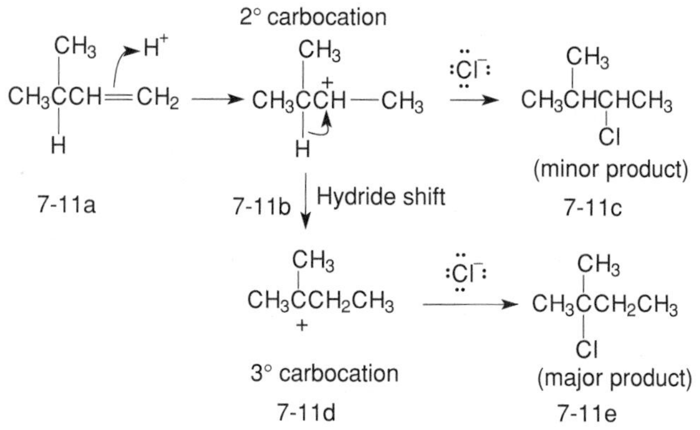

Fig. 7-11. Hydride shift.

Rearrangement Reactions of Carbocations

Sometimes the products of electrophilic addition to alkenes are not the expected products. Figure 7-11 shows the products of a reaction of an alkene, 3-methyl-1-butene (Structure 7-11a), with HCl in an inert solvent. One would expect the 2° carbocation intermediate 7-11b to result from the addition of H$^+$ to the carbon atom containing the most hydrogen atoms (Markovnikov addition). Reaction of intermediate 7-11b with chloride anion gives chloride 7-11c.

HYDRIDE SHIFT

However, the major product observed for the addition of HCl to 3-methyl-1-butene is 7-11e. If the tertiary hydrogen atom (attached to the tertiary carbon atom) in Structure 7-11b transfers, with its bonding electrons, to the adjacent 2° carbon atom, a 3° carbocation (Structure 7-11d) results. Since 3° carbocations are more stable than 2° carbocations, there is a thermodynamic reason for the hydrogen transfer to take place. A hydrogen atom with two electrons is called a hydride anion. The transfer of a hydride anion between adjacent carbon atoms is called a *1,2-hydride shift*. The 1,2 shift refers to a shift between adjacent carbons atoms and not necessarily between carbon atoms C-1 and C-2. The chloride anion reacts with 3° carbocation 7-11d to give the observed product (7-11e).

QUESTION 7-7

What product would you expect from the addition of HBr to 3-methyl-1- pentene?

2° carbocation   3° carbocation

CH₃–C(–CH₃)(–CH₃)–CH=CH₂ $\xrightarrow{H^+}$ CH₃–C(–CH₃)(–CH₃)–CH⁺–CH₃ $\xrightarrow[\text{shift}]{\text{Methyl}}$ CH₃–C(–CH₃)–CH–CH₃

7-12a   7-12b   7-12c

:Cl̈:⁻

CH₃–C(–CH₃)–CH–CH₃
                Cl CH₃

7-12d

Fig. 7-12. Methyl shift.

ANSWER 7-7

3-Bromo-3-methylpentane. The initially formed carbocation undergoes a hydride shift to give a more stable carbocation before being attacked by the choride anion.

METHYL SHIFT

Another example of a rearrangement reaction is shown in Fig. 7-12, when HCl reacts with 3,3-dimethyl-1-butene (7-12a). Structure 7-12a undergoes Markovnikov addition with HCl to form 2° carbocation 7-12b. The methyl group and its bonding electrons on carbon atom C-3 transfers to the adjacent carbon atom C-2. This results in a 3° carbocation (Structure 7-12c) being formed. The thermodynamic (energy) driving force is the formation of a 3° carbocation from a 2° carbocation. A CH_3^- group is called a methide anion and the reaction should be called a 1,2-methide shift. Most authors refer to this as a 1,2-methyl shift. The chloride anion reacts with carbocation 7-12c to give alkyl chloride 7-12d.

A key point: whenever a carbocation is involved in a reaction, one has to be aware of possible rearrangement reactions. The extent of rearrangement reactions depends upon reaction conditions: temperature, solvent, etc. Rearrangement reactions also suggest that carbocations may be involved as reaction intermediates.

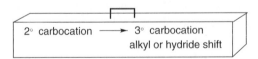

2° carbocation $\longrightarrow$ 3° carbocation
alkyl or hydride shift

Toolbox 7-2.

Quiz

1. Thermodynamics is the study of
    (a) hot dynamite
    (b) the speed of a reaction
    (c) energy changes
    (d) containers that keep drinks cold

2. Kinetics is the study of
    (a) pilates
    (b) molecular orbitals
    (c) energy at equilibrium
    (d) the rates of reactions

3. Exergonic reactions
    (a) are at equilibrium
    (b) give off energy
    (c) take up energy
    (d) remove demons

4. Gibbs free energy is
    (a) always greater for reactants than products at equilibrium
    (b) always greater for products than reactants at equilibrium
    (c) a measure of the change in the energy of a reaction
    (d) available from Mr. Gibbs at no charge

5. Transition states are _____ in energy than/as the reactants.
    (a) higher
    (b) lower
    (c) the same

6. A carbocation is an example of a
    (a) transition state
    (b) reactive intermediate
    (c) negatively charged species
    (d) neutral species

7. The order of stability of carbocations is
    (a) $3° > 2° > 1°$
    (b) $2° > 1° > 3°$
    (c) $1° > 3° > 2°$
    (d) $1° > 2° > 3°$

8. Carbocations are
   (a) neutral species
   (b) electrophilic
   (c) nucleophilic
   (d) stable molecules

9. Hyperconjugation
   (a) decreases stability
   (b) has no effect on stability
   (c) involves sharing electrons in overlapping orbitals
   (d) refers to jumping up and down

10. A transition state represents the _____ in a reaction.
    (a) final product
    (b) starting material
    (c) least stable species
    (d) most stable species

11. Markovnikov addition
    (a) gives the most stable carbocation
    (b) gives the least stable carbocation
    (c) is addition to a carbon atom containing the least hydrogen atoms
    (d) is a new Russian vodka

12. Carbocations rearrange to give the _____ carbocation.
    (a) most stable
    (b) least stable
    (c) same

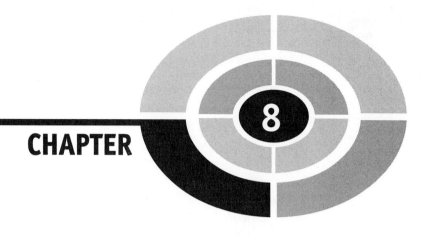

Reactions of Alkenes

Alkenes readily share their π electrons with electron-deficient (electrophilic) reagents. Thus alkenes act as Lewis bases. They are also called nucleophiles since they are nucleus-loving or nucleus-seeking reagents. Alkenes do *not* react with the nucleus of an atom (which is positively charged), but share their electrons with electron-deficient species. It will help to look at the structures in the various figures as you read this chapter for a better understanding of the reactions and reaction mechanisms.

Reaction with Hydrogen Halides in Inert Solvents

Hydrogen halides (HCl, HBr, and HI) undergo Markovnikov addition with alkenes to give the corresponding alkyl halides, as shown in Fig. 8-1. Hydrogen fluoride (HF) is not used as it reacts too vigorously. A halogen or halide refers to the elements in Group 7B in the periodic table (see Appendix A). The symbol X^- is often used to represent a halide. Reactions of alkenes with HX can be

$$RCH{=\!=}CH_2 \xrightarrow[\text{inert solvent}]{H^+} \overset{+}{R}CHCH_3 \xrightarrow{X^-} \overset{\overset{X}{|}}{R}CHCH_3$$

Carbocation
intermediate

X = Bromine, chlorine, or iodine

Fig. 8-1. Reaction of alkenes with hydrogen halides.

carried out in a solvent that does not react with the alkene, the hydrogen halide, or any intermediate species. Such a solvent is called an *inert solvent*. Examples of inert solvents are carbon tetrachloride (CCl_4) and methylene chloride (CH_2Cl_2).

QUESTION 8-1
What is the product of the reaction of HI with 2-methylpropene?

ANSWER 8-1
2-Iodo-2-methylpropane.

Alkene $\xrightarrow{\text{HX}}$ Alkyl halide

Markovnikov addition
possible rearrangement

Toolbox 8-1.

Reaction with Hydrogen Halides in Protic Solvents

A different product results if HCl, HBr, or HI is reacted with an alkene in protic (proton-containing) solvents such as water (H_2O) or alcohols (ROH). The reactions are shown in Figs. 8-2 and 8-3. In these cases the solvent is not inert. It has nonbonding electrons on the oxygen atom that can be shared with an electrophile (the carbocation in this example). The reaction medium (the solvent) also contains the nucleophilic halide anion (from HX) but the solvent is present in much greater concentration than is the halide anion. The higher concentration of solvent competes very successfully with the halide anion for reaction with the carbocation.

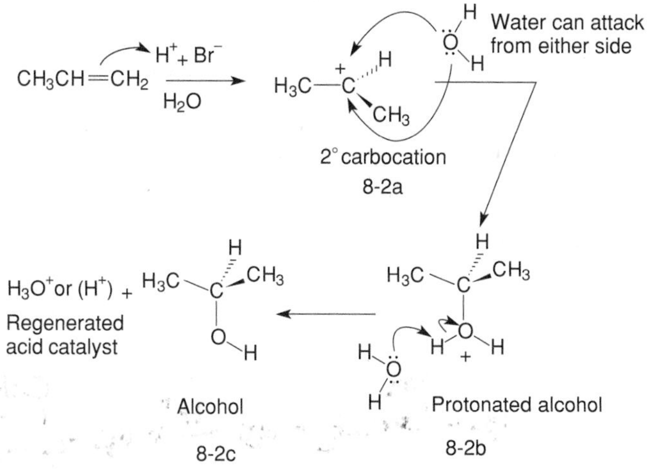

Fig. 8-2. Acid-catalyzed addition of water to propylene.

WATER AS THE SOLVENT

The reaction of HBr with propylene in water is shown in Fig. 8-2. Markovnikov addition of a proton to propylene gives a reactive intermediate, a planar 2° carbocation (8-2a). The nonbonding electrons on the oxygen atom in a water molecule attack the carbocation. Since carbocation 8-2a is planar, water can attack it from either side as shown in Fig. 8-2. The reaction of a positively charged carbocation with a neutral water molecule produces a cationic species,

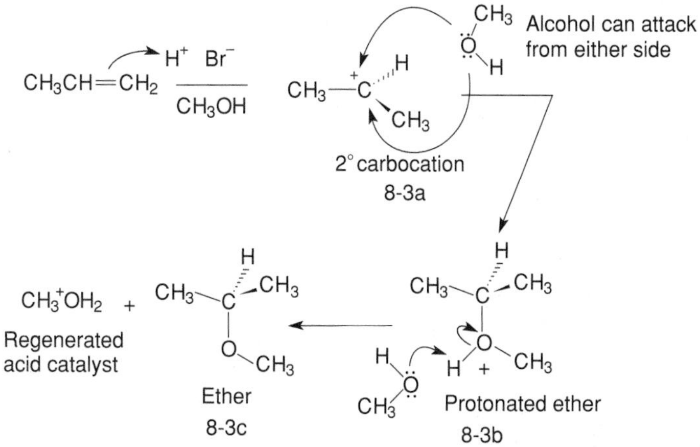

Fig. 8-3. Acid-catalyzed addition of an alcohol to propylene.

8-2b. This cation is called a *protonated alcohol* (an alcohol with an additional proton). Protonated alcohols are very strong acids with pK_a values of about −2.5. (For a review of K_a and pK_a, see Chapter 3.)

The acidic proton in 8-2b is transferred to a water molecule forming a hydronium ion (H_3O^+, $pK_a = -1.7$). This is an equilibrium reaction favoring the weaker acid (survival of the weakest acid) and regenerating the acid catalyst. Alcohol 8-2c results when 8-2b loses a proton. *The net result is the Markovnikov addition of one molecule of water to one molecule of alkene to form an alcohol.*

QUESTION 8-2

What product is formed by the acid-catalyzed addition of water to 2-methylpropene?

ANSWER 8-2

2-Methyl-2-propanol (*tert*-butyl alcohol).

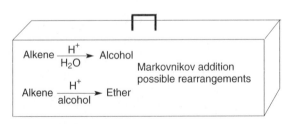

ALCOHOL AS THE SOLVENT

Figure 8-3 shows a reaction mechanism similar to the one shown in Fig. 8-2. The first step is protonation of the alkene by a strong acid (HBr) to give a planar carbocation, 8-3a. The planar carbocation is attacked by a nucleophilic alcohol molecule, such as methanol (CH_3OH). The attack can be from either side of the planar carbocation. The oxygen atom in methanol has nonbonding electron pairs it readily shares with the carbocation. The resulting protonated ether 8-3b is a very strong acid (pK_a of about −3.5). Another methanol molecule removes the proton from 8-3b, giving an ether (8-3c) and a protonated methanol molecule ($pK_a = -2.5$). Since a strong acid is regenerated, only catalytic amounts of HBr (or another strong acid) are used. *The net result is the Markovnikov addition of one molecule of an alcohol to one molecule of alkene to form an ether.*

The two reactions discussed above involve the formation of a carbocation, and therefore rearrangement reactions (hydride and alkyl shifts) are possible. Also, if a stereocenter is formed two enantiomeric products can be formed.

Alkene $\xrightarrow[H_2O]{H^+}$ Alcohol

Alkene $\xrightarrow[\text{alcohol}]{H^+}$ Ether

Markovnikov addition
possible rearrangements

Toolbox 8-2.

Oxymercuration-Demercuration Reactions

The acid-catalyzed addition of water or an alcohol to an alkene may require high temperatures that may result in undesirable reactions in other parts of the molecule. Also, rearrangement reactions (hydride and methyl shifts) are observed when a carbocation intermediate is involved. Another method of adding water or an alcohol to an alkene in a Markovnikov manner involves the use of mercury(II) salts. Mercury(II) acetate is commonly used as the mercury salt. These reactions can be carried out under mild conditions (although mercury compounds are toxic) and no rearrangement reactions occur. An example of this reaction is shown in Fig. 8-4. The mercury(II) cation is electron deficient and reacts with the π electrons in the double bond of an alkene (8-4a) to form a three-membered ring called a *mercurinium cation* (8-4b). The three-membered ring has a positive charge that is shared among the three atoms in the ring.

REGIOSELECTIVE ADDITION OF WATER

When nucleophilic water is used as the solvent in a mercury(II)-catalyzed reaction, it can share its nonbonding electrons on the oxygen atom with the electron-deficient carbon atoms in the three-membered mercurinium ring. Water could react with either of the two carbon atoms in the three-membered ring. Figure 8-5 shows three resonance structures for the mercurinium ion. Structure 8-5a has a full positive charge on the secondary carbocation and Structure 8-5b has a full positive charge on the primary carbocation. The actual structure is some hybrid of the three structures shown, although 8-5a would be expected to contribute

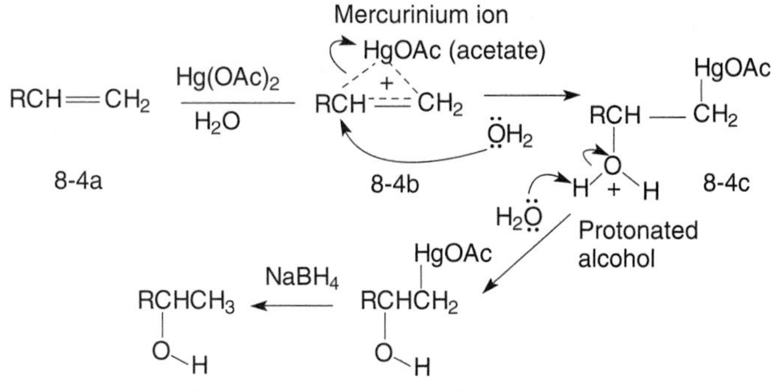

Fig. 8-4. Mercury(II) salt-catalyzed addition of water to an alkene.

$$
\underset{\substack{\text{HgOAc (acetate)} \\ \overset{+}{} \\ RCH \text{---} CH_2}}{} \quad \longleftrightarrow \quad \underset{\substack{\text{HgOAc} \\ RCH \text{---} \overset{+}{C}H_2 \\ 2° \text{ carbocation} \\ \text{8-5a}}}{} \quad \longleftrightarrow \quad \underset{\substack{\text{HgOAc} \\ \overset{+}{R}CH \text{---} CH_2 \\ 1° \text{ carbocation} \\ \text{8-5b}}}{}
$$

Fig. 8-5.   Resonance forms of the mercurinium ion.

more to the resonance hybrid than 8-5b since a secondary carbocation (8-5a) is more stable than a primary carbocation (8-5b).

Water attacks the secondary carbon in the mercurinium ion since it contributes more to the resonance hybrid than does the primary carbocation. Attack by water opens the ring, giving Structure 8-4c. Mercury is a fairly large atom and prevents the water molecule from attacking the secondary carbon atom from the same side of the molecule occupied by the mercury atom. The easiest approach for water is from the side opposite the mercury atom. This is called *anti addition* of water.

DEMERCURATION REACTIONS

Reaction of the mercurinium ion with water opens the three-membered ring and yields a protonated alcohol (8-4c). Water removes the proton from the protonated alcohol, giving the weaker acid, H_3O^+, and mercury compound 8-4d. A special reducing reagent, sodium borohydride ($NaBH_4$), is added to remove the mercury atom (demercuration). The product (8-4e) is an alcohol that results from the Markovnikov addition of one molecule of water to one molecule of an alkene. If the reaction is carried out in an alcohol solvent rather than water, the final product is an ether that results from the Markovnikov addition of one molecule of an alcohol to one molecule of an alkene. *A key point: mercuration-demercuration does not involve a carbocation intermediate and no rearrangement reactions (hydride or alkyl shift) are observed.*

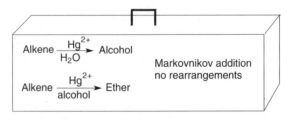

Toolbox 8-3.

QUESTION 8-3

What product results from the mercury-catalyzed addition of water to 4-methyl-1-pentene followed by reaction with sodium borohydride?

ANSWER 8-3

4-Methyl-2-pentanol.

4/14

Hydroboration-Oxidation

The preceding section gave examples of preparing an alcohol via Markovnikov addition in which the alcohol functionality (—OH) becomes attached to the carbon atom in a double bond that is bonded to the least number of hydrogen atoms. The concept of Markovnikov addition of water to an alkene can be restated as the alcohol functionality becomes bonded to the *most highly substituted* carbon atom. A 3° carbon atom is more highly substituted than a 2° carbon atom, which is more highly substituted than a 1° carbon atom. Can alcohols be made so that the alcohol functionality is bonded to the least substituted carbon atom? Yes, but the mechanism must not involve a carbocation intermediate or potential rearrangement reactions could occur. Addition of —OH to the least substituted carbon atom in a double bond is called *anti-Markovnikov addition*.

ANTI-MARKOVNIKOV ADDITION

Borane (BH_3) or substituted boranes (BHR_2) are used to accomplish anti-Markovnikov addition to an alkene. R is often some large, bulky alkyl group. The Lewis structure of BH_3 (Fig. 8-6, Structure 8-6a) shows a boron atom

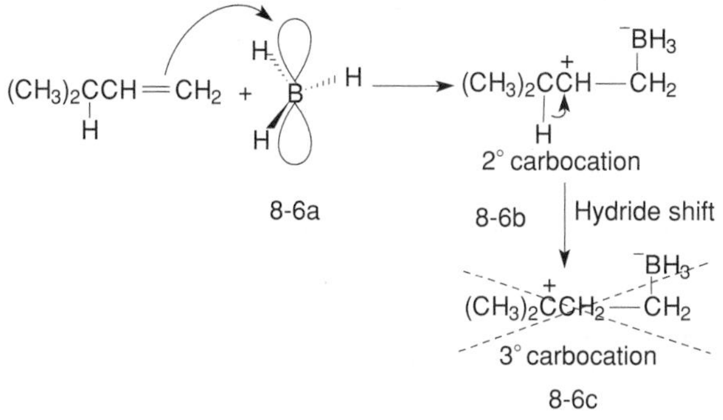

Fig. 8-6.   Markovnikov addition of borane.

with only six valence electrons in the three B—H bonds. Since an alkene has π electrons to share, it would be reasonable to expect Markovnikov addition of BH_3 to an alkene to give a secondary carbocation as shown in Structure 8-6b. Boron now has eight valence electrons. If this carbocation is formed, one might expect a rearrangement reaction (a hydride shift) to occur to give a tertiary carbocation (8-6c). No rearrangements were observed in a series of reactions of BH_3 with various alkenes where rearrangement was expected. Therefore it appears that a carbocation *is not* involved in these reactions.

A concerted reaction?

A mechanism has been proposed that does not involve a carbocation. Figure 8-7 shows the reaction of an alkene (8-7a) with BH_3 to give a four-membered transition state (8-7b). Transition states represent a bond-breaking and/or bond-making process. The broken lines shown in 8-7b indicate partial bonds. A C—H bond is being formed, an H—B bond is being broken, a B—C bond is being formed, and a C—C π bond is being broken. All bond making and breaking can take place at the same time and not in a stepwise manner. The simultaneous making and breaking of bonds is called a *concerted step* (or *reaction*). You can think of an analogy of a *concert* where all the players have to play their appropriate parts at the same time. It is also possible that boron may attack the terminal carbon atom and a hydride shift from BH_3 occurs to form 8-7c before a rearrangement reaction can take place. Since a transition state is a transitory species, it is difficult to fully characterize these species.

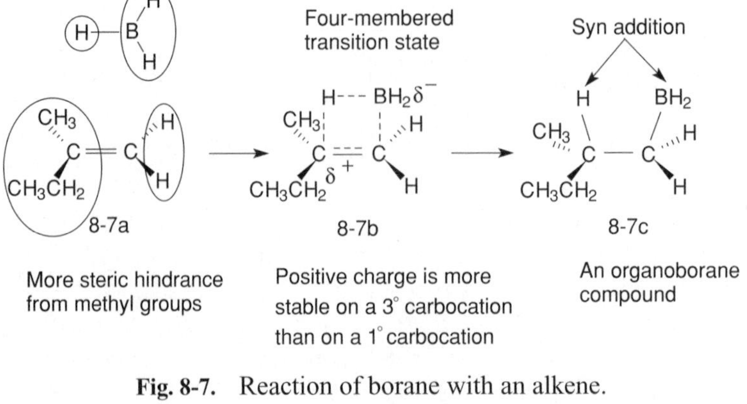

Fig. 8-7.   Reaction of borane with an alkene.

SYN ADDITION

If the reaction proceeds by the mechanism shown in Fig. 8-7, the lack of carbocation formation would explain the absence of rearrangement reactions. The transition state results when BH_3 approaches the planar alkene bond from one side of the molecule. One H from BH_3 begins to form a bond to the tertiary carbon atom. The boron begins to form a bond to the terminal carbon atom. Regardless of whether the reaction is concerted or the C—H bond forms immediately after the B—C bond forms, both bond-forming processes occur on the same side of the alkene. This is called *syn addition*. BH_3 could react on the "top" side or the "bottom" side of the planar alkene. In either case, *syn addition* would result.

One molecule of BH_3 reacts with one molecule of an alkene (R) to form the alkylborane, RBH_2. RBH_2 has two remaining B—H bonds that can react with two additional molecules of alkene to form R_3B. The reaction mechanism will be explained with the addition of just one alkene, to simplify the structures drawn.

REGIOSELECTIVITY

The reaction with boron is a *regioselective* reaction. In a regioselective reaction one constitutional isomer is formed in preference to all other isomers. The major product results from anti-Markovnikov addition. One reason for this selectivity is a steric (space-filling, bulky) effect. An example of this steric effect is shown between BH_3 and alkene 8-7a. There is more steric interaction between the boron atom and the two alkyl groups on the alkene than between the boron atom and the two hydrogen atoms on the alkene. It is easier for the boron atom to approach the carbon atom containing the less bulky hydrogen atoms. Boron compounds with larger (more bulky) groups than hydrogen, HBR_2, give a greater degree of regioselectivity, which supports the steric effect theory.

ANTI-MARKOVNIKOV FORMATION OF ALCOHOLS

The value of organoboron compounds (8-8a) is that they can be converted to other compounds with desirable functional groups as shown in Fig. 8-8. The reaction of organoborane 8-8a in basic hydrogen peroxide solution gives an alcohol. The mechanism will not be discussed here, except to state that the —OH group replaces the boron group from the *same side of the molecule* as shown in Structure 8-8b. The net result of this reaction is the *syn addition* of one molecule of water to one molecule of alkene via anti-Markovnikov addition. *Two methods to make alcohols, by Markovnikov and anti-Markovnikov addition of water to an alkene, have now been discussed.*

Syn addition of water

H_2O_2

^-OH

^-OH replaces $^-BH_2$ on the same side of the molecule

8-8a

8-8b

Br_2

CH_3O^-

CH_3OH

Syn addition of HBr

8-8c

$CH_3CH_2CO_2H$

Syn addition of H_2

8-8d

Fig. 8-8. Reaction of organoboranes.

ANTI-MARKOVNIKOV FORMATION OF BROMIDES

Organoborane 8-8a reacts with bromine in basic alcoholic solutions to replace the boron-carbon bond with a bromine-carbon bond. The bromide anion replaces the boron from the same side of the molecule as shown in Structure 8-8c. The net result is anti-Markovnikov, syn addition of HBr to an alkene. *Two methods of making alkyl bromides, by Markovnikov and anti-Markovnikov addition to an alkene, have now been discussed.*

Organoborane 8-8a reacts with propionic acid, $CH_3CH_2CO_2H$, to replace the B—C bond with an H—C bond. *The net reaction is to convert an alkene to an alkane,* 8-8d.

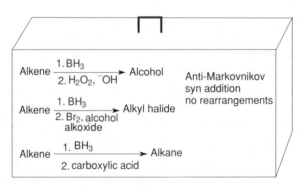

Toolbox 8-4.

OPTICALLY INACTIVE PRODUCTS

Products 8-8b and 8-8c in Fig. 8-8 contain a chiral carbon atom. However, the product mixtures are optically inactive. Borane adds in a syn manner to alkene 8-7a. Since the alkene function in 8-7a is planar, borane can add to the "top" or "bottom" side of the alkene molecule with equal probability. If a stereocenter is generated, equal amounts of enantiomers are formed giving a racemic mixture. Racemic mixtures are optically inactive. The starting alkene, 8-7a, is achiral and optically inactive. *A key point: if starting materials in a chemical reaction are optically inactive or achiral, the products will be optically inactive because they are either achiral or a racemic mixture.*

QUESTION 8-4
3-Methyl-2-butene is reacted with HBR_2 to form an organoborane which is subsequently reacted with aqueous basic hydrogen peroxide. What are the products? Is the product solution optically active?

ANSWER 8-4
The product is a racemic, optically inactive mixture of $(2R)$- and $(2S)$-3-methyl-2-butanol $((2R)$- and $(2S)$-2-hydroxy-2-methylbutane).

Halogenation in Inert Solvents

Bromine (Br_2) and chlorine (Cl_2) readily react with alkenes. Fluorine (F_2) and iodine (I_2) are not discussed here because F_2 reacts explosively with alkenes while I_2 reacts to only a limited extent. Figure 8-9 shows the mechanism of

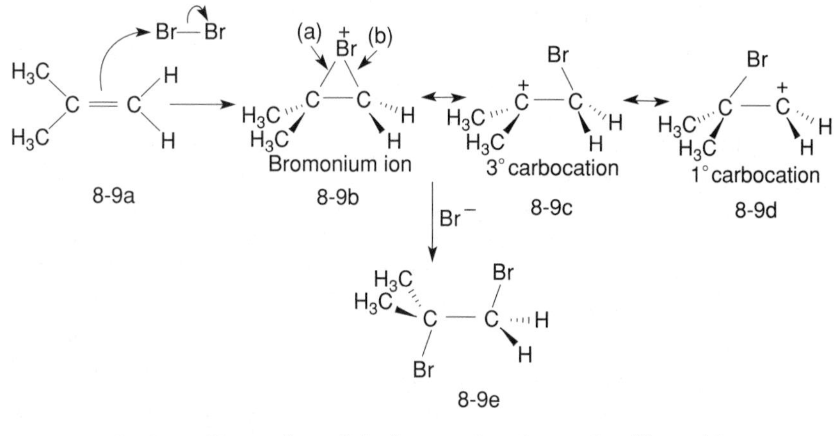

Fig. 8-9.   Formation of the bromonium ion and a dibromide.

Br$_2$ reacting with an alkene (8-9a) in an *inert solvent*. Bromine is not a polar molecule but it has a large electron cloud that is easily polarized. As a bromine molecule approaches an alkene, the electron-rich alkene induces polarity in the bromine molecule. The alkene π electrons "push" the bromine electrons to the opposite side of the bromine molecule. This gives the side of the bromine molecule approaching the alkene a partial positive charge. The alkene then donates its π electrons to the bromine atom with the partial positive charge. The bromine-bromine bond breaks releasing a bromide anion, Br$^-$. A bromine atom becomes bonded to the alkene by syn addition, forming a three-membered ring, called a *bromonium ion* (8-9b). Since the alkene functionality is planar, the bromine molecule can approach the alkene from either the "top" or the "bottom" side. The bromonium ion is a positively charged electrophile that seeks electrons and the bromide anion is available to share its nonbonding electrons.

RESONANCE STRUCTURES

The bulky bromine atom in the three-membered bromonium ion ring prevents the bromide anion from attacking on the same side of the molecule. The least sterically hindered approach for the bromide anion is from the side opposite of the bromine atom in the three-membered ring. This is called *anti addition*. There are two carbon atoms in the three-membered ring. Figure 8-9 shows three resonance forms for the bromonium cation: 8-9b, 8-9c, and 8-9d. Resonance forms result by moving bonding electron pairs between the bromine and the carbon atoms. Resonance structure 8-9c has a 3° carbocation and 8-9d has a 1° carbocation. Since 3° carbocations are more stable than 1° carbocations, the resonance structure with the 3° carbocation contributes more to the true structure (resonance hybrid) of the bromonium ion than does 8-9d. If 8-9c is a greater contributor to the resonance hybrid than 8-9d, bond (a) in 8-9b is expected to be weaker and more easily broken than bond (b). Attack by the bromide anion at the 3° carbon atom results in bond (a) being broken.

VICINAL DIHALIDES

If one just considered steric effects, one would predict the bulky bromide ion would attack the less sterically hindered primary carbon atom, which does not happen. Attack occurs at the 3° carbon atom since 8-9c contributes more to the resonance hybrid than does 8-9d. Attraction of the bromide anion to the 3° carbon atom is apparently more important than the steric hindrance effect. The resulting product is a vicinal dibromide 8.9e. Vicinal (vic) groups are bonded to adjacent (next to) carbon atoms.

QUESTION 8-5
What product results from the reaction of chlorine with ethylene (ethene)?

ANSWER 8-5
1,2-Dichloroethane.

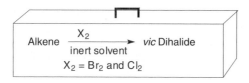

Alkene $\xrightarrow[\text{inert solvent}]{X_2}$ *vic* Dihalide

$X_2 = Br_2$ and Cl_2

Toolbox 8-5.

Stereochemistry of Halogenation

Since the alkene functional group is planar, the reaction with a bromine molecule can occur with equal probability from the "top" or "bottom" side of the alkene. Figure 8-10 shows the reaction of an alkene (8-10a) with bromine to give two bromonium ions, 8-10b and 8-10c in equal amounts. The bromide anion then attacks each bromonium ion at the most substituted carbon atom, giving vicinal dibromides 8-10d and 8-10e. The dibromides are produced in equal amounts and each contains one stereocenter. Structure 8-10d has an S configuration and Structure 8-10e has an R configuration. They are enantiomers. Rotating the entire molecule 180° around the carbon-carbon single bond axis as shown in Structure 8-10d gives Structure 8-10f. Comparing 8-10e and 8-10f confirms that

Fig. 8-10.   Stereochemistry of bromine addition to an alkene.

8-10d and 8-10e are enantiomers (the ethyl and isopropyl groups are reversed). Equal amounts of the dibromides are produced by both reactions. The final solution is a racemic mixture and is optically inactive. Repeating what was mentioned above, *optically inactive starting materials (the alkene) produce optically inactive achiral products or a racemic mixture.* In this case a racemic mixture results.

QUESTION 8-6
Is the product of the reaction of bromine with 2-methylpropene a racemic mixture?

ANSWER 8-6
No, the product contains no chiral carbon atom.

Halogenation in Reactive Solvents

The reaction of bromine or chlorine with alkenes in "reactive" solvents, such as water or an alcohol, does *not* yield the dihalide as the major product.

HALOHYDRIN FORMATION

The reaction of bromine with propylene gives, initially, a bromonium ion. Figure 8-11 shows the reaction of a water molecule with the initially formed bromonium ion (8-11a) to give a protonated alcohol (8-11b). The water molecule

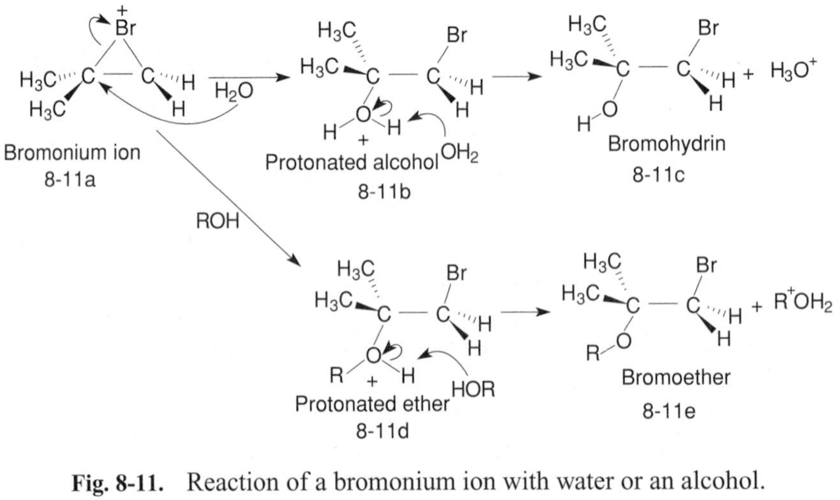

Fig. 8-11.   Reaction of a bromonium ion with water or an alcohol.

attacks the more highly substituted carbon atom of the cyclic bromonium ion. The protonated alcohol is a very strong acid and a water molecule takes a proton from 8-11b to give alcohol 8-11c. Compounds that contain adjacent halogen and alcohol (hydroxyl) functions are called *halohydrins*. In this case, the product is a *bromohydrin*.

BROMOETHERS

When the reaction of bromine (or chlorine) with an alkene is carried out in an alcohol as the solvent, a *haloether* is formed. Figure 8-11 also shows the reaction of a generic alcohol, ROH, with the initially formed bromonium ion (8-11a) to give a protonated ether (8-11d), a very strong acid. An alcohol molecule will attack the most highly substituted carbon atom of the cyclic bromonium ion. The protonated ether (8-11d) transfers a proton to an alcohol molecule, giving a *bromoether* 8-11e and a protonated alcohol.

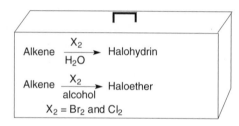

Toolbox 8-6.

Radical Bromination

A method of using BH_3 to add HBr to an alkene in an anti-Markovnikov manner was previously discussed. The anti-Markovnikov addition of HBr to an alkene can also be accomplished by carrying out the reaction in the presence of peroxides. The mechanism of this reaction is shown in Fig. 8-12. Peroxides (8-12a) have an oxygen-oxygen σ bond, which is relatively weak compared to carbon-carbon σ bonds. Moderate heat or ultraviolet light can provide sufficient energy to break the oxygen-oxygen bond. The bond breaks *homolytically*; that is, the bond breaks so one electron (of the two-electron bond) remains with each alkoxy fragment (8-12b). Each fragment now has an unpaired electron. Species with unpaired electrons are called *radicals* (or *free radicals*). The single-headed arrow indicates the movement of a single electon.

RO—OR ⟶ 2 RO·
8-12a   8-12b

RO· + H—Br ⟶ ROH + Br·
8-12b   8-12c

Fig. 8-12. Radical bromination of alkenes.

A CHAIN REACTION

Radicals are very reactive species. Radical 8-12b reacts with HBr to produce a bromine radical (8-12c). The bromine radical is also very reactive and reacts with alkene 8-12d to give a new radical species (8-12e). Radical 8-12e is another very reactive species and reacts with HBr to form bromoalkane 8-12f, and a bromine radical (8-12c). The new bromine radical can react with another alkene molecule and the reaction sequence repeats itself. This type of "circular" reaction is like a chain where the two ends are connected giving a circle and is thus called a *chain reaction*.

Anti-Markovnikov addition

When a bromine radical reacts with the alkene (8-12d) it could react with either of the two alkene carbon atoms. Reacting with C-1 results in the unpaired electron being on a 3° carbon atom (8-12e). If the bromine radical had reacted with C-2, the unpaired electron would have been on a 1° carbon atom. The stability of radicals parallels that of carbocations: 3° radicals > 2° radicals > 1° radicals > methyl radicals (least stable). Even though HBr is added to an alkene in an anti-Markovnikov manner, the reaction actually occurs by Markovnikov addition via a radical mechanism. The bromine radical adds to the carbon atom with the greatest number of hydrogen atoms, giving a tertiary radical (8-12e). This chain reaction does not occur with other halogens for thermodynamic (energy) reasons.

Toolbox 8-7.

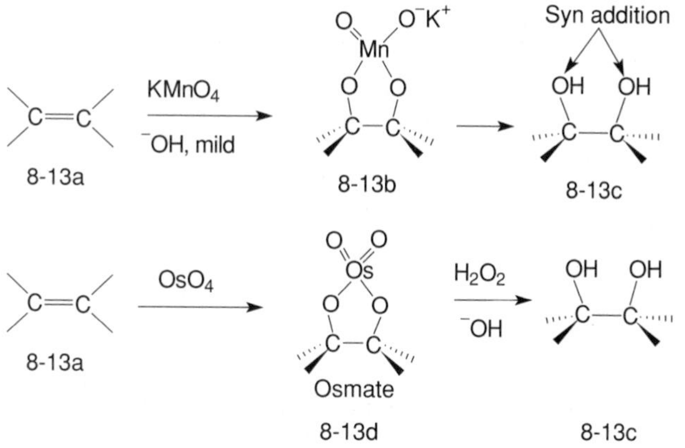

Fig. 8-13. Formation of diols.

Formation of Diols

Two common methods for converting alkenes into alcohols containing two alcohol (—OH) groups on adjacent carbon atoms are shown in Fig. 8-13. The resulting molecules are called *dialcohols*, *1,2-diols*, *vicinal diols*, or *glycols*. Alkene (8-13a) can be oxidized with potassium permanganate (KMnO$_4$) under mild, basic conditions to give *vicinal* diols (8-13c). The reaction proceeds through a cyclic intermediate (8-13b). Alkenes also react with osmium tetroxide (OsO$_4$) in an inert solvent to give a cyclic osmate (8-13d). The osmate is converted to a diol (8-13c) by treatment with aqueous basic hydrogen peroxide. The peroxide regenerates osmium tetroxide and thus only small amounts of this expensive, toxic material are needed.

SYN ADDITION

Both reactions for making diols involve a cyclic intermediate. The cyclic structure necessarily attaches to both oxygen atoms on the same side of the alkene molecule. This is called *syn addition* giving a cis-1,2-diol. In these reactions the π bond of the alkene is broken and two new carbon-oxygen sigma bonds are formed.

QUESTION 8-7
What product is formed when cyclopentene is reacted with osmium tetroxide followed by treatment with basic hydrogen peroxide?

ANSWER 8-7
cis-1,2-Dihydroxycyclopentane.

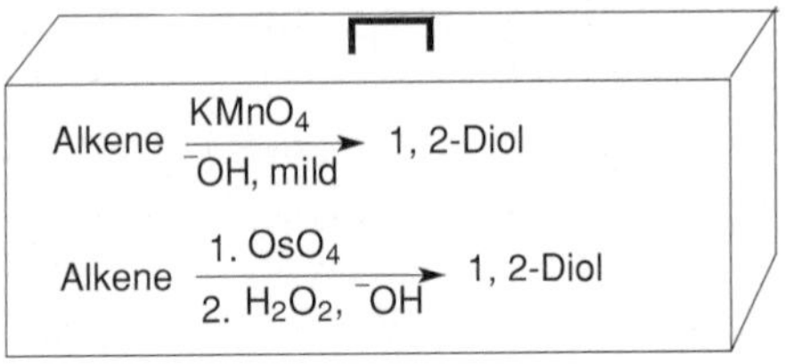

Alkene $\xrightarrow[\text{$^-$OH, mild}]{\text{KMnO}_4}$ 1, 2-Diol

Alkene $\xrightarrow[\text{2. H}_2\text{O}_2, \text{ }^-\text{OH}]{\text{1. OsO}_4}$ 1, 2-Diol

Toolbox 8-8.

Double Bond Cleavage

Alkenes react with strong oxidizing agents that break the alkene double (sigma and π) bond, giving aldehyde, ketone, and/or acid functional groups. Figure 8-14 shows the reaction of alkenes with ozone (O_3) and permanganate anion (MnO_4^-).

OZONE OXIDATION

Reaction of a tetrasubstituted alkene (8-14a) with ozone gives ketones 8-14b and 8-14c. The reaction of ozone with a trisubstituted alkene (8-14d) gives a ketone (8-14e) and an aldehyde (8-14f). The reaction of ozone with a disubstituted terminal alkene (8-14g) gives a ketone (8-14h) and formaldehyde (8-14i). Ozone is easy to use but is very toxic and care must be taken when using this reagent. Ozone reacts with an alkene to initially form a cyclic ozonide. In a subsequent step the ozonide is reduced with dimethyl sulfide ($(CH_3)_2S$) to a carbonyl-containing ($C=O$) compound. Ozone is a very reactive molecule

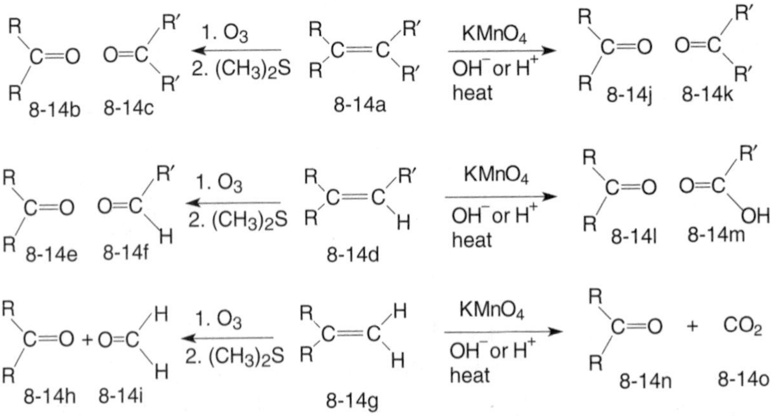

Fig. 8-14. Oxidation of alkenes.

(an oxidizing agent) and is able to break the σ and π bonds in alkenes and alkynes, but it does not break isolated C–C single σ bonds.

Similar products are obtained when alkenes are reacted with $KMnO_4$ under more vigorous conditions than used to make vicinal diols. Follow these reactions in Fig. 8-14. The reaction of $KMnO_4$ with a tetrasubstituted alkene (8-14a) gives ketones 8-14j and 8-14k. The reaction of $KMnO_4$ with a trisubstituted alkene (8-14d) gives a ketone (8-14l) and a carboxylic acid (8-14m) under acidic reaction conditions and the carboxylate salt under basic reaction conditions. Reaction of $KMnO_4$ with a disubstituted terminal alkene (8-14g) gives a ketone (8-14n) and carbon dioxide (8-14o). Under these conditions, $KMnO_4$ can break C–H bonds and σ and π carbon-carbon bonds in alkenes but does not break isolated C–C single σ bonds.

In summary, if one hydrogen atom is attached to a carbon atom in a double bond (8-14d), treatment with ozone converts that carbon atom into an aldehyde and potassium permanganate converts that carbon atom into a carboxylic acid. If one of the carbon atoms in the double bond contains two hydrogen atoms (8-14g), ozone converts this carbon atom into formaldehyde (8-14i) and permanganate converts this carbon atom into carbon dioxide (8-14o). If two alkyl groups are attached to each carbon atom in the double bond (8-14a), only ketones are obtained with both oxidizing agents.

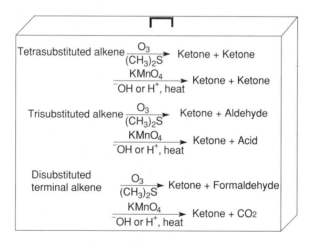

Toolbox 8-9.

QUESTION 8-8

What functional groups result from the reaction of 2-methylcyclopentene with (a) ozone followed by reaction with dimethyl sulfide and (b) $KMnO_4$ in acid with heating?

ANSWER 8-8

(a) A ketone and an aldehyde. (b) A ketone and a carboxylic acid.

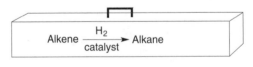

Fig. 8-15. Hydrogenation of alkenes.

Hydrogenation

Alkenes are converted to alkanes by reaction with hydrogen, H_2. These reactions have large activation energies, so a catalyst must be used to obtain reasonable reaction rates. Figure 8-15 shows the reaction of hydrogen with an alkene. Typical catalysts are nickel, palladium, platinum, or platinum dioxide. These metals are ground into fine powders to give a large surface area per unit weight of the metal. The powered metals are then coated on inert particles (silica/sand) for the hydrogenation reaction (reaction with hydrogen). The exact mechanism is not known in detail but it is proposed that the metal surface "activates" the hydrogen molecule and reaction with the alkene takes place on the metal surface. The hydrogen atoms are added in a *syn* manner even though it is not necessarily a concerted reaction. Reactions often take place at moderate temperatures but require high pressures. Specialized high-pressure equipment is needed to run these reactions.

$$\text{Alkene} \xrightarrow[\text{catalyst}]{H_2} \text{Alkane}$$

Toolbox 8-10.

Epoxides

The reaction of alkenes with peroxyacids gives epoxides, as shown in Fig. 8-16. Peroxyacids contain an additional oxygen atom in the carboxyl group (RCO_3H vs. RCO_2H). Peroxyacids are good oxidizing agents, donating oxygen atoms to other molecules. Peroxyacids break the π bond in an alkene forming a three-membered ring, called an *epoxide* or an *oxirane*. The oxygen atom bonds

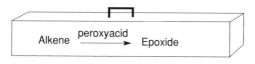

Fig. 8-16. Epoxide formation.

to an alkene in a *syn* manner. Epoxide compounds are very reactive cyclic ethers. Epoxide chemistry is discussed in Chapter 18.

Alkene →(peroxyacid)→ Epoxide

Toolbox 8-11.

Cyclopropanes

Alkenes react with *carbenes* to give cyclopropanes, rings consisting of three carbon atoms. This reaction is shown in Reaction (a) in Fig. 8-17. Carbenes are carbon compounds with two σ bonds and a pair of nonbonding electrons. A generic carbene structure is shown as 8-17a. Carbenes have six valence electrons and are electron deficient. They want eight valence electrons to be "octet happy", so they are electrophilic and react with nucleophilic alkenes. Carbenes react in a *syn* manner to give cyclopropanes [Reaction (a) in Fig. 8-17].

(a) $C=C$ + $R_2C:$ ⟶
         8-17a

(b) H_2CN_2 →(UV or heat)→ $H_2C:$ + N_2
                                    carbene

(c) $CHCl_3$ →(^-OH)→ $Cl_2C:$ + Cl^- + H_2O
                        Dichlorocarbene

Fig. 8-17. Formation of cyclopropanes and carbenes.

Carbene (H_2C:) is formed by the decomposition of diazomethane, CH_2N_2, with heat or UV radiation [Reaction (b) in Fig. 8-17]. Treatment of chloroform, $HCCl_3$, with base gives dichlorocarbene, Cl_2C: [Reaction (c) in Fig. 8-17].

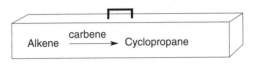

Toolbox 8-12.

Quiz

1. Alkenes react with strong halogen acids (HX) in inert solvents to give
    (a) alkanes
    (b) alcohols
    (c) alkyl halides
    (d) alkenyl halides

2. Alkenes react with aqueous acids to give
    (a) alkanes
    (b) alcohols
    (c) alkyl halides
    (d) alkenyl halides

3. Alkenes react in acidic alcoholic solutions to give
    (a) alcohols
    (b) halohydrins
    (c) alkyl halides
    (d) ethers

4. Mercury(II)-catalyzed addition of water to alkenes gives
    (a) alkanes
    (b) alcohols
    (c) ethers
    (d) hydride shifts

5. The reaction of an alkene with borane followed by reaction with basic hydrogen peroxide gives
    (a) Markovnikov addition of water
    (b) anti-Markovnikov addition of water
    (c) a dialcohol
    (d) a methyl shift

6. The addition of borane to an alkene involves a
    (a) radical intermediate
    (b) carbocation intermediate
    (c) four-membered transition state
    (d) rearrangement reaction

7. The reaction of an alkene with borane followed by addition of bromine
    in basic alcohol gives a(n)
    (a) alkene
    (b) alkyl bromide
    (c) bromohydrin
    (d) dibromide

8. Borane is used for the —— of HBr from/to an alkene.
    (a) replacement
    (b) removal
    (c) anti-Markovnikov addition
    (d) Markovnikov addition

9. The reaction of chlorine with an alkene in an inert solvent gives
    (a) an alkyne
    (b) a dichloride
    (c) a chlorohydrin
    (d) a trans alkene

10. The reaction of bromine with an alkene in water as the solvent gives a
    (a) diol
    (b) dibromide
    (c) halohydrin
    (d) haloether

11. The addition of HBr to an alkene in the presence of a peroxide
    gives
    (a) Markovnikov addition
    (b) anti-Markovnikov addition
    (c) a dibromide
    (d) a dialcohol

12. The reaction of an alkene with osmium tetroxide gives
    (a) anti addition
    (b) trans addition
    (c) syn addition
    (d) Markovnikov addition

13. The reaction of $KMnO_4$ with an alkene under mild conditions gives
    (a) an acid
    (b) an osmate
    (c) a *vicinal* diol
    (d) a *vicinal* dihalide

14. Ozone reacts with tetrasubstituted alkenes to give
    (a) aldehydes
    (b) ketones
    (c) aldehydes and ketones
    (d) acids

15. The reaction of an alkene with hydrogen in the presence of a catalyst gives
    (a) trans alkenes
    (b) cis alkenes
    (c) syn addition
    (d) anti addition

16. Carbenes react with alkenes to give
    (a) alcohols
    (b) aldehydes
    (c) acids
    (d) cyclopropanes

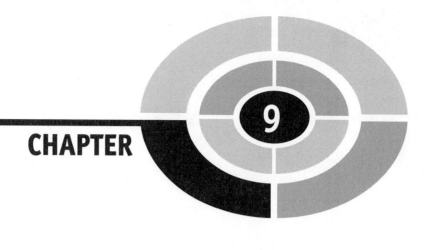

CHAPTER 9

Alkynes

Introduction

Alkynes are compounds that contain a carbon-carbon triple bond. The general formula for an acyclic, monoalkyne is C_nH_{2n-2}. There are two main classifications of alkynes: terminal ($RC\equiv CH$) and internal ($RC\equiv CR$). The alkyne group is very reactive and few molecules with a triple bond are found in nature. Acetylene ($HC\equiv CH$) is the most widely used alkyne and is probably best known for its use in oxyacetylene torches.

Structure

The triple bond ($-C\equiv C-$) is drawn with three identical lines suggesting all three bonds are identical. This is not the case however, as one bond is a σ bond and the other two bonds are π bonds. Each carbon atom in a triple bond is sp hybridized according to hybrid atomic orbital theory. (For a review of hybrid atomic orbitals see the section entitled *Wave Equations* in Chapter 1.) The alkyne carbon atom

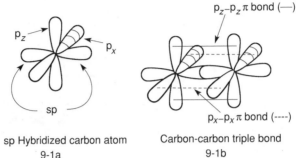

sp Hybridized carbon atom
9-1a

Carbon-carbon triple bond
9-1b

Fig. 9-1. The sp hybridized carbon atom and π bonds.

has two sp orbitals that are 180° apart, which minimizes repulsion between electrons in these two orbitals. Each carbon atom also has two unhybridized p orbitals, shown in Fig. 9-1, Structure 9-1a. The p orbitals on one carbon atom (p_x and p_z) are in a common plane and perpendicular to each other. Structure 9-1b shows the orbital overlap forming the three bonds between two adjacent alkyne carbon atoms. Two sp orbitals, one from each carbon atom, overlap to form a σ bond. The two p orbitals on one carbon atom overlap with the two p orbitals on the adjacent carbon atom to form two π bonds. The overlapping p orbitals must be in the same plane for maximum overlap and bond strength.

QUESTION 9-1
Why isn't an alkyne bond three times as strong as a single bond?

ANSWER 9-1
An alkyne bond consists of one σ and two π bonds. π bonds have less orbital overlap than do σ bonds and are therefore weaker bonds.

Alkynes have four electrons in their π bonding system, two in each π bond. Alkynes are willing to share these electrons with electrophiles (a species that wants electrons). Thus alkynes are nucleophilic. They seek positively charged or electron-deficient species.

Nomenclature

The rules for naming alkynes are similar to those for naming alkanes. An example of naming an alkyne is given in Fig. 9-2. As you read the rules, see the compound in Fig. 9-2 for an explanation of each rule.

1. Choose the longest chain containing the triple bond(s).

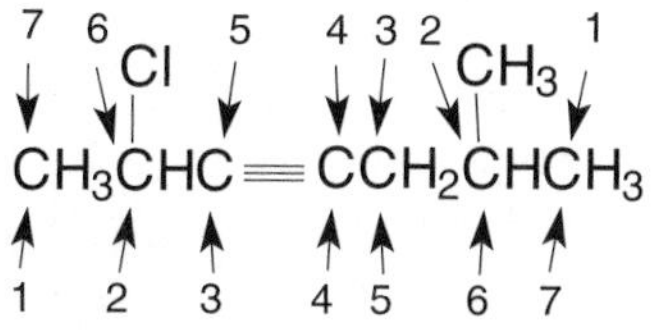

$$\overset{7\ \ \ 6\ \ \ \ \ \ 5\ \ \ \ \ 4\ 3\ 2\ \ \ \ \ \ \ 1}{\underset{1\ \ \ \ 2\ \ \ \ 3\ \ \ \ \ 4\ 5\ \ \ 6\ \ \ 7}{CH_3CHC \equiv CCH_2CHCH_3}}$$

Fig. 9-2.   Naming compounds: (1) The longest chain containing the triple bond contains seven carbon atoms. (2) A seven-carbon atom chain is called a heptyne. (3) Numbering the chain from the left (bottom numbers) makes the first carbon atom in the triple bond C-3. Starting from the right (top numbers) makes the first carbon atom in the triple bond C-4. So one starts from the left since 3 < 4. (4) There is a chlorine atom on carbon C-2 and a methyl group on carbon C-6. Chloro comes before methyl, since c precedes m in the alphabet. The name is 2-chloro-6-methyl-3-heptyne or 2-chloro-6-methylhept-3-yne.

2.   Name this longest chain as you would an alkane, changing the -ane suffix to -yne.
3.   Number the carbon atoms starting at the end of the chain that gives the lowest number to the first carbon atom in a triple bond.
4.   List the substituents alphabetically and indicate the point of attachment to the parent chain. Substituents should have the lowest parent chain numbers possible.

QUESTION 9-2
Name the following compound.

$$CH_3CHC \equiv CCH_2CH_2Br$$
$$\underset{CH_3}{|}$$

ANSWER 9-2
1-Bromo-5-methyl-3-hexyne (or 1-bromo-5-methylhex-3-yne). If the chain is numbered from the other end, the substituents will not have the lowest number combination.

COMMON NAMES

The simplest alkyne is $HC \equiv CH$. Its formal (IUPAC) name is ethyne. Its common name is acetylene. This common name is unfortunate since the -ene suffix implies it is an alkene, which it is not. Many compounds have common names based on acetylene. For example, $CH_3C \equiv CH$ is methylacetylene and $CH_3C \equiv CCH_2CH_3$ is ethylmethylacetylene. Even the common name alphabetizes the substituents (ethyl before methyl).

When the alkyne group is at the end of a molecule ($RC \equiv CH$) it is called a *terminal*, or monosubstituted, alkyne. Methylacetylene ($CH_3C \equiv CH$) is a terminal

Table 9-1.   Bond length and strength of carbon-carbon bonds

		Bond length, pm	Bond strength, kJ/mol (kcal/mol)
Ethane	$H_3C\text{—}CH_3$	154	368 (88)
Ethylene	$H_2C\text{=}CH_2$	133	606 (145)
Acetylene	$HC\text{≡}CH$	120	836 (200)

alkyne. An alkyne that has a group (not hydrogen) bonded to each alkyne carbon atom ($RC\equiv CR$) is called an *internal*, or disubstituted alkyne. The R groups may be the same or different. Ethylmethylacetylene, $CH_3CH_2C\equiv CCH_3$, is an internal alkyne.

Physical Properties

The physical properties of unsubstituted alkynes are similar to alkenes and alkanes. Alkynes are less dense than water, hydrophobic (insoluble in water), soluble in organic solvents such as benzene and carbon tetrachloride, and their boiling points increase with increasing molecular weight. Alkynes of four or fewer carbon atoms are gases at room temperature.

BOND STRENGTH

Triple bonds are shorter and stronger than double or single bonds. Table 9-1 gives the values of bond lengths in picometers (or 10^{-12} m) and bond strengths in kilojoules per mole (and kilocalories per mole) for single, double, and triple bonds in ethane, ethylene, and acetylene. Note that a triple bond is not three times as strong as a single bond. It takes about 230 kJ/mol to break one π bond in acetylene. A slightly greater amount of energy (238 kJ/mol) is required to break the π bond in ethylene. It requires 368 kJ/mol to break the carbon-carbon σ bond in ethane. This comparison is qualitative since the σ and "second" π bond in acetylene may not have the same bond strengths as the π bond in ethylene and the σ bond in ethane.

Chemical Properties

Chemical properties of alkynes are similar to those of alkenes. Alkynes react with many of the same reagents that react with alkenes. (Reactions of alkenes are discussed in Chapters 6 and 8.) With care, alkynes can undergo reactions

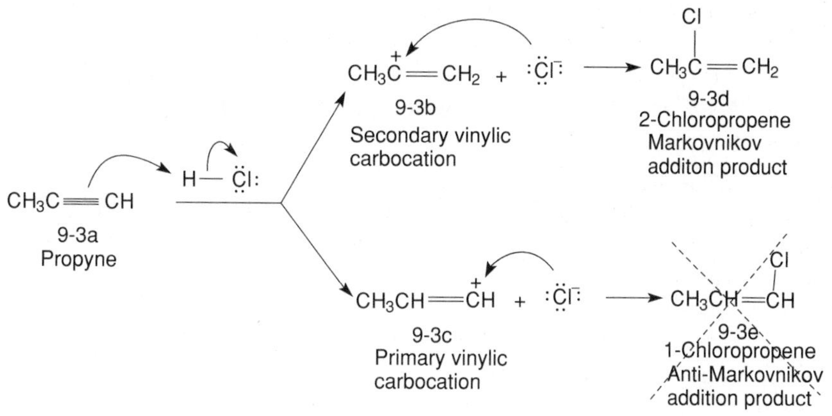

Fig. 9-3.   Reaction of HCl with propyne.

that involve only one of the π bonds. When one π bond of an alkyne reacts, an alkene is formed. The π electron density of alkynes is twice that of alkenes but some alkynes tend to be less chemically reactive than alkenes.

Reactions with Brønsted-Lowry Acids

Alkynes undergo Markovnikov addition reactions with HCl, HBr, and HI. Figure 9-3 shows the reaction of HCl with propyne (Structure 9-3a). The nucleophilic alkyne initially reacts with the electrophilic proton (H$^+$). Markovnikov addition gives the secondary vinylic carbocation (also called a vinyl cation) 9-3b while anti-Markovnikov addition gives the primary vinylic carbocation 9-3c. A reaction will occur predominantly by the lowest energy pathway. The order of stability of vinylic carbocations follows the order of stability of alkyl carbocations; that is, a secondary vinylic carbocation is more stable than a primary vinylic carbocation. Figure 9-4 lists the relative stability of vinylic and alkyl carbocations.

$$R_3\overset{+}{C} \; > \; R_2\overset{+}{C}H \; > \; R\overset{+}{C}{=}CH_2 \; > \; R\overset{+}{C}H_2 \; > \; RHC{=}\overset{+}{C}H \; > \; \overset{+}{C}H_3$$

3° alkyl 2° alkyl 2° vinylic 1° alkyl 1° vinylic Methyl

(most stable)   (least stable)

Fig. 9-4.   Relative stability of carbocations.

THE HAMMOND POSTULATE

According to the Hammond postulate (for a review see the section entitled *The Hammond Postulate* in Chapter 7) the transition state leading to a carbocation resembles (in structure and energy) that carbocation. The greater the stability of the carbocation, the greater the stability (the lower the energy barrier/hill) of the transition state leading to that carbocation. One would then predict, correctly, that the reaction of HCl with propyne proceeds via the more stable secondary vinylic carbocation giving 2-chloropropene (9-3d) and not 1-chloropropene (9-3e).

QUESTION 9-3
Why don't tertiary vinylic carbocations exist?

ANSWER 9-3
A tertiary carbocation must be bonded to three other carbon atoms. If one of these carbon atoms is bonded with a double bond (to make the alkene structure) there would then be four bonds to the carbocation. Draw the structure of a protonated internal alkyne to convince yourself that a maximum of only two alkyl groups are attached to an vinylic carbocation.

CARBOCATION STABILITY

One explanation why alkynes are less reactive than alkenes in reactions with hydrogen halides (protic acids) is based on the order of stabilities of the carbocation intermediates. Figure 9-5 shows an energy profile of transition states and carbocations for vinylic and alkyl carbocations. The relative energy of these species is the basis for the following discussion.

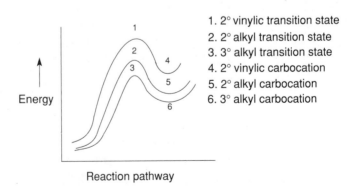

1. 2° vinylic transition state
2. 2° alkyl transition state
3. 3° alkyl transition state
4. 2° vinylic carbocation
5. 2° alkyl carbocation
6. 3° alkyl carbocation

Fig. 9-5. Transition states of carbocations.

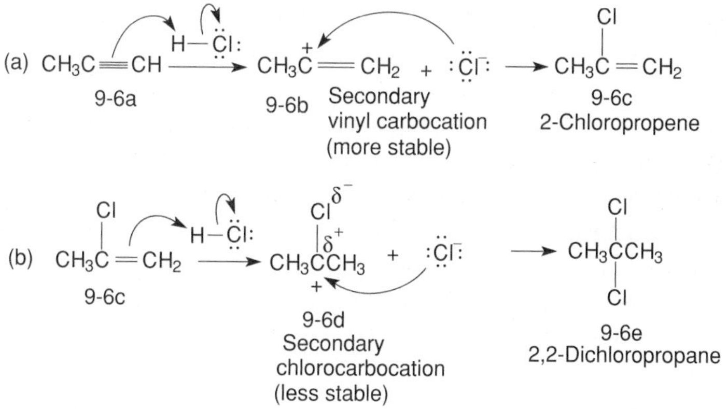

Fig. 9-6. Reaction of HCl with propyne and 2-chloropropene.

The addition of HCl to propyne [Fig. 9-6, Reaction (a)] gives 2-chloropropene (Structure 9-6c). In Reaction (b), 2-chloropropene can react with an additional equivalent of HCl to give 2,2-dichloropropane (Structure 9-6e). The ratio of products formed, 9-6c/9-6e, depends upon the rates of both reactions. The rate of a reaction is directly proportional to the activation energy (height of the energy hill) for that reaction.

AN UNEXPECTED RESULT

The reaction of alkyne 9-6a with HCl proceeds via the secondary vinylic carbocation intermediate 9-6b [Reaction (a)]. The reaction of chloroalkene 9-6c with HCl proceeds via a secondary alkyl carbocation, 9-6d [Reaction (b)]. As shown in Fig. 9-4, a secondary alkyl carbocation is more stable than a secondary vinylic carbocation. According to the Hammond postulate, the transition state leading to the secondary alkyl carbocation should then be lower in energy than the transition state leading to the secondary vinylic carbocation. One would therefore predict monochloroalkene 9-6c reacts with HCl faster than alkyne 9-6a reacts with HCl. Structure 9-6a reacts with HCl—but *Structure 9-6c does not react faster than Structure 9-6a.* The monochloroalkene (9-6c) is the major product when *equal molar (equivalent) quantities* of propyne and HCl react.

Competing reactions

To follow this discussion you need to carefully follow the reactions in Fig. 9-6. Why does the addition of HCl to propyne result primarily in 2-chloropropene (9-6c) and not 2,2-dichloropropane (9-6e)? Consider the two carbocation intermediates (9-6b and 9-6d) in the competing reactions shown in Fig. 9-6.

Addition of HCl to propyne 9-6a results in a secondary vinylic carbocation 9-6b. Reaction of HCl with 2-chloropropene (9-6c) results in a *chlorinated* secondary alkyl carbocation 9-6d.

Chlorine is more electronegative than carbon and the chlorine atom will withdraw the Cl–C bonding electrons away from the positively charged carbon atom in Structure 9-6d. An electron-withdrawing atom or group, such as chlorine, bonded to a carbocation gives that carbocation an even greater positive charge, making it even less stable. In this case, chlorine decreases the stability (increases the energy) of the secondary alkyl carbocation (Structure 9-6d) so it is *less stable* than the nonchlorinated secondary vinylic carbocation (Structure 9-6b). Since the vinylic cation (Structure 9-3b) is lower in energy than is the chlorinated secondary alkyl carbocation (Structure 9-6d) the transition state leading to Structure 9-6b will also be lower in energy than is the transition state leading to Structure 9-6d. Therefore Reaction (a) will be faster than Reaction (b). With equal molar quantities of alkyne and HCl, most of the HCl will have reacted with the alkyne (Structure 9-6a) before it can react with the chloroalkene (Structure 9-6c).

ANTI ADDITION

Figure 9-7 shows the reaction of HCl with dimethylacetylene (9-7a) to give the *anti addition* product 9-7c. A three-membered transition state (Structure 9-7b) involving the hydrogen atom has been proposed. The hydrogen atom bridge blocks attack by the chloride ion from the "bridge side" of the molecule and chloride ion attack occurs from the opposite (anti) side. In this example the reaction is *stereospecific*. However syn addition is also observed in reactions run under different conditions, suggesting a planar vinylic sp^2 hybridized cation (Structure 9-7d) is involved. In the second case, the reaction is not *stereospecific*. This is an example where reaction conditions determine the reaction mechanism and the stereochemistry of the products.

When 2 mol of HCl reacts with 1 mol of propyne (9-6a), HCl will add in a Markovnikov fashion, first to the alkyne and then to the resulting alkene (9-6c) to give geminal 2,2-dichloropropane (9-6e). This is a regiospecific reaction. Geminal (gem) means two groups or atoms are bonded to the same atom.

1 Alkyne + 2 HCl ⟶ Gem dichloroalkane
Markovnikov addition

Toolbox 9-1.

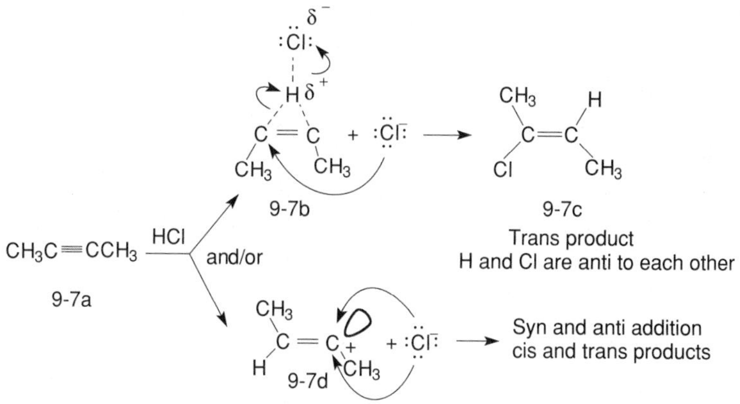

Fig. 9-7.   Addition of HCl to 2-butyne.

Reactions with HBr and Peroxides

Alkynes react with HBr, *in the presence of peroxides*, to give the anti-Markovnikov addition product. The reaction mechanism is shown in a series of reactions in Fig. 9-8. The mechanism can be explained in terms of the more general definition of Markovnikov addition—the reaction will take place by way of the most stable intermediate(s)/transition state(s).

RADICAL MECHANISM

The mechanism is best understood by following the reaction sequence in Fig. 9-8 as the text is read. In Reaction (a), peroxide (9-8a) undergoes homolytic bond cleavage to give two radicals (9-8b). Each radical reacts with HBr to produce a bromine radical (9-8c). In Reaction (b), the bromine radical can react with a terminal alkyne (9-8d) to give a secondary vinylic (9-8e) or a primary vinylic (9-8f) radical. The stability of the radical species follows the same order as the stability of carbocations. Secondary radicals (9-8e) are more stable than primary radicals (9-8f). According to the Hammond postulate, the transition state leading to a secondary radical will be of lower energy than the transition state leading to a primary radical. The reaction thus proceeds by the most energy efficient pathway, the secondary radical intermediate. Radical 9-8e then reacts with HBr [Reaction (c)] to abstract (remove) the hydrogen atom from HBr giving monobromoalkene 9-8g and a bromine radical, 9-8c. The reaction is not stereospecific as a mixture of cis and trans bromoalkenes (9-8g) is produced.

(a) RO—OR $\xrightarrow{\text{heat}}$ 2 RO· $\xrightarrow{2H—Br}$ 2 ROH + 2 Br·
         9.8a   or light   9-8b   9-8c
      Peroxide   Radical   Radical

(b) Br· + RC≡CH ⟶ RĊ=CHBr   or   RC=ĊH
                                      9-8e |
        or   2° vinylic radical   Br
     9-8c   9-8d   9-8f
                                                  1° vinylic radical

(c) RĊ=CHBr + H—Br ⟶ RCH=CHBr + Br·
        9-8e   9-8g   9-8c
                                 Cis and trans

(d) RCH=CHBr + Br· ⟶ RĊH—CHBr$_2$
        9-8g   9-8c   9-8h

(e) RĊH—CHBr$_2$ + H—Br ⟶ RCH$_2$CHBr$_2$ + Br·
        9-8h   9-8i   9-8c

Fig. 9-8. Reaction of HBr with an alkyne in the presence of peroxide.

In Reaction (d), bromine radical (9-8c) reacts with the bromoalkene (9-8g) to give the 2° alkyl dibromo radical intermediate 9-8h. Intermediate 9-8h abstracts a hydrogen atom from HBr [Reaction (e)] to give the geminal dibromo compound 9-8i and another bromine radical. The overall reaction is regiospecific. This is another example of a radical chain reaction. HBr is the only halogen acid that reacts by this radical mechanism to give the anti-Markovnikov product. HCl and HI do not give the anti-Markovnikov product in the presence of peroxides for thermodynamic reasons.

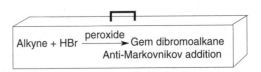

Alkyne + HBr $\xrightarrow{\text{peroxide}}$ Gem dibromoalkane
                        Anti-Markovnikov addition

Toolbox 9-2.

Bridged bromonium ion

$$RC\equiv CH + Br-Br \longrightarrow$$

9-9a

$$\overset{R}{\underset{Br}{}}C=C\overset{Br}{\underset{H}{}}$$

9-9b

Anti-addition
trans product

$$\underset{Br}{\overset{Br}{H-C}}-\underset{Br}{\overset{Br}{C-H}} \xleftarrow{\quad Br_2 \quad}$$

9-9c
Tetrabromoalkane

Fig. 9-9. Reaction of bromine with an alkyne.

Reactions with Halogens

Alkynes react with bromine (Br_2) and chlorine (Cl_2) to give the dihalo- and tetrahalo-substituted products, as shown in Fig. 9-9. Bromine reacts with an alkyne in an anti addition fashion, giving the trans dibromoalkene (9-9b) as the major product. This suggests a bridged bromonium ion intermediate (9-9a) is involved. This bromonium bridge intermediate prevents the bromide anion from attacking the "bridge side" of the molecule. Attack by the bromide anion from the side opposite to the bridge gives the anti (trans) addition product (9-9b). However, the reaction is not as stereospecific as is the addition of bromine to an alkene. (For a review see the section entitled *Halogenation of Alkenes in Inert Solvents* in Chapter 8.) The dibromo compound readily reacts with bromine to form the tetrabromoalkane, (9-9c). Under controlled conditions, such as low temperatures and slow addition of bromine, one can obtain the vicinal dibromo adduct as the major product.

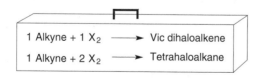

Toolbox 9-3.

Hydration Reactions

Alkynes react with water in the presence of protic acids and mercury(II) (Hg^{2+}) salt catalysts to form a molecule containing alkene (C=C) and alcohol (OH)

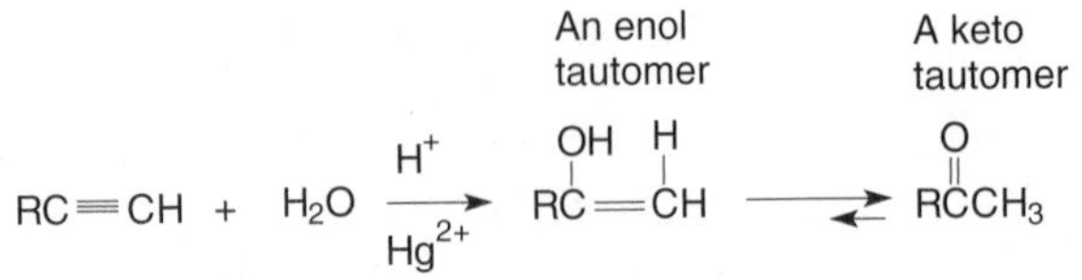

$$\text{RC} \equiv \text{CH} + \text{H}_2\text{O} \xrightarrow[\text{Hg}^{2+}]{\text{H}^+} \underset{\substack{\text{An enol} \\ \text{tautomer}}}{\overset{\substack{\text{OH} \quad \text{H}}}{\text{RC}=\text{CH}}} \rightleftharpoons \underset{\substack{\text{A keto} \\ \text{tautomer}}}{\overset{\text{O}}{\text{RCCH}_3}}$$

Fig. 9-10. Acid-catalyzed hydration of an alkyne and tautomerization.

group bounded to an alkene carbon atom, shown in Fig. 9-10. The resulting molecule is called an *enol, en* from the alk*ene* and *ol* from the alcoh*ol*. The net reaction is Markovnikov addition of water to an alkyne.

QUESTION 9-4
Why does acid-catalyzed addition of water to a terminal alkyne give the Markovnikov product?

ANSWER 9-4
Markovnikov addition results in a secondary vinylic carbocation intermediate which is more stable than a primary vinylic carbocation intermediate obtained by anti-Markovnikov addition. The reaction will proceed predominately via the transition state of lowest energy and the corresponding carbocation intermediate.

TAUTOMERIZATION

The enol compound is in equilibrium with a carbonyl (C=O, e.g., aldehyde or ketone) compound. The equilibrium, called an *enol-keto equilibrium*, is shown in Fig. 9-10. The equilibrium is strongly shifted toward the keto form. (The keto form refers to a ketone or an aldehyde.) The enol and carbonyl compounds are called *tautomers*. The process of interconverting tautomers is called *tautomerization*.

Addition of water

Figure 9-11 shows the catalyzed addition of water to three different alkynes: acetylene (9-11a), 1-butyne (9-11b), and 2-butyne (9-11c). Acetylene reacts with water [Reaction (a)] to give an aldehyde, acetaldehyde (9-11d). This is a special case where the keto-enol equilibrium involves an aldehyde and not a ketone. Terminal (monosubstituted) alkynes (9-11b) react with water [Reaction (b)] to give methyl ketones (9-11e). Internal alkynes (9-11c) react with water [Reaction (c)] to give two products from nonsymmetrical alkynes and one product from symmetrical alkynes (9-11c). Symmetrical alkynes have the same

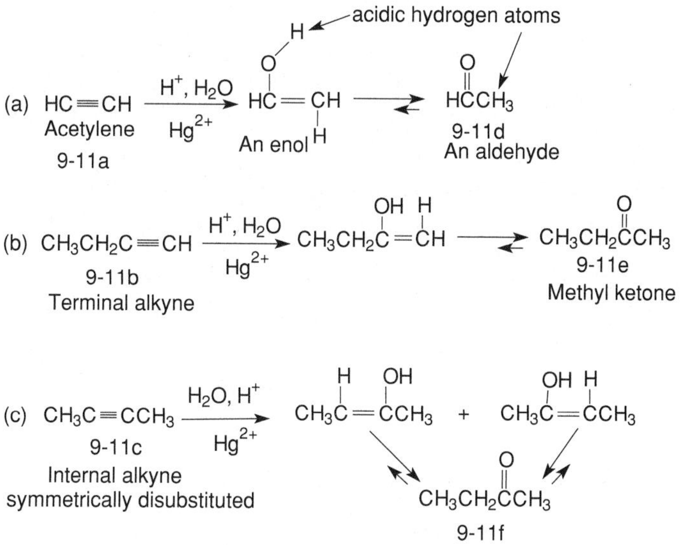

Fig. 9-11. Acid-catalyzed hydration of alkynes.

groups bonded to each side of the alkyne functional group while nonsymmetrical alkynes have different groups bonded to each side of the alkyne functional group.

Since each alkyne carbon atom in an unsymmetrical alkyne is bonded to a different alkyl group, H and OH can bond to either carbon atom resulting in two different ketones. The hydration of unsymmetrical internal alkynes results in two products that would have to be separated.

QUESTION 9-5
What product(s) would you expect from the acid-catalyzed addition of water to 2-pentyne, $CH_3C{\equiv}CCH_2CH_3$?

ANSWER 9-5

$$CH_3\overset{\overset{\displaystyle O}{\|}}{C}CH_2CH_2CH_3 \text{ and } CH_3CH_2\overset{\overset{\displaystyle O}{\|}}{C}CH_2CH_3.$$

SURVIVAL OF THE WEAKEST

The equilibrium between the enol and keto form is usually shifted strongly toward the keto form as shown in the reactions in Figs. 9-10 and 9-11. The keto form must therefore be more thermodynamically stable (lower in energy).

This equilibrium can be explained in terms of an acid-base reaction. In acid-base reactions, the general rule is survival of the weakest (acid or base). The hydrogen atom of an alcohol (—OH, the enol form) is more acidic than an alpha hydrogen atom attached to α carbon atom adjacent to the carbonyl group (the keto form). These hydrogen atoms are indicated in the structures in Figure 9.11, Reaction (a). Survival of the weakest acid would favor the keto form since the hydrogen atom bonded to the carbon atom is less acidic than the hydrogen atom bonded to the oxygen atom.

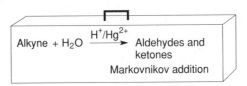

Alkyne + H_2O $\xrightarrow{H^+/Hg^{2+}}$ Aldehydes and ketones

Markovnikov addition

Toolbox 9-4.

Hydroboration-Oxidation Reactions

Borane, BH_3, reacts with alkynes by a mechanism similar to its reaction with alkenes. Figure 9-12 shows a disubstituted borane, HBR_2, reacting in a syn manner (Structure 9-12a) with an alkyne. *After the reaction of one molecule of borane with one molecule of alkyne, the resulting adduct is an alkene* (9-12b). This alkenylborane compound can react with another molecule of HBR_2 to give the diborane adduct (9-12c). The addition of a second molecule of HBR_2 can be minimized by having two large (bulky) R groups attached to the borane atom. The major product is then monoborane adduct 9-12b. When Compound 9-12b reacts with basic, aqueous hydrogen peroxide (an oxidation reaction), boron is replaced by an —OH group giving Structure 9-12d, an *enol*. This represents

Fig. 9-12. Hydroboration-oxidation of an alkyne.

the anti-Markovnikov addition of a molecule of water to an alkyne. The enol undergoes tautomerization to give an aldehyde (9-12e).

Hydroboration-oxidation of terminal alkynes gives aldehydes while the acid-catalyzed addition of water to terminal alkynes gives ketones. Hydroboration-oxidation of internal alkynes gives ketones, as does the acid-catalyzed addition of water to internal alkynes.

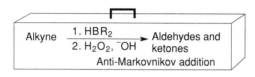

Toolbox 9-5.

QUESTION 9-6

What product would you expect from the hydroboration-oxidation of 2-butyne, $CH_3C{\equiv}CCH_3$?

ANSWER 9-6

Since this is a symmetrical alkyne, only one product is obtained.

$$CH_3\overset{\overset{\displaystyle O}{\|}}{C}CH_2CH_3$$

Hydrogenation/Reduction Reactions

Alkynes react with hydrogen in the presence of a finely divided (powdered) metal catalyst such as platinum (Pt), palladium (Pd), or nickel (Ni), to give the corresponding alkane. It is difficult to stop the reaction at the alkene stage since the same reaction conditions are used to convert alkenes to alkanes. (For a review see the section entitled *Hydrogenation of Alkenes* in Chapter 8.) The addition of hydrogen atoms to a carbon compound is called a reduction reaction.

LINDLAR'S CATALYST—CIS ALKENE FORMATION

Hydrogenation of alkynes can be stopped at the alkene stage with the use of special catalysts. A common catalyst is *Lindlar's catalyst*. The catalyst, palladium, is made less reactive (it is deactivated or poisoned) by treating it with lead (Pb^{2+}) salts and an organic base. Like the hydrogenation of an alkene, hydrogenation of an alkyne occurs in a syn (same side) manner. The syn addition to internal alkynes gives cis alkenes, as shown in Fig. 9-13, Structure 9-13a. A more recently developed catalyst, nickel boride, Ni_2B, can also be used to synthesize cis alkenes from alkynes.

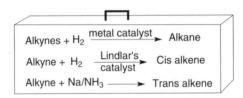

Fig. 9-13. Hydrogenation of an alkyne to an alkene.

TRANS ALKENE FORMATION

Anti (opposite side) addition of hydrogen to alkynes to give trans alkenes can also be accomplished with special reagents. The reaction of sodium (Na) or lithium (Li) metal in liquid ammonia results in the addition of hydrogen to internal alkynes in an anti manner to give trans alkenes (9-13b).

Alkynes + H_2 $\xrightarrow{\text{metal catalyst}}$ Alkane

Alkyne + H_2 $\xrightarrow{\text{Lindlar's catalyst}}$ Cis alkene

Alkyne + Na/NH_3 $\longrightarrow$ Trans alkene

Toolbox 9-6.

QUESTION 9-7
Methylacetylene (propyne) reacts with hydrogen in the presence of Lindlar's catalyst. Is the cis or trans alkene obtained?

ANSWER 9-7
Neither. Addition of 1 mol of hydrogen to 1 mol of a terminal alkyne results in an alkene with two hydrogen atoms on the terminal carbon atom, $CH_3CH=CH_2$. The compound cannot be cis or trans.

Oxidation Reactions

Alkynes can be oxidized (the addition of oxygen to a carbon compound) with a variety of reagents. Only two reagents, shown in Fig. 9-14, will be discussed. One reagent is the permanganate ion, MnO_4^-. Under mild conditions — neutral aqueous solutions and moderate temperatures — oxidation of an alkyne gives a diketone (9-14a). Under more rigorous conditions — strong base and high temperatures — alkynes are oxidized to salts of carboxylic acids (9-14b).

Ozone, O_3, is also used to oxidize alkynes to carboxylic acids (9-14c) at room temperature. Ozone is relatively easy to use but quite toxic and care is needed

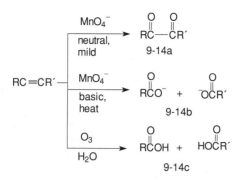

Fig. 9-14. Oxidation of alkynes.

when using this reagent. The choice of an oxidizing agent depends upon the sensitivity of other functional groups in an alkyne. Oxidation with ozone or permanganate may cause an undesirable reaction in other parts of a molecule and therefore the oxidizing reagent must be chosen with care.

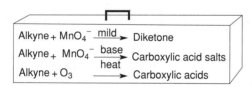

Toolbox 9-7.

Acidity of Alkynes

The Brønsted-Lowry definition of acids are compounds that donate hydrogen atoms (protons). Terminal alkynes ($RC{\equiv}CH$), alkenes ($R_2C{=}CH_2$), and alkanes (R_3CH) each have a hydrogen atom bonded to an sp, sp^2, or sp^3 carbon atom, respectively. Each of these molecules has the potential to act as an acid. The strength of an acid depends on how willing the compound is to give up (donate) a proton to a base.

THE pK_a SCALE

Strong acids such as HCl donate a proton readily and have small pK_a values (approx. -1 to -10). Carboxylic acids are weaker acids and have pK_a values of about 5. Water and alcohols are still weaker acids with pK_a values of 16 to 18. Alkanes are acids, but very weak acids with pK_a values of about 50. Alkenes

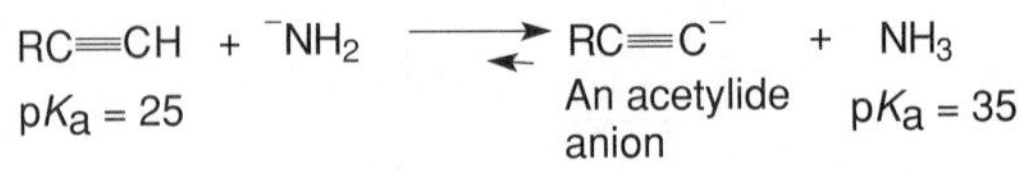

$$RC{\equiv}CH + {^-}NH_2 \rightleftarrows RC{\equiv}C^- + NH_3$$

$$pK_a = 25 \qquad\qquad \text{An acetylide} \quad pK_a = 35$$
$$\text{anion}$$

Fig. 9-15.   Formation of an acetylide anion.

are slightly more acidic, but still very weak acids with pK_a values of about 45. Alkynes are more acidic than alkanes and alkenes with pK_a values of about 25. Alkynes are still very weak acids, a billion times weaker than water but 10^{20} times more acidic than alkenes. (For a review of acidity values see Table 3-1 in Chapter 3.)

SURVIVAL OF THE WEAKEST

Terminal alkynes act as acids in the presence of a strong base. The amide anion, $^-NH_2$, the conjugate base of ammonia ($pK_a = 35$), is a strong enough base to completely remove a proton from a terminal alkyne. Figure 9-15 shows this reaction and the corresponding pK_a values. This is an acid-base reaction where $RC{\equiv}CH$ is the acid on the left side of the equilibrium expression and ammonia, NH_3, is the conjugate acid on the right side of the expression. In acid-base reactions, the rule is survival of the weakest. Since ammonia is a weaker acid ($pK_a = 35$) than the alkyne ($pK_a = 25$), the reaction is shifted strongly to the right. The larger the pK_a value, the weaker the acid. The conjugate base of this reaction is the *acetylide anion,* $RC{\equiv}C^-$.

Alkylation Reactions

Organic chemistry includes the study of synthesizing molecules that are not readily available from natural sources. The acetylide anion is a useful synthetic reagent. These anions are strong bases and good nucleophiles. They have a non-bonding electron pair they are willing to share with an electrophile. Acetylide anions react readily with methyl compounds (CH_3X) and primary alkyl compounds (RCH_2X) where X is a good leaving group like a halide or tosylate anion (Structure 9-16a). An example of this reaction is shown in Fig. 9-16.

The C—X bond is polar when the X atom/group is more electronegative than C. Bond polarity makes C partially positive and electrophilic. The reaction shown in Fig. 9-16 is called a *nucleophilic substitution reaction*. The X group is replaced (substituted) with the acetylide group.

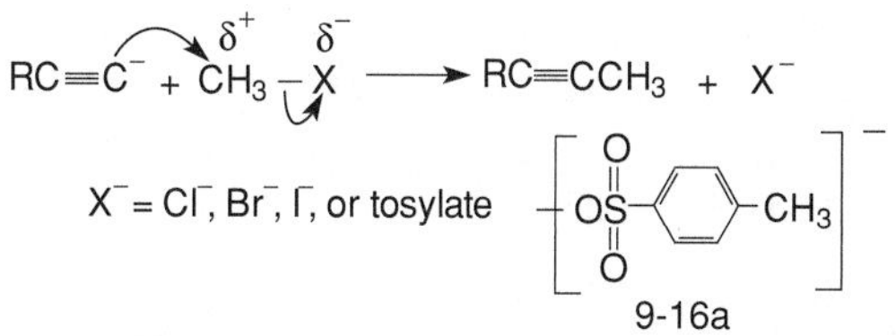

Fig. 9-16.   Alkylation via an acetylide anion.

SYNTHESIS OF ALKYNES

A nucleophilic substitution reaction is a convenient way of making larger molecules containing an alkyne function. Figure 9-17 shows a reaction scheme starting with acetylene. In the first step, a proton is removed from acetylene by the amide anion ($^-NH_2$). The resulting acetylide anion reacts with a primary alkyl halide to give a terminal alkyne (9-17a). In the next step, amide anion is again used to remove the remaining acetylenic proton. The resulting acetylide anion reacts with another alkyl halide molecule to give a disubstituted internal alkyne (9-17b).

   Only methyl and primary alkyl—X reagents can be used in these reactions. Secondary and tertiary alkyl—X compounds tend to give elimination reactions (alkene formation), not substitution reactions.

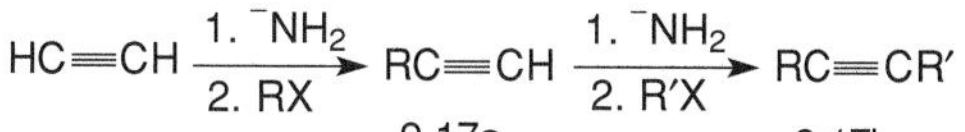

Fig. 9-17.

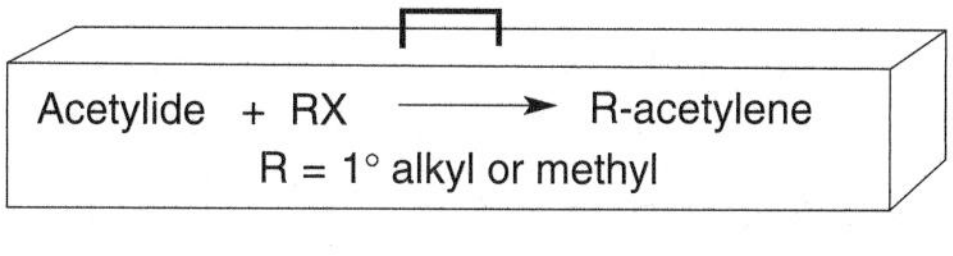

Toolbox 9-8.

QUESTION 9-8

One mole of acetylene is reacted with 1 mol of amide anion ($NH_2{}^-$) to give product A. A is reacted with bromoethane to give B. One mole of B is reacted with 1 mol of $NH_2{}^-$ to give C. C is reacted with 1 mol of bromoethane to give D. What are the structures of A, B, C, and D?

ANSWER 9-8

A is the acetylide anion, $HC{\equiv}C^-$; B is 1-butyne (ethylacetylene), $CH_3CH_2C{\equiv}CH$; C is ethylacetylide, $CH_3CH_2C{\equiv}C^-$; and D is 3-hexyne (diethylacetylene), $CH_3CH_2C{\equiv}CCH_2CH_3$.

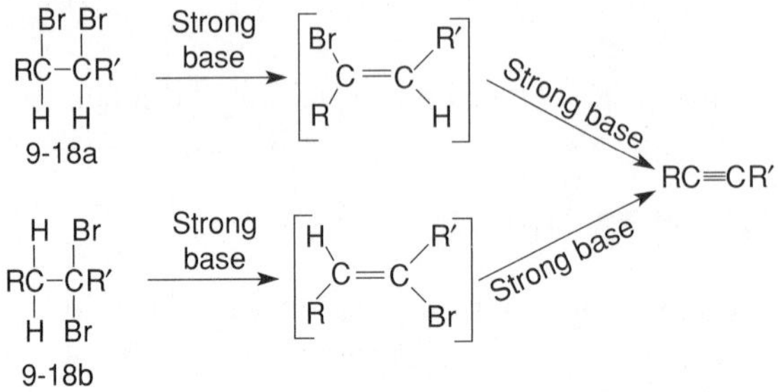

Fig. 9-18.   Preparation of alkynes.

PREPARATION OF ALKYNES

Alkynes can be prepared by *dehydrohalogenation* of vicinal (9-18a) or geminal (9-18b) dihalides by reaction with a strong base. Figure 9-18 shows both reactions. Brackets around a compound indicate it is an intermediate product. Gem is the abbreviation for geminal (twins) and vic is an abbreviation for vicinal (nearby).

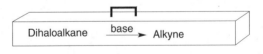

Toolbox 9-9.

Quiz

1. The strength of a triple bond is
   (a) equal to the strength of a double bond
   (b) less than the strength of a double bond
   (c) greater than the strength of a double bond
   (d) less than the strength of a single bond

2. The length of a triple bond is
   (a) equal to the length of a single bond
   (b) less than the length of a single bond

(c) greater than the length of a single bond
(d) greater than the length of a double bond

3. An alkyne's name ends in
    (a) -ane
    (b) -ene
    (c) -yne
    (d) diene

4. Alkynes have a density ——— that of water.
    (a) equal to
    (b) greater than
    (c) less than

5. An alkyne function has ——— pi bond(s).
    (a) one
    (b) two
    (c) three
    (d) four

6. Alkynes react with HCl by a mechanism called
    (a) elimination
    (b) Markovnikov addition
    (c) anti-Markovnikov addition
    (d) substitution

7. A secondary alkenyl carbocation is ——— a secondary alkyl carbocation.
    (a) equal in stability to
    (b) more stable than
    (c) less stable than

8. Excess bromine reacts with an alkyne to give a
    (a) diketone
    (b) tetrabromide
    (c) hexabromide
    (d) tetraalcohol

9. Alkynes react with water in the presence of a catalyst to give
    (a) a dialcohol (diol)
    (b) an alkane
    (c) an enol
    (d) a dibromide

10. An enol-keto equilibrium is usually shifted toward
     (a) the enol form
     (b) the keto form
     (c) equally in both directions
     (d) no equilibrium exists

11. The hydroboration-oxidation of an alkyne gives ——— addition.
     (a) Markovnikov
     (b) anti-Markovnikov
     (c) dehydration
     (d) no

12. The conversion of alkynes to alkanes is an example of
     (a) oxidation
     (b) reduction
     (c) chlorination
     (d) dehydration

13. Lindlar's catalyst is used to convert an alkyne to
     (a) an alkane
     (b) a cis alkene
     (c) a trans alkene
     (d) a cis alkyne

14. Ozone reacts with an alkyne to give
     (a) alcohols
     (b) acids
     (c) chlorides
     (d) water

15. A terminal alkyne is ——— an alkane.
     (a) more acidic than
     (b) less acidic than
     (c) more basic than
     (d) equal in acidity to

16. An alkylation reaction refers to
     (a) adding an alcohol group
     (b) forming an alkylate ion
     (c) removing an alkyl group
     (d) adding an alkyl group

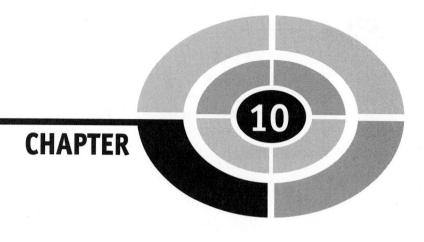

CHAPTER

10

Characterization

Introduction

When a new compound is made in a laboratory or is isolated from some natural source, such as a plant or microorganism, its physical properties and chemical structure must be determined. Chemists have many analytical methods to purify and characterize compounds. Five methods are discussed here: chromatography, infrared spectroscopy, ultraviolet spectroscopy, nuclear magnetic resonance spectroscopy, and mass spectrometry.

Chromatography

Chromatography is a method of physically separating the components in a mixture of compounds. This technique does not necessarily tell what the separated compounds are but can give information about their general structures (polar characteristics). There are several chromatography methods that are all based on the same separation principle. The sample to be analyzed may be a gas, a liquid, or a solid dissolved in a liquid. The sample to be analyzed is passed over a solid, called a *stationary phase*. The stationary phase is typically powdered silica gel (hydrated SiO_2) or alumina (Al_2O_3). These materials are polar. They may be used as is or coated with another material to make their surface more or less polar. The sample being analyzed has some degree of attraction for the stationary phase. *Thin layer chromatography* (TLC) is used to discuss the basic principles of chromatography.

THIN LAYER CHROMATOGRAPHY

A thin layer of the stationary phase is applied to a solid surface/backing. The solid surface is often a glass plate or a thin plastic sheet. The size of this plate/sheet can vary. Typical sizes are 2.5 cm by 10 cm (1 by 3 in.) to 900 cm^2 (1 ft^2). The stationary phase adheres to the solid surface/backing so that the plate/sheet can be placed in a vertical position.

Spotting

A very small amount of sample to be tested is put (spotted) near the bottom of the coated plate/sheet. Typically, the sample is dissolved in a solvent and a very thin capillary is used to place (spot) a small amount of sample on the sheet. The sheet is then placed in a vertical position in a beaker containing a small amount of liquid. The liquid is called the developing solvent or the mobile phase. An example of a TLC experiment is shown in Fig. 10-1. It is critical that the sample (spot) is placed *above the level* of the (developing) solvent in the beaker.

Sample separation

The solvent will wick up (climb up/wet) the stationary phase. As the solvent wicks up the stationary phase it encounters the sample. The solvent dissolves

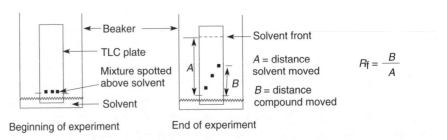

Fig. 10-1.   TLC experiment.

the sample because of the intermolecular attractions between the sample and the solvent. The sample will also be attracted to the stationary phase. As the sample, now dissolved in the solvent, moves up the coated plate/sheet it will *partition* between the solvent and the stationary phase. The sample will take a "zigzag path"—dissolved in the solvent, absorbed on the stationary phase, dissolved in the solvent, absorbed on the stationary phase, dissolved in the solvent, and so on. Since different materials will have different attractions for the solvent and for the stationary phase, they will move up the stationary phase at different rates. In this manner compounds in a sample are separated.

R_f values

The distance the material moves from the place where it was spotted is compared to the distance the solvent moves from the point where the sample was spotted. One can usually visually observe how high the solvent rises (the solvent front). The procedure is stopped before the solvent reaches the top of the plate/sheet. The ratio of how far a component in the sample moves divided by the distance the solvent moves is called the retention factor, or R_f value. Figure 10-1 shows the R_f calculation when B is the distance moved by one compound and A is the distance moved by the solvent. The R_f value will be a specific value under the conditions of the test for each component in the sample.

QUESTION 10-1
Does a polar compound have a larger or a smaller R_f value when the polarity of the stationary phase is increased?

ANSWER 10-1
The R_f value will decrease as the material will "stick" to the stationary phase more tightly and move more slowly toward the solvent front.

Developing

Often the materials of interest are colorless and not visible on the stationary phase at the end of the experiment. The sample can be made visible by staining with iodine (materials turn a blue/purple color) or the sample may fluoresce under a UV light source. This is called *developing* the TLC plate/sheet. It is possible that two different compounds may have the same R_f value. In this case a mixture of three compounds may indicate only two materials (two R_f values). Changing the solvent composition is a way of separating (resolving) R_f values if two different materials happen to have the same R_f value in one solvent system. The sample is usually not recovered since very small quantities of sample are typically used.

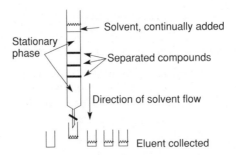

Fig. 10-2.   Column chromatography.

COLUMN OR LIQUID CHROMATOGRAPHY

Larger quantities (milligrams to grams) may be separated by *column chromatography*. The equipment consists of a glass (or plastic) column with a control valve (stopcock) on one end. Columns vary in length from a few centimeters (inches) to a meter or greater (several feet). A column chromatography experiment is shown in Fig. 10-2. Solid support is put (packed) in the column containing the desired solvent (eluent). The sample to be analyzed is added at the top of the column. The solvent exits at the bottom of the column and fresh solvent is continually added at the top of the column. As the solvent moves down the column, the sample also moves down the column. Components in the sample are separated by the same zigzag pathway discussed for TLC. Eluent exiting the column is collected in a series of vials. Each vial is tested for the presence of a component of the sample.

QUESTION 10-2
Which component in a mixture will exit the column first: the most or the least polar?

ANSWER 10-2
The least polar component; the most polar component will "stick" to the stationary phase and move down the column at a slower rate.

HIGH PERFORMANCE LIQUID CHROMATOGRAPHY

The efficiency of separating a complex sample mixture depends on the number of times the zigzag process occurs. The extent of sample separation is proportional to the total surface area of the stationary phase since a greater surface area gives the components of a mixture more opportunity to interact with the stationary

phase surface. If the stationary phase is ground to a fine powder, its surface area is greatly increased. These small particles pack more tightly together in a column than do larger particles and it is more difficult for the solvent (the eluent) to flow through the column. The flow rate is increased by using pressure to force the solvent through the column. This chromatography method gives a more efficient separation than does column chromatography and hence the name *high performance liquid chromatography (HPLC)*.

The sample is injected into the chromatograph with a syringe. The diameter of the column, a metal tube, is much smaller than the diameter of columns used in column chromatography. These tubes have very small (a few millimeters) inner diameters and are typically just a few centimeters long. The column can be heated to assist separation. A detection system is used to analyze the eluent as it leaves the chromatograph.

GAS (VAPOR) PHASE CHROMATOGRAPHY

Samples that can be vaporized can be separated in the gas phase. The instrument used to carry out these separations is similar to an HPLC chromatograph. One difference is that a carrier gas is used to move the sample through the column. A schematic of a *gas phase chromatograph (GC)* unit is shown in Fig. 10-3. A sample of a liquid (or gas) is injected into a heated port of a GC where the sample is vaporized. An inert carrier gas (that replaces the solvent used in other methods) enters the chromatograph in the heated port and flows through the column packed with a stationary phase. The carrier gas moves (carries) the sample through the column. Components in the sample partition between the vapor phase and the stationary phase. Components move in a zigzag pathway through the column and are separated for the same reasons explained for the TLC process. A detection system is used to tell when components exit the chromatograph.

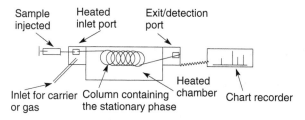

Fig. 10-3. GC chromatograph.

QUESTION 10-3

A sample containing CCl_4 (boiling point 76°C) and CH_3CH_2OH (boiling point 78°C) is analyzed by GC. Which material is expected to exit the column first?

ANSWER 10-3

CCl_4, since it is less polar and would "stick" to a lesser extent to the stationary phase.

Chromatography gives information about how many components are in a sample but does not necessarily give information about what the components are. Injecting a known component with the unknown mixture may help identify (but not confirm) the presence of a known compound in the mixture. Two identical materials in the same sample will necessarily exit the chromatograph at the same time but two nonidentical materials may also exit the chromatograph at the same time. Chromatographs can be connected to other analytical devices to help characterize the components as they exit the chromatograph. These devices include infrared spectroscopy and mass spectrometry instruments, which are discussed below.

Spectroscopy

Spectroscopy is the study of the interaction of electromagnetic radiation with matter. Electromagnetic radiation makes up the electromagnetic spectrum, shown in Fig. 10-4. The electromagnetic spectrum is subdivided into regions: radio waves, microwaves, infrared (IR) radiation, visible light, ultraviolet (UV) radiation, x-rays, and gamma rays.

ELECTROMAGNETIC RADIATION

Electromagnetic radiation can be described as a wave traveling at the speed of light. The wave, shown in Fig. 10-5, can be characterized by a *wavelength* and

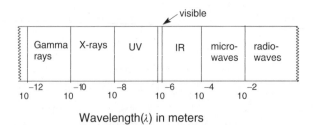

Fig. 10-4. Electromagnetic spectrum.

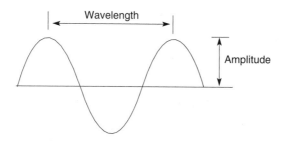

Fig. 10-5.   A standing wave.

a *frequency*. A good analogy is a series of water waves. The wavelength λ is the distance between crests of the wave. The frequency ν, is the number of crests that pass a given point in one second or the number of times a cork floating on water bobs up and down per second. Frequency is expressed in hertz which has units of 1/sec or sec^{-1}.

Frequency times wavelength equals a constant, the speed of light ($\nu\lambda$ = constant). Frequency and wavelength are inversely proportional. If the frequency increases, the wavelength decreases. Electromagnetic radiation can also be described as a stream of photons (a massless packet of energy). The energy of the photons is related to the frequency of radiation by the equation $E = h\nu = h(c/\lambda)$, where h is a constant. High energies are associated with high frequencies and short wavelengths. Low energies are associated with low frequencies and long wavelengths.

QUESTION 10-4
How does the energy of electromagnetic radiation change with increasing wavelength?

ANSWER 10-4
Energy decreases with increasing wavelength.

QUESTION 10-5
Can the frequency increase while the wavelength stays constant?

ANSWER 10-5
No, if the frequency increases, the wavelength decreases.

Molecules are capable of absorbing radiation in the various regions of the electromagnetic spectrum. The energy of the absorbed radiation has different effects on molecules. X-rays have sufficient energy to cause ionization (loss of electrons), UV and visible radiation promote electrons into higher energy levels (excited states), IR radiation excites vibrational modes, and microwave radiation causes the entire molecule to tumble/rotate.

INFRARED SPECTROSCOPY

Infrared spectroscopy measures the interaction (absorption) of infrared radiation with molecules. Bonds between atoms are usually given as specific lengths, implying rigid bonds between atoms. The bond between two atoms is also described as having the properties of a spring. The bond (spring) vibrates and the *average distance* between the two connected atoms (in their lowest energy ground state) is given as the *bond length*. The bond vibrates with a frequency that is characteristic of that specific bond. A C–H bond, a C–C single bond, and a C=C double bond have different vibrational frequencies. Electromagnetic radiation is also described in terms of a frequency. Molecules can absorb electromagnetic radiation if the frequency of the bond vibration matches the frequency of the electromagnetic radiation. Bonds vibrate in the IR frequency range and thus absorb IR radiation.

Energy ranges

Most molecules absorb IR radiation in the wavelength range of 2.5×10^{-4} to 25×10^{-4} cm, which is a small segment of the IR radiation range (8×10^{-5} to 1×10^{-2} cm). This small segment corresponds to a frequency range of 12×10^{13} to 1.2×10^{13} sec^{-1} and an energy range of 4.6 to 46 kJ/mol (1.1 to 11 kcal/mol). This energy range is much smaller than the energy required to break a typical single bond (about 400 kJ/mol or 100 kcal/mol) so the absorption of IR radiation does not cause bonds to break.

Wavenumbers

Radiation in the IR spectroscopy range is usually expressed in units of *wavenumbers* $\overline{\upsilon}$. A wavenumber usually is a frequency measurement and is equivalent to the reciprocal of the wavelength in centimeters: $\upsilon = 1/\lambda$ cm. The wavenumber range of IR spectra is from 400 to 4000 cm^{-1}. Larger values of wavenumbers represent higher energies.

QUESTION 10-6
If a sample absorbed radiation at 1.0×10^{-3} cm, will it appear in an IR spectrum?

ANSWER 10-6
Yes, 1.0×10^{-3} cm is between 2.5×10^{-4} and 25×10^{-4} cm and its reciprocal is between 4000 and 400 cm^{-1}.

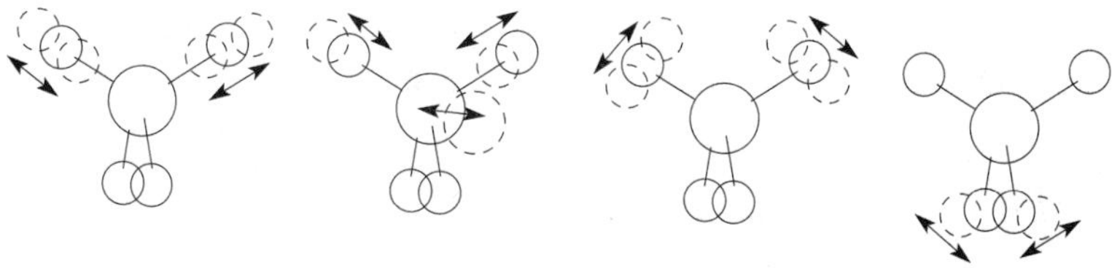

| Symmetrical stretching | Asymmetric stretching | Scissoring bending | Twisting bending |

Fig. 10-6.   Molecular vibrations.

Molecular vibrations

Molecules undergo two types of vibrations: *stretching* and *bending*. More energy is required for a stretching vibration than for a bending vibration. Examples of these vibrations are shown in Fig. 10-6. When a stretching vibration absorbs IR radiation, the amplitude of the vibration changes but the frequency of the vibration does not change. The "spring" stretches twice as far (the amplitude increases) but undergoes the same number of stretches-compressions per second (the frequency) as it did before absorbing the radiation. A bending vibration involves at least three atoms. As an analogy, hold your arms straight out from your shoulders, parallel to the ground. Now bring your hands together, holding your arms stiff. Now return your arms to their original positions. This is analogous to a symmetric bending (scissoring) vibration.

The spectrometer

An *IR spectrometer* is used to obtain spectra of compounds. The spectrometer generates a range of IR radiation from 4000 to 400 cm^{-1}. Radiation is directed through the sample and wavenumbers for the radiation absorbed by the sample are recorded. Older instruments scan through this wavenumber range. Newer instruments generate and expose the sample to all the wavenumbers in this range simultaneously. Newer instruments take much less time to analyze a sample, and give spectra with better resolution. This newer method is called FT-IR spectroscopy. FT stands for Fourier transformation, a mathematical process for analyzing and converting the data into an IR spectrum.

Sample preparation

The sample to be tested can be a gas, a liquid, or a solid. The sample is put in a cell or between plates of a material that does not absorb IR radiation in the 4000 to 400 cm^{-1} wavenumber range. Cells or plates are made from salts like

NaCl, NaBr, and AgCl. Window glass and plastic materials absorb IR radiation in the wavenumber range of interest, and their absorptions would appear on the spectra with those of the sample. Therefore, these materials are not used as cells or plates. Solid samples can be mixed with powdered salt (e.g., NaBr) and pressed into an optically clear pellet. A solid sample can also be ground in mineral oil to give a *mull*. Mineral oil contains C—H and C—C bonds that absorb IR radiation in the IR spectroscopy range and obscure these absorptions in the sample. The sample is placed in the instrument such that the IR radiation from the radiation source passes through the sample.

IR spectra

Four typical *IR spectra* are shown in Fig. 10-7 for ethanol, propionic acid, acetone, and cyclohexene*. The scale on the bottom of the spectrum is in wavenumbers. The left vertical scale is *percent transmittance*. The "baseline" goes across the top of the IR spectrum. A peak (also called an absorption, an absorption band, or a spike) is observed if the sample absorbs IR radiation in the wavelength range of the instrument.

The transmittance varies from 100% (no absorption) to 0% (complete absorption). Transmittance is the opposite of absorption; 0% transmittance equals 100% absorption. The characteristic absorptions for these four compounds are discussed below under the corresponding functional group headings.

Absorption requirements

A molecule will absorb IR radiation if a *polar bond* in that molecule is undergoing a vibration, and the vibration causes a periodic change in the dipole moment of that bond.

A polar bond can be thought of as having a positive "end" and a negative "end." Electromagnetic radiation can also be thought of as an alternating positive and negative charge (the crest and trough of a wave). When the negative "end" of the molecular dipole interacts with the positive charge of the electromagnetic radiation it "stretches" the bond and when the negative charge of the electromagnetic radiation interacts with the negative end of the dipole it compresses the bond. Thus the interaction (absorption) of radiation causes the length (amplitude) of the bond (spring) to increase or decrease. This interaction is shown in Fig. 10-8.

Bonds that are nonpolar do not absorb IR radiation. The nonpolar triple bond in acetylene (HC≡CH) does not show an IR absorption.

---

*IR spectra are reprinted with permission form: SDBSWEB:http://www.aist.go.jp/RIODB/SDBS/ (National Institute of Advanced Industry and Technology, 11-15-05)

Ethanol (liquid film)

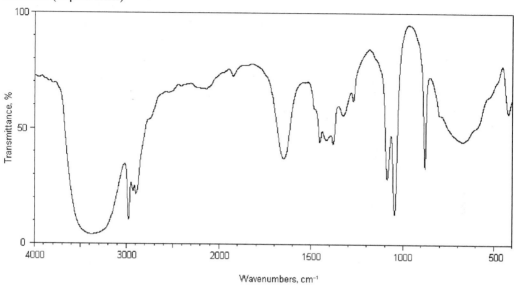

Propionic acid (CCl₄ solution)

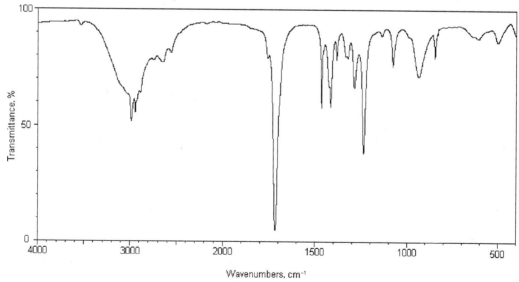

Fig. 10-7. Typical IR spectra of organic compounds.

Acetone (liquid film)

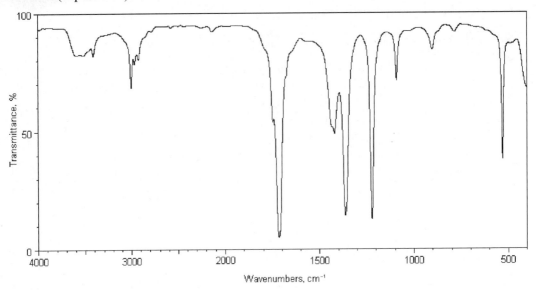

Cyclohexene (liquid film)

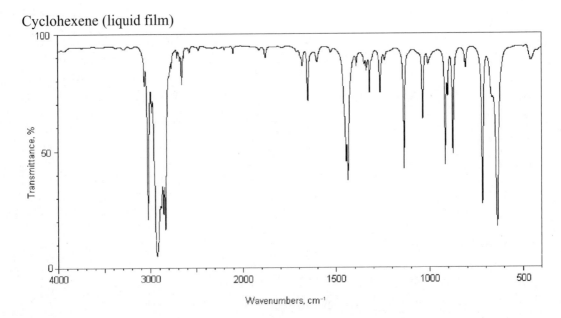

Fig. 10-7. (*Continued*)

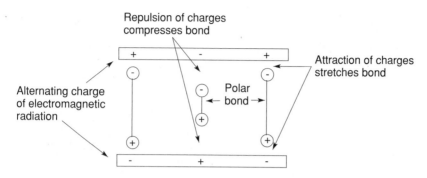

Fig. 10-8.   Interaction of EM radiation with polar molecules.

QUESTION 10-7
Would you expect to see an absorption of the carbon-carbon double bond in $(CH_3)_2C=C(CH_3)_2$?

ANSWER 10-7
No, it is a symmetrical molecule and the carbon-carbon double bond is nonpolar and does not absorb IR radiation (it is infrared inactive).

Absorption ranges

The amount of radiation and frequency of the radiation absorbed depend on several factors. The higher the concentration of the sample, the greater the amount of radiation absorbed. The strength of the bond and the polarity of the bond are also factors. Stronger bonds require more energy for stretching and absorb at higher wavenumber values. Table 10-1 shows the absorptions of carbon-carbon single, double, and triple bonds. The strongest bond, the triple bond, absorbs at the highest wavenumber (energy) value. The single bond absorbs at the lowest wavenumber (energy) value. Heavier atoms vibrate at lower wavenumbers (or frequencies). Values of wavenumbers for bonds between carbon atoms and isotopes of hydrogen atoms are also shown in Table 10-1.

Table 10-2 lists the absorption ranges for the more common functional groups. The positions of the absorption bands for specific functional groups do not change much in different compounds and are usually found within a given range. For example, the carbonyl absorptions in aldehydes, ketones, acids, and esters are usually found between 1780 and 1650 cm^{-1}.

Table 10-1. IR absorptions as a function of bond strength and atomic weight.

Bond	Wavenumber, cm^{-1}	
C—C	1200–800	
C=C	~1650	Bond strength
C≡C	~2100	
C—H	~3000	Atomic weight
C—D	~2200	

Fingerprint region

An IR spectrum (see Fig. 10-7) is divided into two regions. The region from 400 to about 1400 cm^{-1} is called the *fingerprint region*. Most of the bending vibrations and some stretching vibrations occur in this range. Nonlinear molecules

Table 10-2. Characteristic IR absorptions.

Bond of functional group	Absorption range, cm^{-1}	Comments
O—H, stretch, alcohol	3200–3600	m to s, H-bonding
	3590–3600	s, no H-bonding
O—H, stretch, carboxylic acid	2500–3500	b
C—H, stretch		
Alkane	2800–3000	m to s
Alkene =C—H	3000–3100	m
Alkyne ≡C—H	3000	m
C≡N, stretch, nitrile	2210–2260	m to w
C≡C, stretch, alkyne	2100–2260	w or absent
C=O, stretch		
Ester	1730–1755	s
Aldehyde	1720–1735	s
Ketone	1700–1725	s
Acid	1680–1725	b
Amide	1620–1690	s
C=C, stretch		
Alkene	1600–1680	w or absent
Benzene	1400–1500, 1585–1600	m
C—O, stretch, ether, ester, alcohol	1200–1250	s
N—H, stretch, amine	1180–1360	s
C—H, bend alkane,		
methylene	1465	w
methyl	1350–1475, two bands	w

with n atoms have $3n - 6$ fundamental vibrational modes. Isobutane (C_4H_{10}), a relatively simple molecule, potentially has $3(14) - 6 = 36$ vibrational modes (some modes may be infrared inactive). The region from 400 to 1400 cm^{-1} tends to have many absorption bands. Many of these bands overlap and it is often very difficult to identify specific absorptions. This region is used for matching the spectra of unknown compounds with the spectra of known compounds, as one matches fingerprints; hence the name fingerprint region. The region from 1400 to 4000 cm^{-1} has fewer absorption bands and it is relatively easy to correlate specific absorptions with functional groups. Note how the spectra in Fig. 10-7 are more complex in the fingerprint region.

Shapes of absorption bands

Absorption bands are identified by the terms strong, medium, weak, broad, and sharp. These terms refer to the shape of the absorption band. A strong absorption means the bond absorbs most of the incident radiation and the peak will be long (will have a large percent absorption). Weak absorptions will appear as short peaks. Broad absorptions are wide peaks and sharp absorptions are narrow peaks.

Note the shapes of the bands in the spectrum for propionic acid in Fig. 10-7. The absorption at 3000 cm^{-1} is strong and broad, the absorption at 1716 cm^{-1} is strong and sharp, and the absorption at 1080 cm^{-1} is medium and sharp.

Alkenes

The IR spectrum of the alkene cyclohexene is shown in Fig. 10-7. Most organic compounds have carbon-hydrogen bonds and show medium to strong absorptions between 2700 and 3300 cm^{-1}. The alkene (C=C) absorption appearing at 1660 cm^{-1} is medium in length and sharp.

The O—H absorption

The O—H absorption band is present in alcohols (ROH) and acids (RC(O)OH), as both contain an —OH group. The O—H absorption tends to be strong and broad. This absorption is readily apparent between 2500 and 3300 cm^{-1} in the spectra for ethanol and propionic acid in Fig. 10-7. It is quite easy to identify this O—H absorption.

Absorption band width

Some absorption bands, such as the O—H absorption band, are very broad. The position of an absorption band depends upon the bond strength, the bond polarity, and the mass of the atoms making the bond. The O—H functional group undergoes hydrogen bonding with other O—H functional groups. At any given time, some hydrogen bonds are stronger than other hydrogen bonds. A hydrogen bond between a hydrogen atom and an OH group has some influence on the polarity and strength of the O—H bond. If the bond strength varies, the wavenumber where absorption occurs will also vary and a broad peak results. A spectrum of a 100% alcohol sample, where hydrogen bonding occurs to a great extent, will have a much broader O—H absorption band than that of a dilute solution of the alcohol where hydrogen bonding between alcohol molecules is diminished.

The carbonyl absorption

A carbonyl (C=O) absorption is present in ketones, aldehydes, acids, esters, amides, and anhydrides. The carbonyl absorption is strong and narrow. It is readily apparent in the spectra for acetone (1715 cm^{-1}) and propionic acid (1716 cm^{-1}), as shown in Fig. 10-7.

Interpreting IR spectra

IR spectra are used to identify the presence of specific functional groups in compounds. *A key point: in interpreting IR spectra, absorptions that are not present are as important as absorptions that are present.* An absorption band that is not present confirms that either the corresponding functional group is not present in the compound being analyzed or a bond may be nonpolar in which case it does not absorb IR radiation may be present. The spectrum for propionic acid has a carbonyl and an O—H absorption band, ethanol shows only an O—H absorption band, and acetone shows only a carbonyl absorption band. It is quite easy to match these three spectra with these three compounds.

QUESTION 10-8
The spectrum of an unknown compound has an absorption at 3000 cm^{-1} and no absorption between 1680 and 1725 cm^{-1}. Could the compound contain an acid and/or an alcohol group?

ANSWER 10-8
It could only be an alcohol since no peak is seen in the carbonyl absorption region.

UV Spectroscopy

The region in the electromagnetic spectrum containing wavelengths from 4×10^{-7} to 1×10^{-8} m is called the UV region (see Fig. 10-4). Only a small part of the UV spectrum, 2×10^{-7} to 4×10^{-7} m, is used to characterize organic compounds. This region is usually expressed as 200 to 400 nm. The energy of electromagnetic radiation in this region ranges from 300 to 600 kJ/mol (70 to 140 kcal/mol). This is sufficient energy to promote electrons into higher energy levels (excited states). This energy range is also sufficient to break some chemical bonds. (For comparison, the energy of IR radiation is much less, around 5 to 25 kJ/mol, and does not promote electrons into excited states.)

QUESTION 10-9

Does UV spectroscopy require longer wavelength radiation than IR spectroscopy?

ANSWER 10-9

No, UV spectroscopy requires shorter wavelength, higher energy radiation.

CONJUGATED MOLECULES

Organic compounds with *conjugated multiple (double and triple) bonds* absorb UV radiation by promoting an electron from a ground state to an excited state. Conjugated bonds consist of two or more multiple bonds connected alternately by a single bond, e.g., $H_2C=CH-CH=CH_2$. Table 10-3 shows additional examples of conjugated double bonds. Nonconjugated double bonds (isolated double bonds), e.g., $H_2C=CH-CH_2-CH=CH_2$, absorb UV radiation at a wavelength shorter than 200 nm and therefore absorptions for isolated carbon-carbon double bonds are not observed in the 200 to 400 nm range.

THE UV SPECTROMETER

A *UV spectrometer* generates electromagnetic radiation in the 200 to 400 nm range. The instrument scans this range and the absorption spectrum is printed on a chart that records the relative absorbance on the vertical scale (0 to 1.0) and the wavelength on the horizontal scale. The sample is dissolved in a solvent and

Table 10-3.   UV absorptions of conjugated alkenes.

Alkene	$\lambda_{max(nm)}$	$\varepsilon_{max(M^{-1}mL^{-1})}$
==	165	15,000
⩘	217	21,000
⩗⩘	256	50,000
⩗⩗⩘	290	85,000

$$A = \log \frac{I_0}{I_{sample}} = \varepsilon\, l\, c$$

ε = extinction coefficient = 1/M-cm
l = length of sample tube, cm
c = molar concentration, M

Fig. 10-9.   Beer-Lambert law.

placed in a cell that does not absorb in the UV region of interest. A reference cell is also used that contains only the solvent.

The absorbance, A, of the sample is reported as the log of the ratio of the beam intensity of the incident light, I_0, divided by the intensity of the light, I, that is transmitted through the sample. The absorbance is also proportional to the sample concentration, c, the length, (thickness) of the sample cell, and the *molar absorptivity*, ε. The molar absorptivity (also called the *extinction coefficient*) is calculated from the spectrum. The equation for the absorptivity (the Beer-Lambert law) is shown in Fig. 10-9. Absorbance is unitless, molar absorptivity has units of L-mol^{-1}cm^{-1}, concentration is in mol/L, and cell length is in cm.

QUESTION 10-10
What change in total UV absorption would you expect if the concentration of a sample is doubled and the cell length is halved?

ANSWER 10-10
There would be no change in amount of radiation absorbed, as calculated from the Beer-Lambert law.

SAMPLE SPECTRUM

A sample spectrum (trace) is shown in Fig. 10-10. The spectrum usually consists of a very broad "peak." Although energy is quantized (only a specific amount of energy is absorbed), the molecules exist in various vibrational and rotational ground and excited states, and a series of closely spaced quantized energies is absorbed, giving a broad curve. Sometimes more than one absorption peak is observed. The spectrum is characterized by stating just the wavelength at the top of the peak(s). This maximum wavelength is called λ_{max} (*lambda max*). The value of ε is also stated. Since the absorbance is affected by the solvent and the temperature, this information should also be stated (but usually is not). A typical UV absorption is recorded as $\lambda_{max} = 300$ nm ($\varepsilon = 15{,}000$).

ELECTRON EXCITATION

When a conjugated molecule absorbs UV radiation, an electron is promoted from a ground state orbital into an unfilled higher energy orbital. The

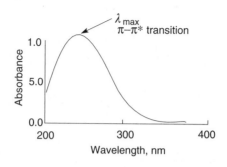

Fig. 10-10. Typical UV spectra.

electrons promoted are usually π (pi) electrons. This transition is shown as $\pi \rightarrow \pi^*$, called a pi to pi star transition. The difference between the π and π^* energy levels decreases as the extent of conjugation increases. Table 10-3 shows the values of λ_{max} and ε_{max} (the value of ε at λ_{max}) for a series of molecules of increasing conjugation. Note how λ_{max} increases and ε increases as conjugation increases.

CHROMOPHORES

Figure 10-11 shows how the energy for promoting an electron from the *highest occupied molecular orbital* (HOMO) into the *lowest unoccupied molecular orbital* (LUMO) *decreases as conjugation increases*. The relative values of ΔE shown in Fig. 10-11 compare the energies required for electron promotion in ethylene (a nonconjugated molecule) and 1,3-butadiene and 1,3,5-hexatriene (conjugated polyenes). The functional group, such as a diene or triene, that absorbs UV radiation is called a *chromophore*. A chromophore does not necessarily consist of only carbon atoms. A carbon-oxygen double bond conjugated with a carbon-carbon double bond is a chromophore that absorbs in the 200 to 400 nm range.

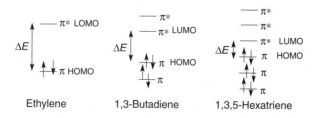

Fig. 10-11. Energy difference between LUMO and HOMO.

QUESTION 10-11
Which would absorb radiation in the 200 to 400 cm range: $CH_2=CHCH_2CH=CH_2$ or $CH_3CH=CHCH=CH_2$?

ANSWER 10-11
$CH_3CH=CHCH=CH_2$, since it is conjugated and the other molecule is not.

Groups attached to a chromophore can cause the absorption to be shifted to longer (lower energy) and shorter (higher energy) wavelengths. Shifts to longer wavelengths are called *red shifts* and shifts to shorter wavelengths are called *blue shifts*.

Nuclear Magnetic Resonance Spectroscopy

Nuclear magnetic resonance (NMR) spectroscopy is one of the most useful tools for determining the structure of organic molecules. It gives information about the number, type, and connectivity of hydrogen and carbon atoms. The technique involves the interaction (*resonance*) of energy from an external energy source with the *magnetic* properties of a *nucleus*. This does not involve a nuclear reaction or a chemical reaction. The material being tested can be recovered unchanged.

Nuclear Properties

Nuclei are charged spinning particles that, like electrons, are considered to have spin states. Nuclei can be compared to a top that spins clockwise or counter-clockwise (two spin states). Charged particles that spin generate a magnetic field. The magnetic field acts like a small bar magnet. The needle of a compass is a small bar magnet that can align with or against an external magnetic field. The needle aligns with the earth's magnetic field and points north. If we want the needle to point south, a small amount of energy would be needed to hold the needle in the opposite direction. The needle will return to pointing north (its "ground state") when it is free to move. The needle would point in random directions if the earth's magnetic field did not exist.

SPIN STATES

Nuclei are very much like a compass needle. Their magnetic component orients in random directions in the absence of a strong external magnetic field. In the presence of a strong external magnetic field, the little magnet (the nuclear spin) can orient with or against the external magnetic field. The lowest energy state exists when the nuclear spin orients with the external field. If the nucleus absorbs the appropriate amount of energy, the magnetic component can align against the external field, like the compass needle forced to point south. The magnetic component of a spin state oriented with an external magnetic field is called the alpha, α, spin state. The beta, β, spin state exists when the magnetic component is oriented against an external magnetic field.

QUESTION 10-12
A proton (a hydrogen nucleus) can exist in how many spin states?

ANSWER 10-12
Two, an α and a β spin state.

SPIN FLIPPING

In order for a nucleus to change from the α to the β spin state, the nucleus must absorb energy. The amount of energy required to go from the α to the β spin state is directly proportional to the strength of the magnetic field felt by the nucleus, as shown in Fig. 10-12. The stronger the magnetic field the nucleus feels, the more energy is needed for the nuclear spin to change directions. Changing spin direction is known as *flipping* or *spin flipping*. The energy required for the nucleus to flip must exactly match the difference in the energy between the two spin states. In Fig. 10-12, two quanta of energy (ΔE_1 and ΔE_2) are shown. Note that more energy is needed for spin flipping in a stronger magnetic field. *Resonance* (in NMR) refers to the absorption of energy from electromagnetic radiation to produce a spin flip. *The relationship between ΔE and the magnetic field the nucleus feels is a critical concept and the basis of the discussion that follows.*

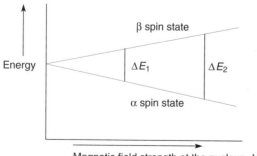

Fig. 10-12.   Energy of spin flipping as a function of magnetic field strength.

The magnetic field felt by a nucleus is also influenced by the electrons surrounding that nucleus. Electrons are moving charged particles. In the presence of an external magnetic field, the electrons generate an induced magnetic field that opposes the external magnetic field. Thus the effective (total) magnetic field, H_{eff}, that reaches the nucleus equals the external magnetic field, H_0, minus the local magnetic field, H_{local}, generated by the electrons; that is, $H_{eff} = H_0 - H_{local}$.

QUESTION 10-13
If H_{eff} increases, is more or less energy needed for spin flipping?

ANSWER 10-13
More energy is needed. A stronger effective magnetic field increases ΔE.

Only certain nuclei interact with a strong external magnetic field in this manner. These nuclei must have an odd atomic number, an odd mass number, or an odd number for both. Hydrogen atoms (^{1}H, or protons) and one isotope of carbon (^{13}C) meet these requirements. Other nuclei also meet these requirements but only ^{1}H and ^{13}C will be discussed since they are the major elements of organic compounds. There are two spin states, α and β, for ^{1}H and ^{13}C. Some atoms have more than two spin states.

NMR Spectrometers

The NMR spectrometer is designed to take advantage of the magnetic properties of nuclei. A diagram of a spectrometer is shown in Fig. 10-13. This instrument contains a very strong magnet. The core of the magnet contains a coil that supplies energy in the radiowave frequency range of the electromagnetic spectrum. The energy, E, of a photon of electromagnetic radiation is directly proportional to its frequency, v; that is, $E = hv$, where h is a constant. Nuclei absorb energy in this radiowave range and flip from the more stable α spin state to the higher energy β spin state, as shown in Fig. 10-12. Another coil senses when the sample absorbs energy and also when energy is released as the nuclei return (relax) to the α spin state and an α-β spin state equilibrium is reestablished.

This energy change information is sent to a computer and analyzed. The computer generates an NMR spectrum and sends the information to a recorder. The recorder prints the spectrum (absorptions) of the sample, giving information about the number, type, and connectivity of ^{1}H and ^{13}C atoms.

ENERGY REQUIREMENTS

The amount of energy required for spin flipping is very small, about 10^{-5} kJ/mol. (The energy range for IR absorptions is 5 to 50 kJ/mol.) Since the energy required

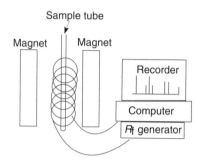

Fig. 10-13. NMR spectrometer.

for spin flipping is so small, one might think there is enough thermal energy at room temperature to cause spin flipping. This is indeed the case. However, a very slight excess (0.0005%) of nuclei are in the lower energy spin state under equilibrium conditions. This small excess is sufficient for the NMR instrument to measure when energy is absorbed or emitted by nuclei changing spin states.

RADIOWAVE FREQUENCY RANGE

NMR spectrometers are usually described in terms of the frequency of radiowaves used for spin flipping. Older spectrometers operate in the 100 MHz range. (Frequency is in units of hertz or sec^{-1}.) The most common spectrometers operate in the 300 MHz range. Larger and much more expensive instruments operate in ranges up to 900 MHz.

In traditional (continuous wave) spectrometers (CW-NMR), the radiowave frequency is held constant and the external magnetic field strength is varied. Newer instruments have a constant external magnetic field strength and apply the entire range of frequencies simultaneously. The intensity and frequency of emitted radiation are analyzed by a mathematical process called Fourier transformation (FT). These spectrometers are called FT-NMR spectrometers. FT-NMR spectrometers offer many advantages over the continuous wave spectrometers in terms of speed and resolution.

SAMPLE PREPARATION

The sample to be tested is put in a thin glass tube, about the size of a pencil, and placed inside the coils (that supply radiowaves) in the core of the magnet (see Fig. 10-13). Only a few milligrams of sample are required. About 0.5 mL of solvent is used. The solvent used does not contain atoms (nuclei) that absorb radiowave energy in the range that protons absorb radiation. Solvents used include carbon tetrachloride (CCl_4), deuterated chloroform ($CDCl_3$), and deuterated water (D_2O). Deuterium (D or ^{2}H), an isotope of hydrogen, does not absorb radiation in the same range as does "normal" ^{1}H.

QUESTION 10-14
Why isn't water (H_2O) used as an NMR solvent?

ANSWER 10-14
The hydrogen atoms in water would give a huge absorption peak that could hide absorptions from the sample compound.

^{1}H NMR Spectroscopy

^{1}H NMR spectroscopy is referred to as proton NMR spectroscopy. A proton, by definition, is a hydrogen atom without an electron, H^+. Proton NMR is a study of hydrogen atoms covalently bonded to another atom, usually carbon, and it is not the study of free protons. However, NMR discussions use the terms proton and hydrogen atom interchangeably.

For NMR to be of value as a diagnostic tool, hydrogen atoms (protons) in different chemical environments need to absorb (resonate) radiowaves at different frequencies. The nucleus of a hydrogen atom is surrounded by an electron cloud (the electron density) of its bonding electrons. The external magnetic field produces a force (torque) on the electrons, causing them to generate a magnetic field, H_{local}, that opposes the external magnetic field, H_0.

The electron density around a hydrogen nucleus is different for protons that exist in different chemical environments. Since the electron density varies, the magnetic field felt by a hydrogen nucleus ($H_{eff} = H_0 - H_{local}$) also varies. As shown in Fig. 10-12, if the magnetic field felt by the nucleus changes, the frequency at which radiation is absorbed also changes. The frequency at which radiation is absorbed gives information about the chemical environment of the protons responsible for those absorptions.

EQUIVALENT AND NONEQUIVALENT PROTONS

Protons in *different* chemical environments are called *nonequivalent protons*. They absorb radiation at *slightly different frequencies* measured in megahertz, MHz, values. It is more convenient to express these frequency differences relative to some standard. The standard chosen in most cases is tetramethylsilane $(CH_3)_4Si$ (TMS). All the protons in TMS exist in an identical chemical environment and are called *equivalent protons*. They all absorb radiation at the same frequency. The difference in absorption frequencies from this standard is in hertz (not megahertz) values.

QUESTION 10-15
Are all the protons in methane, CH_4, equivalent?

ANSWER 10-15
Yes, the symmetrical, tetrahedral methane molecule has four equivalent protons.

CHEMICAL SHIFTS

An example of an ^1H NMR spectrum is shown in Fig. 10-14. The horizontal scale goes from 0 to 10 (right to left). The scale is in parts per million (ppm) or delta (δ) units. One δ unit equals 1 ppm. Nonequivalent protons absorb radiowave radiation at slightly different megahertz values. TMS is arbitrarily given a value of 0 on the ppm (or δ) scale and other absorptions are relative to TMS. Most protons absorb at values to the left (0 to 10) of TMS on this scale. The vertical lines represent proton absorptions. The difference between the absorption of a proton and TMS is called a *chemical shift*. If a chemical shift occurs at 5 ppm, it is said to be *downfield* relative to TMS and TMS absorbs *upfield* relative to the absorption at 5 ppm. The terms upfield and downfield are commonly used in describing spectra.

PPM AND δ VALUES

Chemical shift values in ppm units may sound strange. Where do these units come from? The frequency of radiation used to cause spin flipping in a 300-MHz NMR spectrometer is in the 300 MHz range. Assume a proton absorbs radiowave radiation of 300 Hz greater than TMS. If we take this value of the chemical shift in hertz and divide it by the operating frequency of the spectrometer, 300 MHz

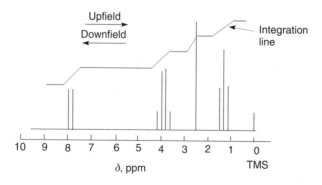

Fig. 10-14. ^1H NMR spectrum.

in this case, this ratio (300 Hz/300,000,000 Hz) is 1 part in 1,000,000 or 1 ppm. Since the energy for spin flipping varies with the strength of the magnetic field felt by the nucleus, more energy (larger MHz values) is required for spin flipping in stronger magnetic fields. The proton in the example above would absorb energy at 600 Hz downfield from TMS in a 600-MHz NMR spectrometer. Dividing the chemical shift (600 Hz) by the operating frequency of the spectrometer (600 Hz/600,000,000 Hz) gives 1 ppm. *A key point: the chemical shift value in ppm (or δ) units is independent of the magnetic field strength of the instrument used.* A spectrum of a sample run on a 100-MHz or 900-MHz spectrometer will have the same chemical shift ppm (or δ) value for corresponding protons.

QUESTION 10-16
The ^1H chemical shift for chloroform ($CHCl_3$) is at 7.2 ppm when run on a 100-MHz spectrometer. What is the ppm value if the sample is run on a 600-MHz spectrometer?

ANSWER 10-16
It will have the same value, 7.2 ppm.

EFFECTIVE MAGNETIC FIELD STRENGTH

The effective magnetic field, H_{eff}, is the magnetic field felt by a nucleus. The energy required for spin flipping is directly proportional to H_{eff}. (See Fig. 10-12 to review this relationship.) There are three factors that effect H_{eff} of protons: the electronegativity of nearby atoms, hybridization of adjacent atoms, and magnetic induction of adjacent π bonds. If the proton is bonded to an atom (or group) that pulls electron density away from that proton, the lower electron density around that proton results in a lower H_{local}. Since the electron's magnetic field opposes the external magnetic field, the net result is a stronger magnetic field felt by the nucleus ($H_{eff} = H_0 - H_{local}$) and more energy (a larger value of MHz) is required for spin flipping.

DOWNFIELD AND UPFIELD SHIFTS

The electron density around a proton in methane, CH_4, is much greater than the electron density around the proton in chloroform, $HCCl_3$. The three electron-withdrawing chlorine atoms decrease the electron density around the hydrogen atom nucleus. A decrease in electron density around an atom is called *deshielding*. Deshielding causes a *downfield chemical shift*. A downfield chemical shift means that absorption is shifted further to the left from TMS. Increasing electron density around a proton is called *shielding* and causes an *upfield chemical shift*.

QUESTION 10-17
Which proton in methanol (CH_3OH) has a chemical shift further downfield?

ANSWER 10-17
The proton bonded directly to the electron-withdrawing oxygen atom.

HYBRIDIZATION

Most of the protons in organic compounds are bonded to carbon atoms. Electronegativity of carbon atoms varies as a function of their hybridization. Carbon atoms that are sp hybridized are more electronegative than sp^2 hybridized carbon atoms and sp^2 hybridized carbon atoms are more electronegative than sp^3 hybridized carbon atoms. (This electronegativity effect is discussed in Chapter 3.) One would, therefore, expect protons attached to sp^2 hybridized carbon atoms to absorb further downfield then protons attached to sp^3 hybridized carbon atoms. This is indeed found. Protons in alkanes (sp^3 hybridized carbon atoms) have chemical shifts from about 0 to 1.5 ppm and protons attached to alkene (vinyl) carbon atoms (sp^2 hybridized) absorb from 4.5 to 6.5 ppm. Protons attached to alkyne carbon atoms (sp hybridized) absorb, unexpectedly, from 2.5 to 3.0, upfield from alkene protons. Table 10-4 shows chemical shifts for protons in different chemical environments.

INDUCTION EFFECTS

Magnetic induction explains the unexpected order of chemical shifts for protons bonded to sp and sp^2 hybridized carbon atoms. An external magnetic field causes the π electrons in an alkene to generate an induced magnetic field as shown in Structure 10-15a in Fig. 10-15. The direction of the induced magnetic field (shown by the arrows) in the area of the vinyl proton (an alkenyl proton) is

Table 10-4. Approximate chemical shift values for ^1H NMR.

Type of proton	Chemical shift, ppm	Type of proton	Chemical shift, ppm
1° alkyl, RCH_3	0.8–1.1	Benzylic $ArCH_3$	2.2–2.7
2° alkyl, R_2CH_2	1.2–1.4	Alkyl halide, RCH_2X	3.0–4.0
3° alkyl, R_3CH	1.5–1.8	Ether, $ROCH_2R$	3.5–4.0
Allylic, $R_2C=CRCH_3$	1.5–2.0	Aromatic, ArH	6.0–9.0
Alkene, $R_2C=CRH$	4.5–6.5	Ketone, $RC(O)CH_3$	2.0–3.0
Alkyne, $RC\equiv CH$	2.5–3.0	Aldehyde, $RC(O)H$	9.0–10
		Carboxylic acid, RCO_2H	10–12

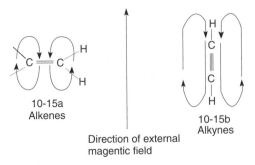

10-15a
Alkenes

10-15b
Alkynes

Direction of external
magentic field

Fig. 10-15. Magnetic induction effects.

in the same direction as the external magnetic field. The effective magnetic field felt by the vinyl proton is the sum of the induced field and the external magnetic field. This increased magnetic field at the nucleus (H_{eff}) results in a greater energy (greater MHz value) needed for spin flipping and the absorption is shifted downfield. This magnetic induction effect causes the downfield shift to be greater than expected from just the electronegativity of the sp^2 hybridized carbon atom.

The induction effect of an alkyne is shown in Structure 10-15b in Fig. 10-15. The alkyne functional group aligns parallel to the external field. The induced magnetic field in the alkyne group, shown by the curved arrows, opposes the external field in the area of the alkyne protons. The magnetic field felt by the nucleus equals the external field less the induced field. Since the magnetic field experienced by the nucleus is less than expected, less energy (a smaller MHz value) is needed for spin flipping. This inductive effect is strong enough that the alkynyl proton absorption is further upfield than the vinyl (alkenyl) proton absorption.

ELECTRON-WITHDRAWING GROUPS

The chemical shift of a proton attached to a carbon atom is dependent on other groups attached to that carbon atom. If the carbon atom is bonded to an electron-withdrawing chlorine atom, electrons will be drawn toward the chlorine atom. Table 10-5 shows the effect of the electron-withdrawing chlorine atom on the chemical shift of a proton in a series of molecules containing an increasing number of chlorine atoms. In this series where no, one, two, and three chlorine atoms are bonded to a carbon atom, the proton is increasingly deshielded and shifted downfield from 0.2 to 7.2 ppm. Electron-withdrawing groups such as the other halogens, oxygen, nitrogen, and sulfur also cause a downfield shift when sigma bonded to a carbon atom to which the hydrogen atom is also bonded.

Table 10-5. Effect of electron-withdrawing groups on chemical shift values.

Compound	ppm
CH_4	0.2
$ClCH_3$	3.0
Cl_2CH_2	5.3
Cl_3CH	7.2

EQUIVALENT AND NONEQUIVALENT PROTONS

Figure 10-16 shows the structures of several hydrocarbons that have equivalent and nonequivalent protons. The protons in Structure 10-16a are all equivalent (identical) to each other, as are the protons in Structures 10-16b and 10-16c. Structures 10-16d and 10-16e each have two groups of nonequivalent protons. Structure 10-16f has three groups of nonequivalent protons and Structure 10-16g has four groups of nonequivalent protons. Circles enclose equivalent protons. Different circles identify groups of nonidentical (nonequivalent) protons.

QUESTION 10-18
How many proton chemical shift values (groups of nonequivalent protons) would you expect for $CH_3C(O)OCH_3$?

ANSWER 10-18
Two, one for each of the nonequivalent methyl groups.

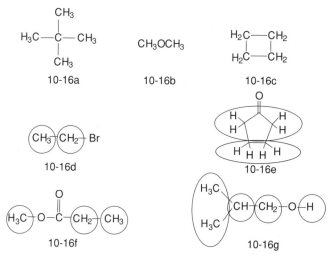

Fig. 10-16. Equivalent and nonequivalent protons.

The X test

One way to tell if protons are equivalent is to substitute hydrogen atoms within the same molecule with an X. First replace any hydrogen atom with an X, giving molecule A. Then replace a different hydrogen atom in the original molecule with an X, giving molecule B. If A and B are superimposable (identical molecules), the hydrogen atoms replaced are equivalent. An example is given in Fig. 10-17. Follow along in Fig. 10-17 as you read this paragraph. An X is substituted for two different hydrogen atoms in Structure 10-17a, giving Structures 10-17b and 10-17c. Rotating around the C—C bond (indicated by the curved arrow) in Structure 10-17c gives conformation 10-17d, which is superimposable on Structure 10-17b. These hydrogen atoms are, therefore, chemically equivalent. In Structure 10-17e, X is substituted for two different hydrogen atoms, giving Structures 10-17f and 10-17g. Rotating Structure 10-17g 180° (top to bottom) in the plane of the page gives Structure 10-17h, which is *not* superimposable on Structure 10-17f. These hydrogen atoms are not chemically equivalent. Nonequivalent hydrogen atoms (protons) have different chemical shifts.

QUESTION 10-19
How many chemical shift values would you expect for $CH_3OC(CH_3)_3$?

ANSWER 10-19
Two peaks, the methoxy methyl protons are equivalent and the methyl protons in the *tert*-butyl group are all equivalent.

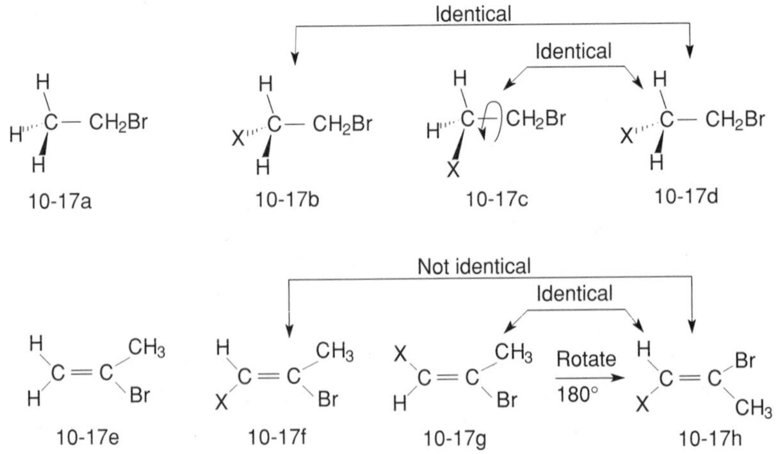

Fig. 10-17. Determining equivalent protons.

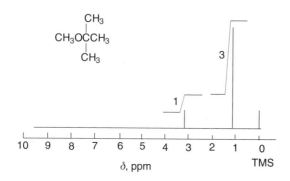

Fig. 10-18. ^{1}H spectrum of *t*-butyl methyl ether.

RELATIVE NUMBER OF HYDROGEN ATOMS

The NMR spectrum for *tert*-butyl methyl ether is shown in Fig. 10-18. The peaks (chemical shift values) are shown as solid lines in spectra in this text. Some spectra will show the peaks as "skinny hills." The NMR spectrometer has software that will integrate (determine the area of) each peak (hill). The area within each peak is proportional to the number of protons that are represented by that peak. A line is drawn on the spectrum that increases in height at each peak. The ratio of the increase in height of the line at the various peaks gives the ratio of the number of protons represented in each peak. Newer spectrometers calculate and print the integration values on the spectra. In Fig. 10-18, the ratio of peak areas at 1.2 and 3.2 ppm is 3:1. The sample contains two groups of equivalent protons: the *tert*-butyl group contains nine equivalent protons and the methyl group contains three chemically equivalent protons, a 3:1 ratio.

SPIN-SPIN SPLITTING

The spectrum for ethyl acetate is shown in Fig. 10-19. There are eight peaks. Four downfield peaks are centered at 4.1 ppm. Three upfield peaks are centered at 1.3 ppm. A single peak is observed at 2.1 ppm. These absorptions represent *three* types of protons: two nonequivalent methyl (−CH$_3$) groups and one methylene (−CH$_2$−) group. One of the methyl groups appears as a single peak and the other methyl group appears as three peaks. The methylene absorption appears as four peaks. Sometimes equivalent protons appear as multiple peaks. This is called *spin-spin splitting*. Although this may appear to make the spectra more complex, it actually gives much information about the structure of the molecule.

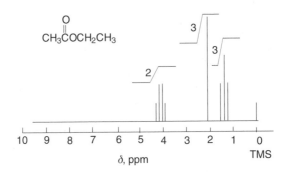

Fig. 10-19. ^1H spectrum of ethyl acetate.

Multiplets

An absorption split into two or more peaks is called a *multiplet*. Two peaks are called a *doublet*, three peaks are called a *triplet*, four peaks are called a *quartet*, five peaks are called a *quintet*, and so on. A single peak is called a *singlet*.

At room temperature, a proton exists in its α and β spin states. Each of these spin states generates a small magnetic field. One of these small magnetic fields aligns with the external magnetic field, increasing the total magnetic field strength. The other spin state generates a magnetic field that aligns against the external magnetic field, decreasing the total magnetic field strength. The net result is two slightly different magnetic field strengths are generated in the vicinity of adjacent protons. The magnetic field strength generated by a proton is much smaller than the external magnetic field strength and thus the two resulting net magnetic fields are only slightly different.

Splitting

Protons that are close to each other (usually within three σ bonds) feel the effect of each other's magnetic fields. The magnetic field of adjacent protons is transmitted through the bonding electrons. The ethyl group ($-CH_2CH_3$) shown in Fig. 10-20 is used to explain spin-spin (or signal) splitting. Proton H_a is within three σ bonds of proton H_c (a nonequivalent proton). The interaction of proton H_c with the external magnetic field generates two different magnetic field strengths (as a result H_c's α and β spin states). Proton H_a will now experience two magnetic fields: the external field plus and minus the two small fields generated by H_c's two spin states. The two magnetic fields felt by H_a will now be a little larger and a little smaller than that of the external magnetic field alone. The energy (frequency) required for an α to β transition for H_a is proportional to the magnetic field strengths H_a experiences. Thus, H_a will have two absorptions

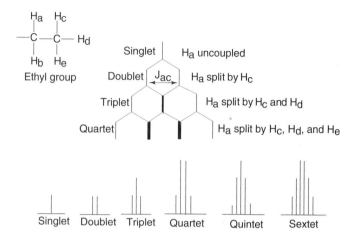

Fig. 10-20. Splitting pattern tree.

in the NMR spectrum. Proton H_c causes H_a to have two absorptions, called a doublet. This is shown in the second line in Fig. 10-20.

Additional splitting

Proton H_d (see Fig. 10-20), like proton H_c, will also generate two magnetic fields. This will cause each of the doublets of proton H_a to again be split, to give "four" peaks. Since H_d is chemically identical to proton H_c, H_c and H_d cause H_a's peaks to be split by the same amount. As shown in Fig. 10-20, only three lines appear. The two center peaks (lines) overlap, as indicated by a bold line. These three lines (peaks) represent a triplet in an NMR spectrum. Proton H_e will also cause each peak in the triplet to be split into a doublet to give "six" peaks. Since H_e, H_d, and H_c are chemically equivalent, H_e causes the lines to be split by the same amount as do H_c and H_d. The center two lines overlap, giving just four lines. The center two lines, shown in Fig. 10-20, are in boldface to indicate that they are formed by overlapping lines. These four lines represent a quartet in an NMR spectrum.

Proton H_b is within three sigma bonds of proton H_a but does not cause H_a to appear as a doublet. H_b is chemically equivalent to H_a and does not cause splitting for reasons that will not be discussed here. *A key point*: *chemically equivalent protons do not split each other*. Since protons H_a and H_b are equivalent, they have the same chemical shift and splitting pattern. They are superimposed in an NMR spectrum and integration of the absorption for H_a and H_b indicates two protons.

(*n* + 1) RULE

If a proton is adjacent to (within three σ bonds of) a single nonequivalent proton, it will appear in the NMR spectrum as a doublet. If the proton is adjacent to two nonequivalent protons, it will appear as a triplet. To generalize, a proton absorption will be split into (*n* + 1) peaks if adjacent to *n* nonequivalent protons. *However, these nonequivalent protons must be equivalent to each other*. In Fig. 10-20, H_c, H_d, and H_e are equivalent to each other and H_a and H_b are equivalent to each other. But H_c, H_d, and H_e are not equivalent to H_a and H_b.

QUESTION 10-20

What is the multiplicity (number of peaks) of each absorption in $CH_3OCH(CH_3)_2$?

ANSWER 10-20

CH_3 is a singlet, CH is a heptet, and $(CH_3)_2$ is a doublet. Draw the Lewis structure of this molecule to visualize the splitting pattern.

The splitting tree

H_a and H_b in Fig. 10-20 will appear as a quartet as they are adjacent to three nonequivalent protons (*n* + 1 or 3 + 1 = 4). H_c, H_d, and H_e will appear as a triplet as they are adjacent to two nonequivalent protons (*n* + 1 or 2 + 1 = 3). One can generate this splitting pattern with a "tree" as shown in Fig. 10-20. This symmetrical tree will only be obtained when H_c, H_d, and H_e are equivalent to each other and H_a and H_b are equivalent to each other.

In this tree, the center peak in the triplet results from two overlapping absorptions and this peak in the NMR spectrum will contain twice the area of the outer peaks that do not overlap. The central peaks in the quartet also result from multiple overlapping absorptions and the areas of the central peaks are three times greater than that of the outer peaks that do not overlap. The resulting multiplets and the relative areas (heights) of each peak are shown in Fig. 10-20 below the "tree."

Pascal's triangle

The number of peaks and their relative areas in a splitting pattern are readily determined by using Pascal's triangle, shown in Fig. 10-21. For example, the relative heights (areas of peaks) for a triplet are 1:2:1 and the relative heights of a quartet are 1:3:3:1.

```
              1    Singlet
           1     1    Doublet
        1     2     1    Triplet
     1     3     3     1    Quartet
  1     4     6     4     1    Quintet
1     5    10    10     5     1    Sextet
```

Fig. 10-21.   Pascal's triangle: multiplicity and intensity.

ASSIGNING CHEMICAL SHIFTS

The splitting patterns and relative areas of the peaks for ethyl acetate are shown in Fig. 10-19. These absorptions (chemical shifts) must still be assigned to the three groups of equivalent protons. The methylene ($-CH_2-$) protons are close to the electron-withdrawing oxygen atom and are expected to be shifted further down-field than either of the methyl groups. The methylene protons should appear as a quartet (as they are adjacent to a methyl group) and can be assigned to the furthest downfield absorption centered at 4.1 ppm. The other two chemical shifts must be the two methyl groups. The methyl group bonded to the electron-withdrawing carbonyl group would be expected to be further downfield than the CH_3 group in the ethyl group. The methyl group bonded to the carbonyl group should appear as a singlet, as there are no adjacent protons to cause splitting. The singlet at 2.1 ppm is assigned to this methyl group. The furthest upfield chemical shift, which is a triplet, can be assigned to the methyl group that is bonded to the methylene group. This methyl group has an absorption centered at 1.3 ppm.

The value for a chemical shift (ppm value) for a singlet can be read directly from the NMR spectrum. The chemical shift of a multiplet is the value for the center of the multiplet, halfway between doublets, the center peak in a triplet, and so on.

Summary of NMR data

Splitting patterns give much useful information about the number of nonequivalent protons adjacent to the proton of interest. Electron-attracting and -donating groups adjacent to the proton of interest will result in a downfield or upfield shift. Integration of the multiplets will confirm the ratio of the number of protons in each multiplet. The relative areas of the peaks of ethyl acetate shown in Fig. 10-19 confirm the assignments for the methylene and methyl groups.

QUESTION 10-21
What is the relative area of each peak in a quartet?

ANSWER 10-21
1:3:3:1.

COUPLING CONSTANTS

When a single peak (a singlet) is split into a doublet, the two resulting peaks appear at slightly different chemical shift (ppm) values. The difference in the ppm values is called the *coupling constant, J.* Protons that cause mutual splitting are said to be coupled to each other. *J* values generally range from 0 to 15 Hz. The symbol J_{ac} represents the coupling constant between protons H_a and H_c, as shown in Fig. 10-20. J_{ca} represents the coupling of H_c and H_a. Coupling between H_a and H_c is reciprocal. If H_a splits H_c by 5 Hz, H_c splits H_a by 5 Hz. In summary, $J_{ac} = J_{ca} = 5$ Hz.

QUESTION 10-22
Two doublets at 1.4 and 4.1 ppm in an ^1H spectrum have *J* values of 3.1 and 1.3 Hz. Are the protons coupled to each other?

ANSWER 10-22
No, if they were coupled to each other, they would have identical *J* values.

NONEQUIVALENT PROTONS

In many cases, one proton is coupled to more than one nonequivalent proton. These nonequivalent protons may also not be equivalent to each other. An example of a vinyl ether is shown in Fig. 10-22. Protons H_a, H_b, and H_c are all not equivalent to each other. (Use the X test to prove that the protons are not equivalent.) Each proton is coupled to the other two protons. Usually, but not

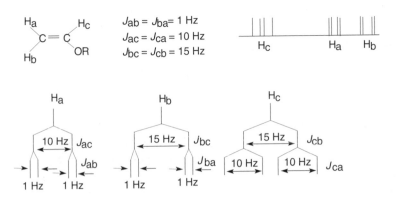

Fig. 10-22. Coupling between vinyl protons.

always, J values differ for different coupling pairs of nonequivalent protons; that is, $J_{ab} \neq J_{ac} \neq J_{bc}$ in most cases.

A NONSYMMETRICAL SPLITTING TREE

When a proton is coupled to more than one nonequivalent proton, the $(n + 1)$ rule *cannot* be used. The splitting pattern is best determined by drawing a tree diagram. The coupling constants given for the vinyl ether in Fig. 10-22 are $J_{ac} = J_{ca} = 10$ Hz, $J_{bc} = J_{cb} = 15$ Hz, and $J_{ab} = J_{ba} = 1$ Hz. The expected splitting pattern is shown in Fig. 10-22 for each proton. There is a coupling constant of 15 Hz in each branch for proton H_b and H_c, suggesting these two protons are coupled to each other. Similar comparisons can be made for coupling constants of 10 and 1 Hz to show which protons are coupled to each other.

It has also been found experimentally that protons that are trans to each other in a carbon-carbon double bond have coupling constants between 11 and 18 Hz, protons that are cis to each other in a carbon-carbon double bond have coupling constants of 6 to 14 Hz, and vinyl geminal protons (protons on the same carbon atom) have coupling constants of 0 to 3 Hz. These coupling constant ranges help in assigning chemical shift values to different protons and identifying their placement about a carbon-carbon double bond.

^{13}C NMR Spectroscopy

Further identification of a compound is obtained by analyzing a ^{13}C spectrum in addition to a ^{1}H spectrum. There are several differences between ^{13}C and ^{1}H NMR spectroscopy. The energy for spin flipping (α to β transition) occurs at a different frequency (energy) range and therefore, ^{13}C absorptions do not appear in ^{1}H spectra. Chemical shifts for ^{1}H occur primarily in a 0–10 ppm range. The range for ^{13}C spectra is about 0 to 220 ppm, which is an advantage as the ^{13}C absorptions are much further apart than ^{1}H absorptions, and there is less chance of nonequivalent peaks overlapping. (When multiplets overlap it is difficult to determine which peaks belong to each overlapping multiplet.) The approximate range for chemical shifts of nonequivalent carbon atoms is shown in Fig. 10-23.

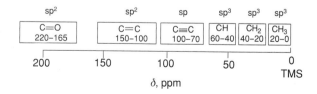

Fig. 10-23. Approximate chemical shift values for ^{13}C NMR.

FT-NMR

Almost all hydrogen atoms consist of the isotope ^1H. About 99% of all carbon atoms are ^{12}C and only 1% are ^{13}C atoms. Thus a ^{13}C signal is much weaker than a ^1H signal in a given sample. The spectrometer also records random, anomalous signals called noise. In a single scan, it is difficult to identify the desired ^{13}C signal from the noise. The advantage of FT-NMR is that each scan takes only a few seconds, and many scans of the same sample can be taken in a few minutes. A computer stores the information from each scan and adds the scans. Since the noise is random ($+$ noise and $-$ noise) it averages to about zero, giving a relatively flat baseline in the spectrum. The signals from ^{13}C are not random (all are positive values) and they are additive for each scan. The final result of many scans and computer averaging gives a spectrum with useful information about the sample.

The chemical shifts for carbon atoms follow the same general pattern discussed for ^1H NMR spectroscopy. Electron-withdrawing atoms or groups adjacent to the ^{13}C atom of interest cause downfield shifts and electron-donating atoms or groups cause upfield shifts. Since there is such a low percentage of ^{13}C atoms, the chance of two ^{13}C atoms being bonded to each other is very low and spin-spin coupling between adjacent ^{13}C atoms is not observed. Spin-spin coupling is however observed between ^{13}C and ^1H atoms bonded directly to each other. Weak coupling is observed between ^{13}C and ^1H atoms separated by two or three σ bonds. This is called long-range coupling. This coupling results in a very complex splitting pattern.

DECOUPLING

A complex splitting pattern can be simplified by spin-spin decoupling. A technique called *broadband decoupling* eliminates all hydrogen-carbon spin-spin coupling interactions. When a spectrum is decoupled, the multiplet is collapsed into a single peak, enhancing the signal-to-noise ratio. The decoupled spectrum for citronella, Spectrum d in Fig. 10-24, shows singlets for each of the 10 nonequivalent carbon atoms. It is more difficult to integrate the areas of ^{13}C peaks (for reasons not discussed here) but newer NMR spectrometers use techniques that give the relative number of carbon atoms in each peak (as is done in proton NMR).

QUESTION 10-23
How many ^{13}C peaks would result in a broadband decoupled spectrum of diethyl ether, $CH_3CH_2OCH_2CH_3$?

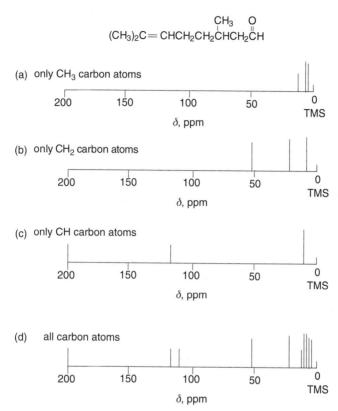

Fig. 10-24. ^{13}C NMR spectra of citronella.

ANSWER 10-23
Two, both methylene carbon atoms are equivalent and both methyl carbon atoms
are equivalent.

Off-resonance decoupling

Selective spin-spin decoupling can be conducted so that coupling occurs only
between carbon and hydrogen atoms bonded directly to each other. Eliminating
long-range coupling (more than one σ bond away) between carbon and hydrogen
atoms is called *off-resonance decoupling*. Carbon atoms now show splitting
patterns conforming to the $(n + 1)$ rule, where n is the number of hydrogen
atoms bonded to that carbon atom when off-resonance coupling is employed.

QUESTION 10-24
What is the splitting pattern of each peak in a ^{13}C off-resonance decoupled
spectrum of diethyl ether, $CH_3CH_2OCH_2CH_3$?

ANSWER 10-24
A quartet and a triplet. The quartet is further downfield.

DEPT

Another technique for analyzing ^{13}C spectra is called *DEPT* (*distortionless enhancement by polarization transfer*). In this process, the spectrometer varies the signal sent to the sample. Three additional spectra of citronella are shown in Fig. 10-24. Spectrum 10.24a only shows carbon atoms bonded to three hydrogen atoms (CH_3), Spectrum 10.24b only shows carbon atoms bonded to two hydrogen atoms (CH_2), and Spectrum 10.24c only shows carbon atoms bonded to one hydrogen atom (CH). Carbon atoms that are not bonded to any hydrogen atoms (quarternary carbon atoms and carbonyl, C=O, carbon atoms) do not appear in DEPT spectra. One can compare these three spectra to a completely decoupled spectrum run at standard conditions (Spectrum 10.24d.) and easily identify primary, secondary, tertiary, and quaternary carbon atoms.

ADVANCED TECHNIQUES

There are even more advanced techniques available for analyzing NMR spectra, called two-, three-, and four-dimensional NMR, that are beyond the scope of this text. These advanced techniques are used to analyze spectra of large complex molecules (proteins) and scans of human tissue. Medical personnel refer to this technique as magnetic resonance imaging (MRI). The fundamental principles of MRI are the same as in NMR.

Mass Spectrometry

Mass spectrometry (MS) is a method for determining the molecular weight of a compound. It also gives information about chemical structure. It is used with other characterization techniques to identify the total structure of a compound.

THE MASS SPECTROMETER

The instrument used is called a *mass spectrometer*. A schematic of a spectrometer is shown in Fig. 10-25. The unit consists of three parts: a chamber, A, where the sample is introduced and ionized, a long "tube", B, that directs the path of the ions, and a detection chamber, C. The sample is heated, if necessary, to put the sample into the vapor (gaseous) state. The spectrometer is run under a high vacuum to help vaporize the sample and remove extraneous gases.

A stream of high-energy electrons, from an external source, is directed at the sample in chamber A. These electrons have energies of about 7600 kJ/mol. This

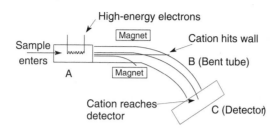

Fig. 10-25.   Mass spectrometer.

corresponds to a wavelength in the far UV/X-ray region of the electromagnetic spectrum. When one of these electrons strikes a molecule of the sample being tested, it knocks an electron out of the molecule. The high-energy electron also "bounces" off the sample. This process is shown in Fig. 10-26 for pentane, C_5H_{12}. It takes about 200 to 400 kJ/mol to knock an electron out of a molecule, and the high-energy electron (from the external source) has plenty of energy to knock an electron out of pentane resulting in a high-energy cation. The resulting high-energy cation is called a *molecular ion*.

THE MOLECULAR ION

Pentane contains 32 electrons in 16 σ bonds. When 1 electron is lost, 31 bonding electrons remain, an uneven (odd) number of electrons. That leaves 15 electron pairs and 1 unpaired electron. A molecule with 1 unpaired electron is called a *radical*, or a *free radical*. The charged molecule produced in chamber A in the mass spectrometer is called a *radical cation*, since it has 1 unpaired electron and a positive charge. Often it is just called a molecular ion, $M^{\overset{+}{\cdot}}$.

The radical cation produced in vacuum chamber A is directed out of A into tube B. The cation will travel in a straight line unless influenced by some external source. Magnets that can be varied in field strength line bent tube B. These magnets can influence the flight path of the cation, changing it from a straight path to a curved path as it moves from A to C. The cation must be directed to a "window" to enter chamber C. Each particle has a charge, usually +1, but the mass of the particles produced varies. By changing the magnetic field strength, particles with different mass-to-charge ratios (m/z, where m represents the mass of and z represents the charge on the particle) can be directed into the "window" in chamber C. Figure 10-25 shows a particle that is directed to chamber C and particles that hit the wall of tube B. If the magnetic field strength is changed in a uniform manner, ions with decreasing mass can be directed into chamber C.

$$C_5H_{12} + e^- \longrightarrow [C_5H_{12}]^{\overset{+}{\cdot}} + 2\,e^-$$

Fig. 10-26.   Formation of the molecular ion.

Particles of decreasing *m/z* ratio and their relative abundance are recorded on a chart, the *mass spectrum*.

QUESTION 10-25
If air is not removed from chamber A in the mass spectrometer, what additional peaks would be expected in the mass spectrum?

ANSWER 10-25
A peak at 28 from nitrogen and one at 32 from oxygen.

MASS SPECTRUM

A plot of the mass spectrum of pentane, C_5H_{12}, is shown in Fig. 10-27. There is a peak at 72 which corresponds to the molecular weight of pentane. This is called the molecular ion peak or parent peak. There are also several other peaks.

QUESTION 10-26
What is the value of the molecular ion peak for methanol, CH_3OH?

ANSWER 10-26
32.

There is a peak at 73 which is greater than the molecular weight of pentane. Some atoms exist as *isotopes*, atoms that have the same number of protons but

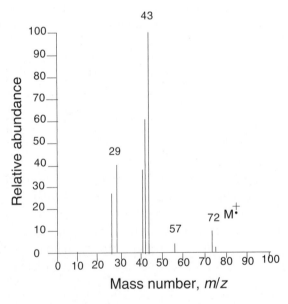

Fig. 10-27. Mass spectrum of pentane.

a different number of neutrons. Most carbon atoms have six protons and six neutrons. This isotope (nuclide) is assigned a mass of 12 amu. About 1% of carbon atoms have six protons and seven neutrons and have a mass of 13 amu. Thus there is a peak at 73 which is about 1% of the peak at 72 and is called the $(M + 1)$ *peak*. Some atoms, such as O, S, Cl, and Br, have isotopes with two additional neutrons and show a significant $(M + 2)$ *peak*. The amount of $(M + 1)$ and $(M + 2)$ peaks gives additional information about the type and number of atoms present in a molecule.

When a high-energy electron collides with pentane, a lot of energy is transferred to pentane. Not only is an electron knocked out, but sufficient energy is transferred to the newly formed cation radical (molecular ion) that σ bonds can also be broken. This process is known as *fragmentation*. In the mass spectrum in Fig. 10-27 there are several peaks with m/z ratios of less than 72. Since the charge on all cation fragments is usually $+1$, the m/z peaks represent the molecular weight (mass) of the cationic fragments. Note that the peak at 72 is not the highest peak. The molecular ion peak is not always the highest peak. The peak at 43 is the highest peak. The highest peak is called the *base peak*, and is arbitrarily set at 100%. All other peak heights are relative to the base peak.

FRAGMENTATION PATTERNS

Fragmentation patterns give additional information about the structure of a molecule, pentane in this example. The peak at 43 is 29 amu less than the molecular ion peak $(72 - 43)$. The peak at 43 could represent the loss of an ethyl $(CH_2CH_3; MW = 29)$ fragment from the molecular (parent) ion. The peak at 29 represents an ethyl carbocation. The peak at 57 $(72 - 15)$ represents the loss of a methyl $(CH_3; MW = 15)$ fragment from the parent ion. Molecules tend to fragment to give the most stable cation. Carbocation stability follows the order: $3° > 2° > 1° >$ methyl (CH_3). Figure 10-28 shows preferred fragmentation patterns for pentane that correspond to the peaks in the spectrum in Fig. 10-27.

Note the methyl carbocation is not formed. Radicals are somewhat more stable than carbocations. A more stable methyl radical, rather than a methyl carbocation, is formed in the fragmentation process but it is not detected by MS since it is not cationic.

Functional groups

Functional groups tend to give common fragmentation patterns. Alcohols (ROH) tend to lose a molecule of water (H_2O), alkenes fragment into allylic cations $(CH_2=CHCH_2^+)$, and aldehydes and ketones give acylium cations $(RC=O^+)$ and enol cations $(R_2C=C(OH)R^+)$. Analyzing a mass spectrum for

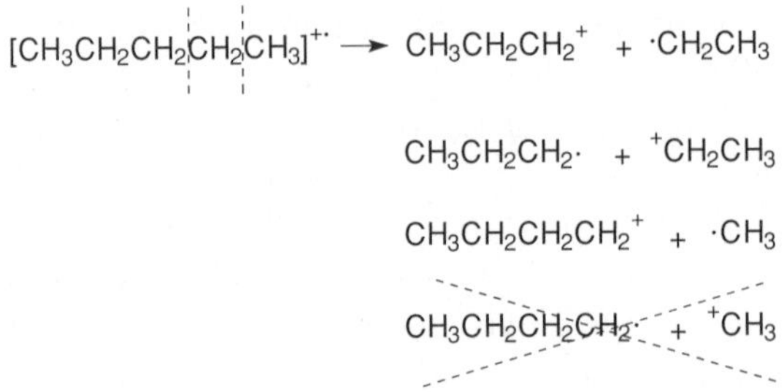

$$[CH_3CH_2CH_2CH_2CH_3]^{+\cdot} \longrightarrow CH_3CH_2CH_2^+ + \cdot CH_2CH_3$$

$$CH_3CH_2CH_2\cdot + {}^+CH_2CH_3$$

$$CH_3CH_2CH_2CH_2^+ + \cdot CH_3$$

$$CH_3CH_2CH_2CH_2\cdot + {}^+CH_3$$

Fig. 10-28. Fragmentation of pentane. All fragmentations give primary carbocations except the last reaction. Methyl carbocations are too high in energy to be formed under typical conditions.

a molecule gives much information about that molecule, but may not be sufficient to identify the material. Other characterization tools are often used in conjunction with MS.

QUESTION 10-27
What fragment peak would be expected in the mass spectrum of ethanol, CH_3CH_2OH?

ANSWER 10-27
Loss of water would give a peak at 28 (46 − 18).

ADVANCED TECHNIQUES

A low-resolution mass spectrometer is the most commonly used instrument and gives nominal molecular weights in whole numbers (43, 72, etc.). A high-resolution mass spectrometer can give molecular weights to the fifth decimal place. For example, C_3H_8, C_2H_4O, CO_2, and CN_2H_4 all have nominal molecular weights of 44. A high-resolution mass spectrometer gives corresponding molecular weight values of 44.06260, 44.02620, 43.98983, and 44.03740, which differentiates among the four materials. The low-resolution mass spectrometer unit is more common because it is more economical.

Traditional MS requires the sample be vaporized and thus samples were limited to those materials that could be vaporized in a high vacuum without decomposition (a molecular weight of up to a few 100 g/mol). Newer techniques with tongue-twisting names like matrix-assisted laser desorption ionization mass spectroscopy (MALDI-MS) and electrospray ionization mass spectroscopy (ESI-MS) allow the analysis of molecules in the 100,000 g/mol range. Proteins and other high molecular weight polymeric materials can now be analyzed by MS.

Quiz

1. The acronym TLC stands for
   (a) total liquid chromatography
   (b) thin layer chromatography
   (c) temperature layer chromatography
   (d) tender loving care

2. Mixtures are separated by TLC based on their
   (a) boiling point
   (b) melting point
   (c) polarity
   (d) density

3. Polar compounds have _____ R_f values on polar stationary phases.
   (a) large
   (b) small
   (c) no

4. Column chromatography uses a _____ as the eluent.
   (a) gas
   (b) liquid
   (c) solid

5. HPLC is used to
   (a) separate mixtures
   (b) separate pure compounds
   (c) cleave chemical bonds
   (d) determine molecular weight

6. IR spectroscopy is used to measure
   (a) ionization energies
   (b) electron excited states
   (c) bond vibrations
   (d) molecular rotations

7. Chemical bonds
   (a) are rigid
   (b) can be described as springs
   (c) have a specific length
   (d) do not absorb IR radiation

8. Carbon-carbon triple bonds absorb IR radiation at _____ wavenumbers than/as do carbon-carbon double bonds.
   (a) higher
   (b) the same
   (c) lower
   (d) exactly 1.5 times longer

9. Heavier atoms vibrate _____ than/as lighter atoms.
   (a) faster
   (b) slower
   (c) the same

10. The absorption of UV radiation results in
    (a) ionization
    (b) electron excited states
    (c) changes in bonding vibrations
    (d) molecular rotational changes

11. Saturated compounds (all sigma bonds) _____ UV radiation in the 200−400 nm range.
    (a) absorb
    (b) do not absorb
    (c) repel
    (d) change the frequency of

12. λ_{max} refers to
    (a) the longest wavelength
    (b) the shortest wavelength
    (c) the highest frequency
    (d) the wavelength of greatest absorption

13. Conjugated molecules
    (a) are saturated compounds
    (b) have a single double bond
    (c) have alternating single and double bonds
    (d) have exactly two double bonds

14. π^* refers to the
    (a) ground state for sigma electrons
    (b) excited state for sigma electrons
    (c) ground state for pi electrons
    (d) excited state for pi electrons

15. Chromophores are
    (a) the results of chromatography

(b) chromium-phosphorus compounds
(c) colored compounds
(d) functional groups that absorb UV radiation

16. NMR stands for
    (a) new molecular reactions
    (b) nuclear material resonance
    (c) nuclear magnetic resonance
    (d) nuclear magnetic resistance

17. Nuclei have properties like
    (a) tiny springs
    (b) tiny magnets
    (c) huge magnets
    (d) chromophores

18. Hydrogen nuclei have _____ spin states.
    (a) one
    (b) two
    (c) three
    (d) four

19. Spin flipping results from absorption of _____ radiation.
    (a) ultraviolet
    (b) visible
    (c) infrared
    (d) radiowave

20. A chemical shift refers to
    (a) hydride shifts
    (b) moving chemicals
    (c) exchanging atoms
    (d) nuclei absorbing radiation

21. Nonequivalent protons
    (a) have the same chemical environment
    (b) have different chemical environments
    (c) are different isotopes of hydrogen
    (d) are different isotopes of carbon

22. The acronym ppm refers to
    (a) impurities in water
    (b) NMR solvents
    (c) chemical shifts
    (d) Peter, Paul, and Mary

23. Downfield chemical shifts result from
    (a) reducing the number of hydrogen atoms
    (b) increasing the number of hydrogen atoms
    (c) adjacent electron-donating groups
    (d) adjacent electron-withdrawing groups

24. Integrating a ^{1}H NMR spectrum gives the
    (a) number of peaks
    (b) number of absorptions
    (c) absolute number of hydrogen atoms
    (d) relative number of hydrogen atoms

25. A triplet has a peak ratio of
    (a) 1:1:1
    (b) 1:2:1
    (c) 1:3:1
    (d) 1:3:3:1

26. A proton split by n equivalent adjacent protons will have
    (a) n peaks
    (b) $(n + 1)$ peaks
    (c) $(n - 1)$ peaks
    (d) $(n + 2)$ peaks

27. A coupling constant refers to
    (a) chemical shift values
    (b) multiplicity values
    (c) spin-spin splitting values
    (d) carbon-carbon bonds

28. The value of J_{ab} _____ that of J_{ba}.
    (a) is less than
    (b) is greater than
    (c) is equal to
    (d) has no relationship to

29. Broadband decoupling
    (a) increases computer speed
    (b) eliminates ^{12}C-^{12}C coupling
    (c) eliminates long- and short-range coupling
    (d) broadens C–H coupling

30. ^{13}C NMR and ^{1}H spectra give _____ information.
    (a) complementary

(b) duplicate
(c) unrelated
(d) conflicting

31. Mass spectroscopy detects
     (a) ionized species
     (b) electron excited states
     (c) changes in bond vibrations
     (d) changes in molecular rotational states

32. The molecular ion peak gives information about
     (a) the height of the peak
     (b) the molecular weight
     (c) fragmentation patterns
     (d) IR radiation absorption

33. MS fragmentation patterns give information about
     (a) bond angles
     (b) functional groups present
     (c) the proton splitting pattern
     (d) crystal structure

CHAPTER

11

Organohalides

Introduction

Man-made organohalides have found uses as solvents, refrigerants, insecticides, polymers, and medicines. Organohalides refer to organic compounds containing a halogen atom: fluorine, chlorine, bromine, or iodine. The organo function may be alkyl, alkenyl (vinyl), or aryl (aromatic). Vinyl polymers and Teflon® coatings are well-known products. In spite of all the useful applications, some fluoroalkanes (Freons®) were determined to be detrimental to earth's upper ozone layer and are banned from use in most parts of the world. Some chlorinated solvents, like carbon tetrachloride, are potential carcinogens. Many tons of chlorinated molecules are produced annually from natural sources. Organohalides are useful intermediates for making compounds containing other functional groups.

Nomenclature

Naming organohalides follow the same general rules discussed for naming alkanes. (For a review see the section entitled *Nomenclature* in Chapter 4.) The formal (IUPAC) system names halides as fluoro-, chloro-, etc., substituents attached to a parent. Common names are accepted and used for smaller molecules. Typical common names for alkyl halides are methyl chloride (CH_3Cl), methylene chloride (H_2CCl_2), chloroform ($HCCl_3$), and carbon tetrachloride (CCl_4). The corresponding formal names are chloromethane, dichloromethane, trichloromethane, and tetrachloromethane.

Properties

Electronegativity of atoms generally increases going from left to right in a row and from bottom to top in a column in the periodic table (see Appendix A). Fluorine is the most electronegative element. The halogens are more electronegative than carbon and halogen-carbon bonds are polar covalent bonds with a partial positive charge on the carbon atom and a partial negative charge on the halogen atom. The carbon atom in carbon-halogen bonds is therefore susceptible to attack by nucleophiles. When a nucleophile attacks and becomes bonded to the carbon atom of an alkyl halide, the carbon-halogen bond is broken. In these reactions the halogen is called the *leaving group* (even though it is an atom). The symbol X is often used to represent any halogen atom.

LEAVING GROUPS

The iodine atom has a much larger diameter than a fluorine atom. The carbon-iodide bond is longer and weaker than a carbon-fluoride bond. For reactions where the carbon-halogen bond is broken, iodide is the best halide leaving group. Bromide and chloride are also good leaving groups but are successively poorer than iodide. Fluoride is a very poor leaving group. Reactions where the halide is the leaving group are discussed in more detail in Chapter 12.

QUESTION 11-1
Why is HI considered a strong acid and HF is a weak acid?

ANSWER 11-1
Even in HX compounds, F^- is a poor leaving group. HI dissociates essentially completely giving high concentrations of protons (H^+), a measure of acidity, while HF dissociates to a much lesser extent.

$$R_3COH \xrightarrow{\text{HCl}} R_3CCl$$

$$RCH_2OH \xrightarrow{\text{SOCl}_2} RCH_2Cl$$

$$R_2CHOH \xrightarrow{\text{PBr}_3} R_2CHBr$$

Fig. 11-1.   Three methods of preparing alkyl halides.

Preparation of Halides

Two methods of preparing alkyl halides from alkenes were previously discussed. The addition of bromine or chlorine to alkenes produces the corresponding dihalide. Addition of HCl, HBr, or HI to alkenes gives alkyl halides. (For a review see the section entitled *Reaction of Alkenes* in Chapter 8.)

Hydrogen halides, with the exception of HF, react with alkyl alcohols to give alkyl halides. Tertiary alkyl alcohols react readily with hydrogen halides at room temperature. Secondary and primary alkyl alcohols are less reactive and heating is often required. Heating may be detrimental to other functional groups in the molecule and therefore other reagents that react at moderate temperatures are used to make alkyl chlorides and bromides from secondary and primary alkyl alcohols. These reagents include thionyl chloride ($SOCl_2$) and phosphorous tribromide (PBr_3). These three halogenation reactions are shown in Fig. 11-1. The mechanisms for these reactions are discussed in Chapter 13.

QUESTION 11-2
What is the product of the reaction of *tert*-butyl alcohol, ((CH_3)$_3$COH), with HI?

ANSWER 11-2
tert-Butyl iodide, (CH_3)$_3$CI.

Radical Halogenation

Alkyl halides can also be prepared by the reaction of an alkane with chlorine or bromine at high temperatures (greater than 300° C) or at room temperature in the presence UV radiation (often just called "light"). This reaction has limited synthetic utility as the product is often a mixture of mono-, di-, tri-, etc. halogenated alkanes. However, it is important to discuss this reaction for a better understanding of the mechanism of this and other similar reactions. This reaction consists of three steps: initiation, propagation, and termination. Fluorine and iodine are generally not used because fluorine is too reactive and iodine is too unreactive.

INITIATION STEP

The reaction mechanism involves radicals (also called free radicals). Species with unpaired electrons are called radicals. The mechanism for the chlorination of propane is shown in Fig. 11-2. In the first step, heat or light (UV) causes the homolytic cleavage of chlorine to give two chlorine radicals (Structure 11-2a). This is called an *initiation step*. Homolytic cleavage means the bond breaks and one electron of the two-electron bond stays with each chlorine atom. The single-headed arrow (fish hook) indicates the movement of just one electron.

PROPAGATION STEP

A chlorine atom reacts with propane to abstract (remove) a hydrogen atom. This reaction also involves homolytic cleavage of a hydrogen-carbon bond. The products are HCl and a propyl radical (Structure 11-2b). The propyl radical reacts with more chlorine to give a monochloropropane (Structure 11-2c) and another chlorine radical. The chlorine radical can react with another propane molecule repeating the prior reaction. This series of reactions (shown in

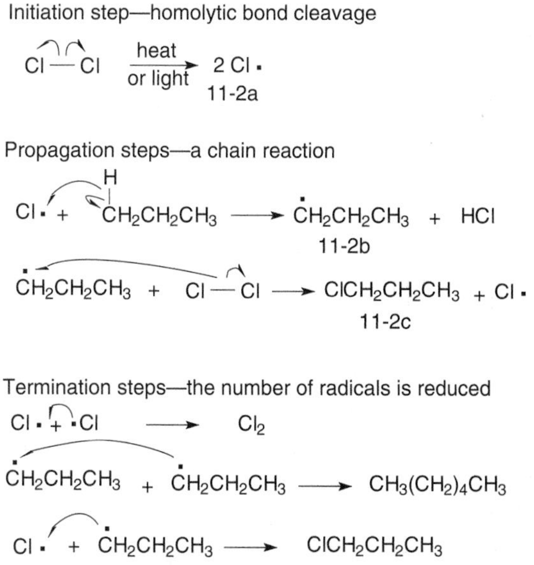

Initiation step—homolytic bond cleavage

$$Cl-Cl \xrightarrow[\text{or light}]{\text{heat}} 2\ Cl\cdot$$
11-2a

Propagation steps—a chain reaction

$$Cl\cdot + CH_2CH_2CH_3 \longrightarrow \overset{\cdot}{C}H_2CH_2CH_3 + HCl$$
11-2b

$$\overset{\cdot}{C}H_2CH_2CH_3 + Cl-Cl \longrightarrow ClCH_2CH_2CH_3 + Cl\cdot$$
11-2c

Termination steps—the number of radicals is reduced

$$Cl\cdot + \cdot Cl \longrightarrow Cl_2$$

$$\overset{\cdot}{C}H_2CH_2CH_3 + \overset{\cdot}{C}H_2CH_2CH_3 \longrightarrow CH_3(CH_2)_4CH_3$$

$$Cl\cdot + \overset{\cdot}{C}H_2CH_2CH_3 \longrightarrow ClCH_2CH_2CH_3$$

Fig. 11-2. Radical chlorination of propane.

$$CI\,CHCH_2CH_3 \;+\; CI\cdot \;\longrightarrow\; CICHCH_2CH_3 \;+\; HCl$$

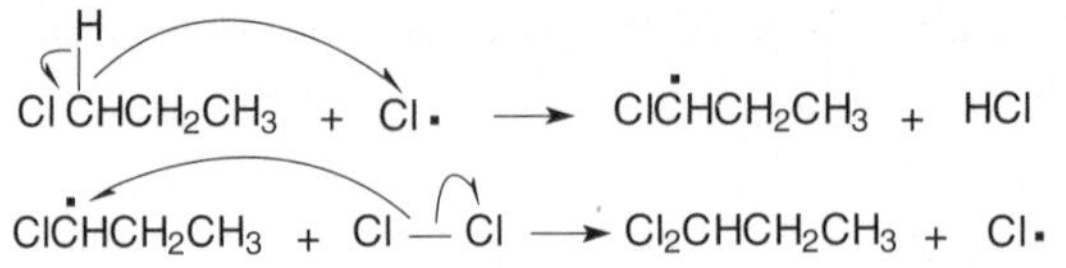

$$CICHCH_2CH_3 \;+\; CI \!-\! CI \;\longrightarrow\; Cl_2CHCH_2CH_3 \;+\; CI\cdot$$

11-3a

Fig. 11-3. Polyhalogenation.

Fig. 11-2) is called the *propagation step*. These two reactions are referred to as a chain reaction since each reaction is like a link of a circular chain that has no beginning or end. The reaction continues (goes in a circle) as long as propane and chlorine are present.

TERMINATION STEP

Radicals are very reactive species that react with almost any molecule they bump into. The reaction between two radicals produces a nonradical species and therefore reduces the total number of radicals in the reaction and slows and ultimately stops the chain reaction. This reaction is known as a *termination step*. Examples of termination steps are shown in Fig. 11-2.

POLYHALOGENATION

Once monochloropropane is formed, it can react by the mechanism shown in Fig. 11-3 to produce a dichloropropane (11-3a). Dichloropropane can react by the same mechanism to give a trichloropropane. In this manner, a mixture of polychlorinated molecules is produced. Poly means more than one. Propane could potentially be converted to octachloropropane. Polyhalogenation is why these reactions are of limited synthetic value.

QUESTION 11-3
How many chloride products could result from the radical chlorination of methane, CH_4?

ANSWER 11-3
Four products: mono-, di-, tri-, and tetrachloromethane.

Isomeric Products

If a large excess of propane is reacted with chlorine, one would expect, on a statistical basis, to get primarily a monochlorinated product. Two monochloropropane isomers could be formed, 1-chloropropane and 2-chloropropane. Since there are six primary hydrogen atoms and two secondary hydrogen atoms in propane that can be replaced by a chlorine atom, one might expect to get three times as much 1-chloropropane as 2-chloropropane (6:2 or 3:1). It is found experimentally that 2-chloropropane is the major product. If bromine is used instead of chlorine, the ratio of 2-bromopropane to 1-bromopropane formed is even greater than the ratio of 2-chloropropane to 1-chloropropane in the chlorine reaction, when run under identical conditions.

THE HAMMOND POSTULATE

The difference of product ratios in bromination and chlorination reactions is explained by considering the mechanism and activation energies of the reaction steps. The first reaction of the propagation steps for each reaction is the rate-determining step. This step is exothermic for the chlorination reaction and endothermic for the bromination reaction. The energy-reaction pathway profiles for the two reactions are shown in Fig. 11-4. The Hammond postulate (for a review see the section entitled *The Hammond Postulate* in Chapter 7) states that the transition state (the top of the energy hill) resembles in structure and energy the reactant or product that is closest in energy to that transition state species.

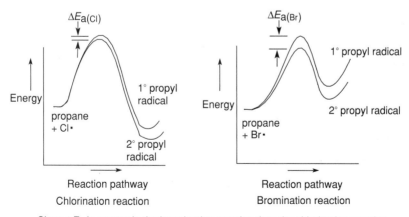

Since ΔE_a is greater in the bromination reaction than the chlorination reaction, the bromination reaction is more selective in forming the secondary radical.

Fig. 11-4. Ratios of isomeric products. Since ΔE_a is greater in the bromination reaction than the chlorination reaction, the bromination reaction is more selective in forming the secondary radical.

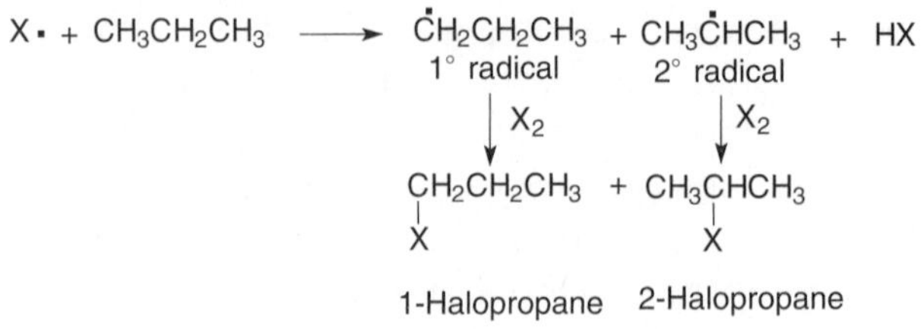

$$X \cdot + CH_3CH_2CH_3 \longrightarrow \overset{\cdot}{C}H_2CH_2CH_3 + CH_3\overset{\cdot}{C}HCH_3 + HX$$

$$\underset{\text{1° radical}}{} \qquad \underset{\text{2° radical}}{}$$

$$\downarrow X_2 \qquad\qquad \downarrow X_2$$

$$\underset{\underset{X}{|}}{CH_2CH_2CH_3} + \underset{\underset{X}{|}}{CH_3CHCH_3}$$

1-Halopropane   2-Halopropane

Fig. 11-5. Competitive radical formation.

RADICAL STABILITY

Propane and halogen radical are the reactants and a propyl radical and hydrogen halide are the "products" in the first reaction of the propagation steps, shown in Fig. 11-5. The "products" are in parentheses since the propyl radical is actually a reactive intermediate in this example. This first reaction controls the overall rate for the propagation steps as it has the highest energy hill (activation energy) to climb.

The stability of the reactants and the "products" must be considered in comparing reaction rates. The order of stability of radicals (electron-poor species) is the same as that for carbocations (also electron-poor species): $3° > 2° > 1° >$ methyl radicals (least stable). The radical halogenation reaction involves two competing reactions, formation of a 1° radical intermediate that subsequently forms 1-halopropane and formation of a 2° radical intermediate that subsequently forms 2-halopropane. These reactions are shown in Fig. 11-5.

Chlorination of propane

The first reaction of the propagation steps, abstraction of a 1° or a 2° hydrogen atom from propane, is exothermic in the radical chlorination of propane. In exothermic reactions, the transition state looks (in energy and structure) more like the reactants than the products. Since the reactants are the same for the competing abstraction steps (and necessarily of equal energy), the difference in energy of the transition states, $\Delta E_{a(Cl)}$, will be small as shown in Fig. 11-4 for the chlorine reaction.

BROMINATION OF PROPANE

The first reaction in the propagation steps of the radical bromination of propane is endothermic. In endothermic reactions, the transition state resembles the products to a greater extent than the reactants. Since there is a difference in the energies of the 1° and 2° radical "products", there will also be a proportional difference in the energies, $\Delta E_{a(Br)}$, of the transition states leading to these products. These energy differences are shown in Fig. 11-4 for the bromine reaction.

Chlorination vs. bromination

Since there is a greater difference in the energies (ΔE_a) of the transition states leading to the 1° and 2° radical "products" for the bromination reaction than the chlorination reaction, one would expect the bromination reaction would give greater selectivity in the product composition. Indeed, the ratio of 2-halopropane to 1-halopropane is greater for the bromination reaction than for the chlorination reaction.

QUESTION 11-4
What constitutional isomers would you expect for the radical monochlorination of pentane, $CH_3CH_2CH_2CH_2CH_3$?

ANSWER 11-4
There are three constitutional isomers: 1-chloro-, 2-chloro-, and 3-chloropentane.

Allylic Halogenation

Allylic halogenation can also give a mixture of products. Allylic compounds refer to compounds that have an alkyl group bonded directly to an alkene function. The hydrogen atoms on the alkyl group are called allylic hydrogen atoms. The allylic hydrogen atoms shown in Structures 11-6a and 11-6d in Fig. 11-6 are indicated with arrows.

Allylic chlorination and bromination reactions follow the same chain reaction mechanism discussed for propane in Fig. 11-2. Abstraction of a hydrogen atom from Structure 11-6a or 11-6d can result in an allylic, alkyl, or vinylic radical. The stability of radical species can now be expanded: 3° allylic > 2° allylic > 1° allylic > 3° alkyl > 2° alkyl > 1° alkyl > vinylic > methyl radicals. According to the Hammond postulate, the stability of the transition states should follow the same order as the stability of the radicals formed. Thus allylic radicals are formed in preference to alkyl or vinylic radicals.

STABILITY FACTORS

The stability of allylic radicals can be explained with resonance structures. (For a review see the section entitled *Resonance Structures* in Chapter 1.) Resonance structures show how a radical can be "spread out" throughout a molecule. The

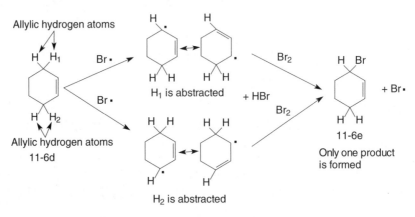

Fig. 11-6. Allylic halogenation.

more "good" resonance structures that can be drawn, the more stable the species. Abstraction of a hydrogen atom from Structure 11-6a gives a radical that can be represented by two resonance structures, 11-6b and 11-6c. These structures have an unpaired electron on different carbon atoms. Each of these radicals can react with halogen to give isomeric monohalogen products. Since the 2° allylic radical (Structure 11-6b) is more stable than 1° allylic radical (Structure 11-6c), the 2° allylic radical contributes more to the resonance hybrid structure. One would then expect that the major product would result from reaction with bromine at the 2° allylic position.

Monohalogenation of cylcohexene (Structure 11-6d) gives one product, Structure 11-6e. Note that the product results from reaction at any of four radical intermediates, all of which are 2° allylic radicals.

Allylic-containing + Br_2 (Cl_2) $\longrightarrow$ Allylic halide
compounds

Toolbox 11-1.

$$\overset{\delta^+}{R}\!\!-\!\!\overset{\delta^-}{X} + Mg \longrightarrow \overset{\delta^-}{R}\!\!-\!\!\overset{\delta^+}{MgX}$$

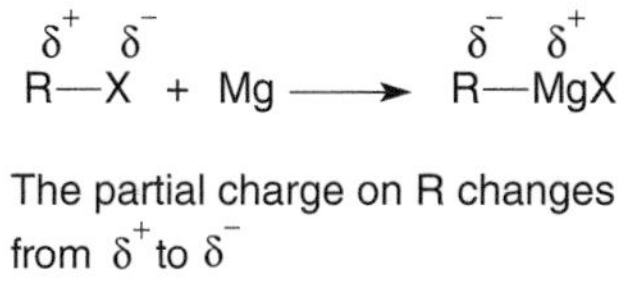

The partial charge on R changes
from δ^+ to δ^-

Fig. 11-7. Preparation of a Grignard reagent. The partial charge on C changes from δ^+ to δ^-.

QUESTION 11-5
What product(s) would result from the monobromination of isobutylene, $(CH_3)_2C=CH_2$?

ANSWER 11-5
Only one product, 3-bromo-2-methylpropene.

Reactions of Organohalides

Organohalides are valuable starting materials in synthetic reactions. Organohalides react with magnesium metal to give organometallic halides. This reaction is shown in Fig. 11-7. These materials are called *Grignard reagents.* The organohalide may be a primary, secondary, or tertiary alkyl halide, a vinylic halide, or an aryl halide. *A key point: the carbon atom bonded to the halide ion has a partial positive charge in the organohalide,* because the halide atom is more electronegative than is the carbon atom. *In the organometallic halide* (the Grignard reagent) *the carbon atom* that was bonded to the halide atom is *now bonded to magnesium* and now *has a partial negative charge* as the carbon atom is more electronegative than is magnesium. These partial charges are shown in Fig. 11-7.

The carbon atom bonded to the halogen in organohalides has a partial positive charge (δ^+) and is susceptible to attack by an electron-rich nucleophile, as discussed above. That same carbon atom in the corresponding Grignard reagent has a partial negative charge (δ^-) and is now a nucleophilic. The reaction

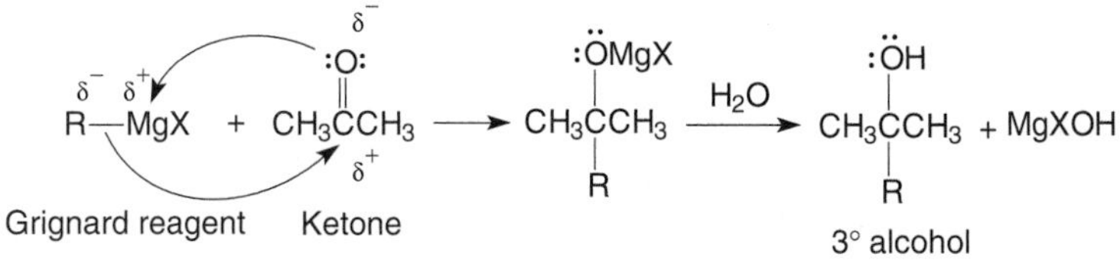

$$\overset{\delta^-}{R}\!\!-\!\!\overset{\delta^+}{MgX} + \underset{\delta^+}{CH_3\overset{\overset{\displaystyle\overset{\cdot\cdot}{O}:}{\|}}{C}CH_3} \longrightarrow \underset{R}{CH_3\overset{\overset{\displaystyle :\overset{\cdot\cdot}{O}MgX}{|}}{C}CH_3} \xrightarrow{H_2O} \underset{R}{CH_3\overset{\overset{\displaystyle :\overset{\cdot\cdot}{O}H}{|}}{C}CH_3} + MgXOH$$

Grignard reagent　　Ketone　　　　　　　　　　　　　　　　　　　3° alcohol

Fig. 11-8. Grignard reaction.

$$RX + Li \longrightarrow RLi \xrightarrow{\text{CuI}} R_2CuLi \xrightarrow{R'X} R-R' + RCu + LiX$$

11-9a   11-9b   11-9c

R is coupled to R′ and
a new C-C bond is formed

Fig. 11-9.   Gilman reagent. R is coupled to R′ and a new C−C bond is formed.

of Grignard reagents with carbonyl compounds to give alcohols is a very important reaction (shown in Fig. 11-8) that will be discussed in more detail in Chapter 13.

Organohalides are also used in coupling reactions, so named because they couple (combine) two molecules forming a new carbon-carbon bond. This reaction is shown in Fig. 11-9. Alkyl halides react with lithium metal to give an alkyllithium compound 11-9a. This material is reacted with copper(I) iodide to give a lithium dialkylcopper compound 11-9b, called a *Gilman reagent*. A Gilman reagent reacts with an alkyl halide (or a vinylic or aryl halide) to give the corresponding coupling compound 11-9c.

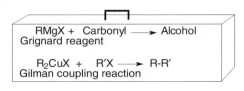

RMgX + Carbonyl ⟶ Alcohol
Grignard reagent

R_2CuX +   R′X ⟶ R-R′
Gilman coupling reaction

Fig. 11-10.

Quiz

1. Which alkyl halide has the most polar alkyl-halogen bond?
    (a) RF
    (b) RCl
    (c) RBr
    (d) RI

2. Which alkyl halide has the strongest alkyl-halogen bond?
    (a) RF
    (b) RCl
    (c) RBr
    (d) RI

3. The reaction of t-butyl alcohol with HCl at room temperature gives
    (a) t-butyl chloride
    (b) t-butyl bromide

(c) isobutyl chloride

(d) an alkene

4. (Free) radicals
    (a) contain no electrons
    (b) contain unpaired electrons
    (c) contain all paired electrons
    (d) cost nothing

5. Homolytic cleavage means a bond
    (a) breaks to form two radicals
    (b) breaks to form ionic species
    (c) is formed from two radicals
    (d) is formed from two ions

6. Monochlorination of butane gives ____ constitutional isomer(s).
    (a) one
    (b) two
    (c) three
    (d) four

7. The ratio of isomeric products formed in the free radical halogenation of
    an alkane depends upon
    (a) the concentration of reactants
    (b) the concentration of products
    (c) the energy differences in the transition states
    (d) the phase of the moon

8. The most stable radical is a/an
    (a) methyl radical
    (b) secondary alkyl radical
    (c) tertiary allylic radical.
    (d) vinylic radical

9. Chlorination of methane catalyzed by light is a/an
    (a) cationic reaction
    (b) anionic reaction
    (c) chain reaction
    (d) rope reaction

10. Allylic radicals are stabilized by
    (a) resonance
    (b) UV light
    (c) halogen atoms
    (d) electron-withdrawing groups

11. Grignard reagents are
   (a) alkyl halides
   (b) alkyl magnesium halides
   (c) magnesium halides
   (d) alkyl manganese halides

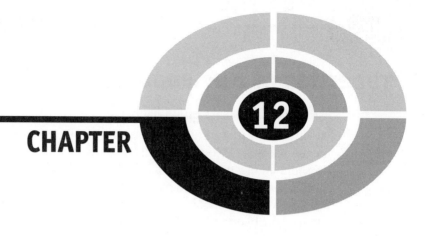

Nucleophilic Substitution and Elimination Reactions

Introduction

The study of organic chemistry involves reactions that convert one functional group into other functional groups. Two important competing reactions are *nucleophilic substitution* and *elimination*. The reactions depend primarily on four factors: the composition of the nucleophile or base, the substrate, the leaving group, and the solvent. Four types of reactions will be discussed: S_N1, S_N2, E1, and E2.

Properties of Nucleophiles

The *nucleophile* is an atom or a functional group with one or more nonbonding electron pairs it is willing to share with an electron-deficient, electrophilic atom in the substrate. Good (strong) nucleophiles are defined by their willingness to share an electron pair. Nucleophiles can also be described in terms of their properties as bases. Good bases tend to be good nucleophiles. The nucleophile may be a neutral species such as water (H_2O) or a negatively charged species such as the hydroxide ^-OH or bromide (Br^-) anions.

A detailed mechanism for a nucleophilic substitution reaction is shown in Reaction (a) in Fig. 12-1.

Five observed trends can be used in predicting the strength of a nucleophile.

(a) Bases are better nucleophiles than their conjugate acids: $H_2N^- > NH_3$, $HO^- > H_2O$, and $F^- > HF$. (For example, H_2N^- is a better nucleophile than its conjugate acid, NH_3.)

(b) For atoms of about the same size (N, O, and F), the more basic species is a better nucleophile: $H_2N^- > HO^- > F^-$. (H_2N^- is the strongest base as well as the best nucleophile in this series.)

(c) For a common atom (O in this case), stronger bases are better nucleophiles: $HO^- > CH_3CO_2^- > H_2O$. (HO^- is the strongest base and best nucleophile.)

(d) For atoms in a family (a column in the periodic table), the larger atoms (those with larger diameters) are better nucleophiles in polar protic

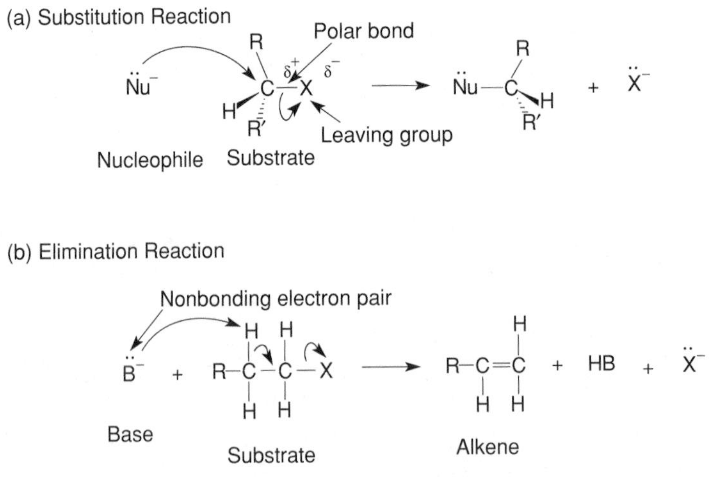

(a) Substitution Reaction

(b) Elimination Reaction

Fig. 12-1. Nucleophilic substitution and elimination reactions.

Table 12-1.   Nucleophiles.

Good	Moderate	Poor
RSe^-	Br^-	F^-
RS^-	N_3^-	ROH
I^-	RSH	H_2O
NC^-	NH_3	
HO^-	Cl^-	
RO^-	RCO_2^-	

solvents: $I^- > Br^- > Cl^- > F^-$. (This relationship is however solvent-dependent, as explained below.)

(e) Bulky alkoxides react more slowly with substrates, since they must approach each other in the transition state. The less bulky ethoxide anion $(CH_3CH_2O^-)$ is a better nucleophile than the more bulky t-butoxide anion $[(CH_3)_3CO^-]$.

QUESTION 12-1
Which has more nonbonding electron pairs, the hydroxide anion or water?

ANSWER 12-1
The hydroxide anion has three nonbonding electron pairs and water has only two.

A list of nucleophiles classified as good, moderate, or poor is given in Table 12-1. The order of nucleophilic strength is highly dependent on the solvent but generally follows the order shown. Nucleophilic strength also depends upon the bulkiness of the nucleophile and that of the substrate.

Properties of Bases

The species with an electron pair to share can alternatively act as a base instead of a nucleophile. If the species acts as a base, it abstracts (removes) a proton (H^+) from the substrate. The leaving group is also cleaved and the end product is an alkene (in most cases). This is shown in Reaction (b) in Fig. 12-1. Bases may have a negative charge like hydroxide and alkoxide anions or may be neutral species like water and ammonia.

STRONG BASES

Base strength is inversely related to the strength of its conjugate acid. Strong bases have weak conjugate acids. The hydroxide and alkoxide ions are strong bases since their conjugate acids, water and alcohols, are weak acids. Water and iodide anion are weak bases as their conjugate acids, H_3O^+ and HI, are strong acids.

QUESTION 12-2
If H_2S is a stronger acid than H_2O, which acid has the weaker conjugate base?

ANSWER 12-2
H_2S has the weaker conjugate base, HS^-.

Properties of Substrates/Electrophiles

The *substrate*, an electrophile, is a molecule that contains a carbon atom bonded to a leaving atom or group that has a greater electronegativity than the carbon atom. The carbon atom bears a partial positive charge (δ^+) and the leaving group a partial negative charge (δ^-), as shown in Fig. 12-1, Reaction (a). The electrophilic atom need not necessarily be a carbon atom, but this discussion will be limited to carbon atoms.

Properties of Leaving Groups

The *leaving group* (the "group" may be a single atom) may be a negative or neutral species and is that part of the substrate which is cleaved in substitution and elimination reactions. The leaving group is more electronegative than the carbon atom it is bonded to, and, when this bond breaks, the bonding electron pair remains with the leaving group. This type of bond cleavage is called *heterolytic cleavage*, since both electrons of the bond remain with the leaving group. Typical atoms of the leaving group are the halogens, oxygen, nitrogen, and sulfur. The best leaving groups are stable, weak bases. Good leaving groups are the halides $(X^-$, except fluorine), sulfonates (RSO_3^-), water (H_2O), alcohols (ROH), and tertiary amines (R_3N).

POOR LEAVING GROUPS

Ions that are strong bases, such as HO^-, RO^-, and H_2N^-, are poor leaving groups. They are not "stable" in the sense that they do not want to be alone. If a proton is present, these anions will want to react with it. If a carbocation is present, these anions also want to react with it to form a C—O or C—N bond. Consequently, if they form a bond with a carbon atom, this bond is not easily broken even though it is a polar bond. A list of leaving groups is given in Table 12-2. The reactions of halides have been studied to a greater extent than other leaving groups and, therefore, alkyl halides will be discussed in this chapter.

QUESTION 12-3
Can an ion that is made up of a single atom, like I^-, be a leaving group?

ANSWER 12-3
Yes, the entity that leaves the substrate is called the leaving group whether it is a monoatomic ion such as I^- and Br^- or a group of atoms such as H_2O.

QUESTION 12-4
Which molecule is a better leaving group, H_2O or NH_3?

ANSWER 12-4
Water is a stronger acid than NH_3 and therefore NH_3 is a stronger base than H_2O. Weaker bases are better leaving groups, therefore, H_2O is a better leaving group.

POLARIZABILITY

Good leaving groups are readily polarizable. That is, they can accommodate a partial negative charge in the transition state species in substitution and elimination reactions. Dispersing the charge on the atoms involved in the transition state (rather than concentrating the charge on one atom) lowers the energy of the transition state. Lowering the energy of the transition state species increases the rate of reaction. (See the section entitled the *Hammond Postulate* in Chapter 7.)

FORMING BETTER LEAVING GROUPS

Poor leaving groups such as HO^-, RO^-, and H_2N^- can be made into better leaving groups in the presence of a strong protic acid. The O or N atom of the leaving group is protonated and the leaving group becomes H_2O, ROH, or NH_3, respectively. Figure 12-2 shows the reaction mechanism for converting

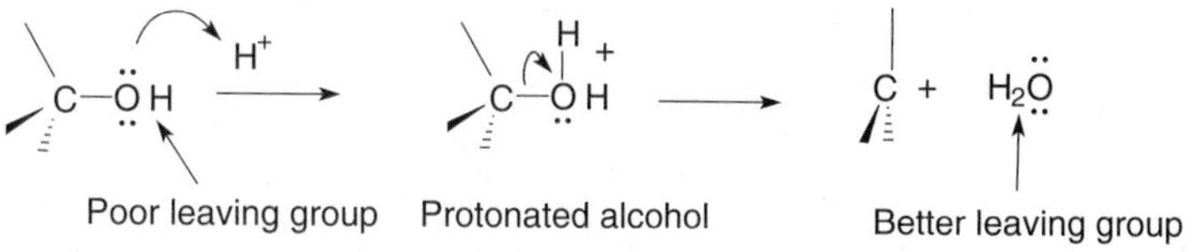

Poor leaving group   Protonated alcohol   Better leaving group

Fig. 12-2. Conversion to a better leaving group.

the $-OH$ substituent into the better leaving group H_2O. Hydrogen halides (HX) are typical strong acids used in this reaction.

Properties of Solvents

The solvent plays a very important role in nucleophilic substitution and elimination reactions. In addition to acting as the reaction medium, it also stabilizes the transition states and intermediates. Solvents are classified as polar and nonpolar. Nonpolar solvents are generally not used, since ionic or polar species in these reactions have limited solubility in nonpolar solvents.

POLAR PROTIC SOLVENTS

Polar solvents are divided into two categories: *aprotic* and *protic*. A protic solvent is one that has an "active" hydrogen atom. An active hydrogen atom is involved in hydrogen bonding or has a strong dipole interaction with other species. Compounds with active hydrogen(s) contain functional groups such as $-OH$, $-NH_2$, $-CO_2H$, and $-SH$.

QUESTION 12-5
Which is a better polar protic solvent, H_2O or H_2S?

ANSWER 12-5
H_2O, since it can hydrogen-bond to a solute. Only hydrogen atoms bonded to O, N, or F form hydrogen bonds to other molecules.

POLAR APROTIC SOLVENTS

Examples of the polar aprotic solvents—acetone, dimethylformamide, and dimethylsulfoxide—are shown in Fig. 12-3. Although they all contain hydrogen atoms, the hydrogen atoms are bonded only to carbon atoms, not to O, N, or F. The carbon–hydrogen bond is considered a nonpolar bond. An interesting property of these solvents is that they *stabilize cations* to a greater extent than they stabilize anions. The partial negative charge on the oxygen atom is available to interact with (stabilize) a positive or partial positive charge in another species. The partial positive charge on the carbon or sulfur atom in these polar aprotic solvents is somewhat shielded, or "buried inside" the solvent molecule, and is not readily available to interact with (stabilize) a negative or partial negative charge on another species. This steric property is demonstrated in Fig. 12-3.

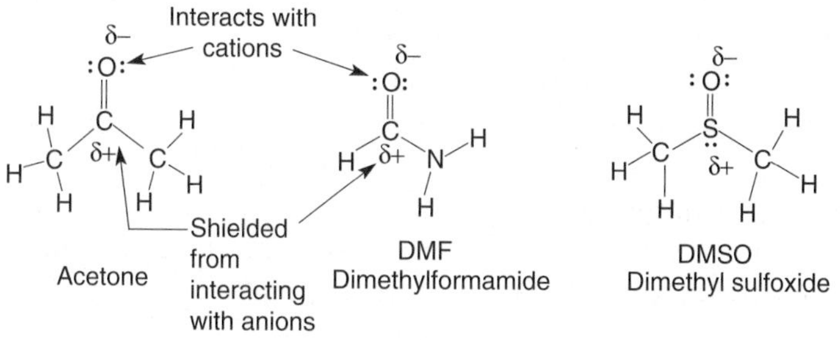

Fig. 12-3. Polar aprotic solvents.

POLAR PROTIC SOLVENTS AND NUCLEOPHILES

Polar solvents can affect the strength of nucleophiles. Consider the halogen series, I^-, Br^-, Cl^-, and F^-. It is stated above that I^- is the best nucleophile in this series. This order is observed in polar protic solvents. The halide nucleophile is initially associated with its countercation, e.g., Na^+X^- or K^+X^- (where X is any halogen). These salts are soluble in polar protic solvents. The halide anion is hydrogen-bonded to many solvent molecules when dissolved in the solvent. The solvent molecules form a solvent shell, or cage, around the halide anion. This makes the halide ion more stable (due to the interaction of the negative and partial positive charges) and less reactive as a nucleophile, since it would have to shed its solvent shell to react with the substrate.

The negative charge on the fluoride atom is more "concentrated" (it has a higher negative charge density) than is the negative charge on the large iodide anion and, therefore, more solvent molecules are attracted to the fluoride anion than the iodide anion, as shown in Fig. 12-4. Thus, electrons in an iodide anion are more available to attack a substrate, and iodide is the best nucleophile in the halide series in polar protic solvents.

Highly solvated fluoride anion.   Slightly solvated iodide anion.

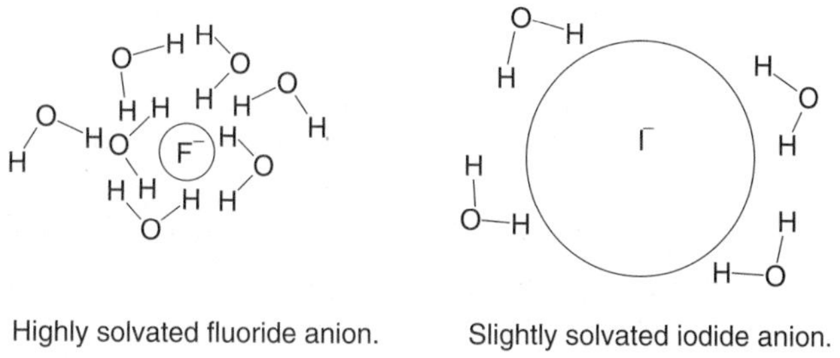

Fig. 12-4. Protic solvent shell around halide anions.

Polar aprotic solvents and nucleophiles

Polar aprotic solvents, with their exposed partial negative charge and hidden partial positive charge, interact more strongly with the cations, such as Na^+ and K^+, than with anions. This helps separate the Na^+X^- or K^+X^- ion pairs and makes the X^- anion more available to act as a nucleophile toward some substrate. The polar aprotic solvent molecules surround the cation and form a solvent shell around it. The anion does not interact with the polar aprotic solvent to any great extent and is said to be "naked." The anion is free to attack the substrate. Since the fluoride anion has a higher charge density in its small electron cloud than does the larger iodide anion, the fluoride anion is the best halogen nucleophile in polar aprotic solvents. The order of nucleophilicity ($F^- >$ $Cl^- > Br^- > I^-$) is *reversed* from that found in polar protic solvents.

QUESTION 12-6
Which will be surrounded by a larger solvent shell in water, Cl^- or Br^-?

ANSWER 12-6
Cl^-, since it has a smaller diameter and has a higher electron charge density.

Second-Order Nucleophilic Substitution (S_N2) Reactions

A variety of substitution reactions were studied and divided into two categories depending upon similar reaction characteristics. One category was given the name *second-order nucleophilic substitution, S_N2, reactions*: S stands for substitution, N for nucleophilic, and 2 for bimolecular (or second order). The effect that the four variables—nucleophile, substrate, leaving group, and solvent—have on S_N2 reactions is discussed below.

KINETICS

Kinetics is the study of how fast (the rate) a reaction occurs. The rate of all S_N2 reactions is dependent on the concentration of two species, the nucleophile and the substrate. This means that both species are involved in the rate-determining step. Doubling the concentration of either species doubles the rate of the reaction according to the following rate-law expression:

$$\text{reaction rate} = k[\text{nucleophile}][\text{substrate}]$$

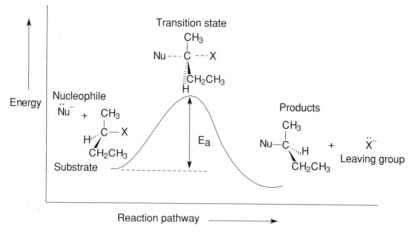

Fig. 12-5. Energy–reaction pathway for a substitution reaction.

The rate constant k is a function of the activation energy E_a (sometimes represented as $\Delta G^{\ddagger}$). (For a review of activation energies, see Chapter 7.) The magnitude of E_a depends on the difference in energy between the transition state species and the starting materials. A energy–reaction pathway diagram is shown in Fig. 12-5.

QUESTION 12-7
Is the rate constant k a constant under all reaction conditions?

ANSWER 12-7
No, k varies with temperature and value of E_a.
(See the section entitled The Rate Constant in Ch. 7)

SUBSTRATE STRUCTURE

The rate of an S_N2 reaction depends on the *bulkiness* of the substrate. Since the transition state involves the reaction of the nucleophile and the substrate, bulky nucleophiles (as stated in item (e) on page 271) and bulky substrates decrease the rate of S_N2 reactions. It becomes more difficult for the reacting centers to interact. This is called a *steric effect*. This steric effect is shown in Fig. 12-6. The rate of reaction of HO^- with a series of alkyl bromides follows the decreasing order: $CH_3Br > CH_3CH_2Br > (CH_3)_2CHBr > (CH_3)_3CBr$. In general, the rate of an S_N2 reaction in a homologous series of alkyl halides is methyl > 1° > 2° > 3° halide. In fact, most 3° carbon compounds do not undergo S_N2 reactions, as it is too difficult (energetically unfavorable) for the nucleophile to closely approach the electrophilic carbon atom in the substrate.

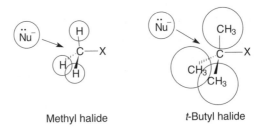

Methyl halide   *t*-Butyl halide

Fig. 12-6. Steric effects in S_N2 reactions.

NUCLEOPHILE

The strength of a nucleophile depends on its willingness to share an electron pair. Strong bases make good nucleophiles. Good nucleophiles are also readily polerizable. The electron cloud of a large atom is more easily distorted (polarized) than the electron cloud of a smaller atom. Thus RSe^- is a better nucleophile than RS^- and I^- is a better nucleophile than the other halogens. See Table 12-1 for the classification of nucleophiles.

STEREOCHEMISTRY

Both the nucleophile and the substrate are involved in the rate-determining step in S_N2 reactions. The most logical mechanism is for the nucleophile to approach the substrate from the side opposite of the leaving group. As the nucleophile approaches the substrate, a bond begins to form and the bond to the leaving group begins to break. As the reaction proceeds, the nucleophile–carbon bond forms and the carbon–leaving group bond breaks. The leaving group takes the bonding electrons. This is called *heterolytic bond cleavage* as shown in Fig. 12-7.

Walden inversion

If an S_N2 reaction takes place at a stereocenter in the starting substrate, the product of the reaction has the opposite configuration at that stereocenter. This is called the *Walden inversion*. As shown in Fig. 12-7, the transition state has

Nucleophile   Substrate   Transition state   Products

Fig. 12-7. Walden inversion.

a trigonal pyramidal structure. The three groups that were oriented toward the "left" in the starting material are oriented to the "right" in the product. This is like an umbrella being inverted on a very windy day. This is a *stereospecific reaction*. If one started with an *S* stereoisomer, the resulting product would have an *R* configuration.

QUESTION 12-8

If a pure stereoisomer (the *R* stereoisomer) undergoes Walden inversion, would the product consist of both *R* and *S* stereoisomers or only one isomer?

ANSWER 12-8

The product will consist (ideally) of only one stereoisomer.

Activation energies

The energy–reaction pathway for an S_N2 reaction in polar protic and polar aprotic solvents is shown in Fig. 12-8. In a polar protic solvent, the solvated anion is more stable and lower in energy than is the naked anion in a polar aprotic solvent. Since the solvated ion is lower in energy, it takes more energy (E_{a_2}) to reach the transition state (the top of the energy hill) than is required for the naked ion (E_{a_1}). This explains why reaction rates for S_N2 reactions are faster in polar aprotic solvents.

This mechanism also explains why 3° alkyl substrates react more slowly (if at all) than 1° alkyl substrates in S_N2 reactions. It is more difficult (a higher activation energy E_a is required) for the nucleophile to approach and react with

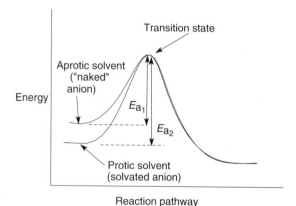

Fig. 12-8. Solvent stabilization of a nucleophile. Solvated anion has a higher activation energy, E_{a_2}.

a bulky, sterically hindered tertiary substrate than to react with the relatively unhindered primary carbon atom of a methyl halide (see Fig. 12-6).

First-Order Nucleophilic Substitution (S$_N$1) Reactions

Another series of substitution reactions carried out between nucleophiles and substrates has different characteristics from those of S$_N$2 reactions. This type of reaction is called a *first-order nucleophilic substitution, S$_N$1*: S stands for substitution, N for nucleophilic, and 1 for unimolecular (or first order).

KINETICS

The rate of an S$_N$1 reaction in dependent only on the substrate and not on the nucleophile, as is the case for S$_N$2 reactions. Doubling the concentration of the substrate doubles the reaction rate (not the rate constant). The rate-law expression for an S$_N$1 reaction is

$$\text{reaction rate} = k[\text{substrate}]$$

NUCLEOPHILE

The structure of the nucleophile does not affect the reaction rate of S$_N$1 reactions since the nucleophile is not involved in the rate-determining step. Typical nucleophiles are the same as those discussed previously for S$_N$2 reactions (see Table 12-1).

QUESTION 12-9
When the concentration of the nucleophile is doubled in an S$_N$1 reaction, will the rate of reaction double?

ANSWER 12-9
No, the reaction rate is independent of nucleophile concentration.

SUBSTRATE

The rate of an S$_N$1 depends on the structure and concentration of the substrate. Here, the order of reactivity of alkyl halides is opposite to that observed for S$_N$2 reactions. In a homologous series, tertiary alkyl halides react the fastest

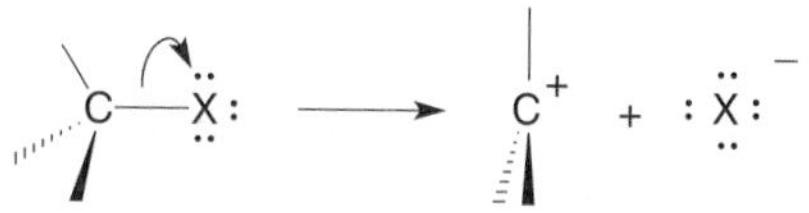

Heterolytic
bond cleavage

Fig. 12-9.   Bond cleavage in an S_N1 reaction.

and methyl halides react the slowest. The order of reactivity for alkyl halides is $3° > 2° > 1° >$ methyl. Methyl and $1°$ halides react so slowly that there is essentially no reaction by an S_N1 mechanism. This order of reactivity is rationalized by a rate-determining step that involves carbocation formation via heterolytic cleavage of the bond between the carbon atom and the leaving group. This reaction is shown in Fig. 12-9. The resulting carbocation is either $3°$ or $2°$, depending on the structure of the starting substrate.

Since $3°$ carbocations are more stable than $2°$ carbocations, the transition state for $3°$ carbocation formation is lower in energy (is more stable) than is the transition state for $2°$ carbocation formation. (For a review of the Hammond postulate, see the section entitled *The Hammond Postulate* in Chapter 7.)

QUESTION 12-10
Which carbocation is more stable, $CH_3CH_2^+$ or $(CH_3)_2CH^+$?

ANSWER 12-10
$(CH_3)_2CH^+$ is more stable since it is a $2°$ carbocation and $CH_3CH_2^+$ is a $1°$ carbocation. A primary carbocation is very unstable and seldom, if ever, observed in solvent-based reactions.

LEAVING GROUP

Factors defining good leaving groups were given previously for S_N2 reactions. These same factors define good leaving groups in S_N1 reactions (see Table 12-2).

Table 12-2.   Leaving groups.

Good	Moderate	Poor
I^-	F^-	HO^-
Br^-	RCO_2^-	RO^-
Cl^-		NH_2^-

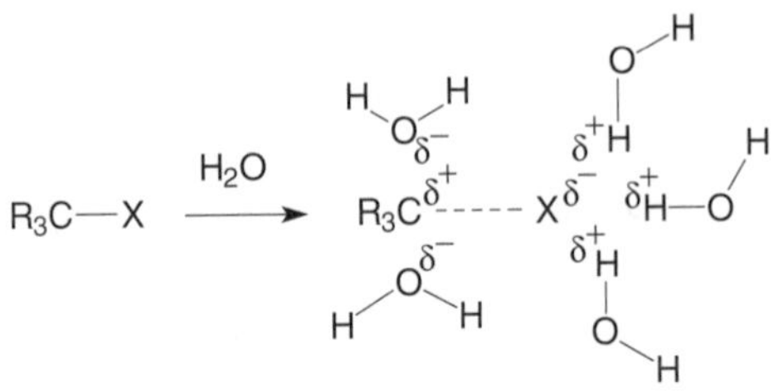

Fig. 12-10. Solvent stabilization of the transition state.

SOLVENT

Polar solvents increase the rate of S_N1 reactions by stabilizing the transition state (carbocation formation) in the rate-determining step. Tertiary and secondary alkyl halide substrates are usually neutral species. Partial charges increase on the carbon atom and on the leaving group as the leaving group begins to leave. This is shown in Fig. 12-10. The transition state has a partial positive charge on the carbon atom and a partial negative charge on the leaving group. A polar solvent will interact with the polar transition state and stabilize it. Stabilizing the transition state lowers the magnitude of E_a. Polar protic solvents hydrogen-bond to the departing negatively charged leaving group. The negative dipole of a solvent molecule interacts with the partial positive charge on the carbon atom in the substrate. These interactions help to stabilize the transition state.

Polar protic solvents stabilize transition states to a greater extent than do polar aprotic solvents. Thus polar protic solvents are better solvents for S_N1 reactions than are polar aprotic solvents, since polar protic solvents lower E_a, increasing the rate of reaction by increasing the rate constant k. Since the nucleophile is not involved in the rate-determining step in S_N1 reactions, the interaction of the solvent with the nucleophile does not affect the rate of reaction. However, the solvent must be sufficiently polar to dissolve the nucleophile.

QUESTION 12-11
Why do polar solvents increase the rate of S_N1 reactions?

ANSWER 12-11
They solvate the transition state, lowering the energy of this species, lowering E_a.

STEREOCHEMISTRY

When an S_N1 reaction takes place at a chiral center of one isomer (e.g., the R isomer), the reaction occurs with *racemization* at that stereocenter. This would suggest that an intermediate in the reaction is a planar species such that attack by the nucleophile can occur in a manner that gives two isomeric (mirror image) products.

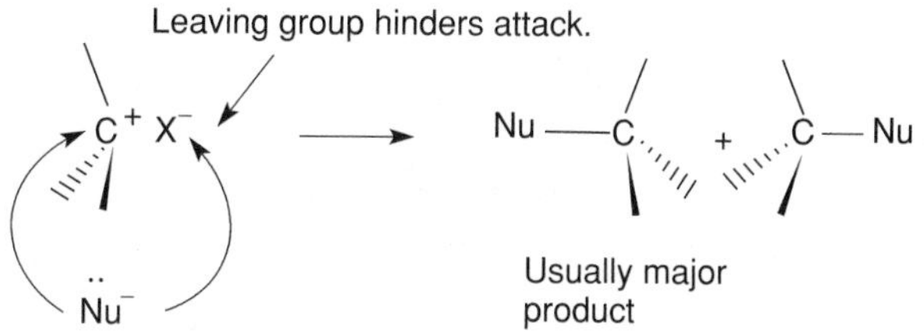

Leaving group hinders attack.

Usually major product

Fig. 12-11.   Hindered nucleophilic attack in an S_N1 reaction.

Carbocation intermediates

A carbocation intermediate would explain racemization. If the leaving group left the substrate in the rate-determining step, the reaction rate would be dependent only on the substrate. A carbocation would result from heterolytic cleavage of the bonding electrons when the leaving group departs.

  Carbocations are planar species. Once formed, the nucleophile could attack from either the front- or backside, giving both stereoisomeric products. Actually, there is usually not an equal probability of nucleophilic attack from both sides. The leaving group may remain in the vicinity of the carbocation and hinder the nucleophile from attacking from that side. This is shown in Fig. 12-11. There is a net inversion of configuration, since it is easier for the nucleophile to attack from the unhindered side.

QUESTION 12-12
If an R stereoisomer undergoes an S_N1 reaction, will equal amounts of R and S isomers be formed?

ANSWER 12-12
Ideally, equal amounts of both isomers would be formed but often there is, net inversion of configuration since the leaving group blocks the incoming nucleophile from the "back-side".

SPECIAL PRIMARY HALIDES

There are two types of "1° like" halides that do react in S_N1 reactions. These are *primary allyl halides and primary benzyl halides*. The reason they undergo S_N1 reactions is that these carbocations are stabilized by resonance. Resonance delocalizes (spreads out) the positive charge, making these carbocations about as stable as 2° alkyl carbocations. Resonance delocalization for these carbocations is shown in Fig. 12-12.

$$H_2C = C - \overset{+}{C}H_2 \longleftrightarrow H_2\overset{+}{C} - C = CH_2$$

Resonance-stabilized allyl carbocation.

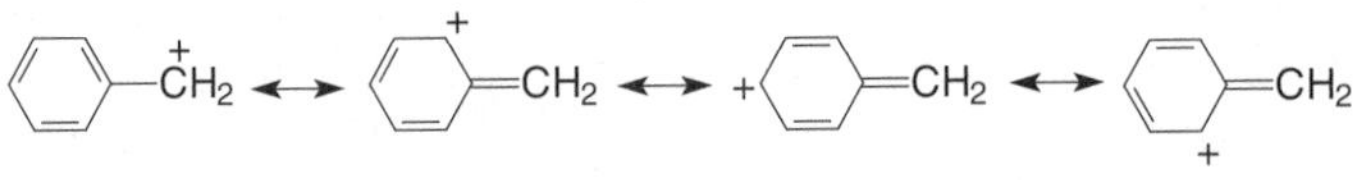

Resonance-stabilized benzyl carbocation.

Fig. 12-12. Primary allyl and benzyl carbocations.

CARBOCATION REARRANGEMENTS

Whenever a carbocation is an intermediate in a reaction, one has to consider *rearrangement* reactions that give a more stable carbocation. These rearrangements involve hydride or alkyl shifts. Tertiary carbocations generally do not undergo these shifts. Secondary carbocations may rearrange to more stable tertiary carbocations. (For a review of these shifts, see the section entitled *Rearrangement Reactions of Carbocations* in Chapter 7.)

QUESTION 12-13
What rearrangement products would you predict for the *tert*-butyl carbocation?

ANSWER 12-13
None, this 3° carbocation cannot rearrange to a more stable species.

Summary of S$_N$1 and S$_N$2 Reactions

S$_N$1 reactions occur with 3° and 2° alkyl halides. S$_N$2 reactions occur with methyl halides and 1° and 2° alkyl halides. We can now predict the substitution reaction mechanism for 3°, 1°, and methyl alkyl halides, but how does one predict the mechanism for 2° alkyl halides since they can undergo either S$_N$1 or S$_N$2 reactions?

Several factors must be considered that influence reaction rates. Good nucleophiles and polar aprotic solvents favor S$_N$2 reactions. Polar protic solvents favor S$_N$1 reactions. A poor nucleophile makes the S$_N$2 reaction less competitive S$_N$1 reactions. One has to consider each of the four factors—nucleophile, substrate, leaving group, and solvent—when predicting the mechanism for a given substitution reaction.

Second-Order Elimination (E2) Reactions

Species that react as nucleophiles can also react as bases. Strong bases are defined by their ability to react with protons. Bases can react with substrates to remove a proton that is on a carbon atom adjacent to the carbon atom containing the leaving group. The leaving group is attached to the α carbon atom and the hydrogen that is removed is attached to a β (an adjacent) carbon atom. These elimination reactions are called *second-order elimination, E2, reactions*. They are also referred to as β-elimination or 1,2-elimination reactions. If a halogen atom is the leaving group, the reaction is called a *dehydrohalogenation reaction*.

REACTION KINETICS

The rate of an E2 reaction is second order: first order in substrate and first order in base. Doubling the concentration of either of these reagents doubles the rate of reaction. The rate-law expression is the same as for the S_N2 reaction:

$$\text{reaction rate} = k[\text{base}][\text{substrate}]$$

The mechanism of an E2 reaction is shown in Fig. 12-13. This reaction is a *concerted* reaction. The rate-determining step is the simultaneous making and breaking of bonds in the transition state. No carbocation intermediate is formed. The C—H and C—X bonds are broken and a π bond is formed in the substrate, giving an alkene product.

QUESTION 12-14
How many steps are there in an E2 reaction?

ANSWER 12-14
One, all bond breaking and making is done in a single concerted step.

Fig. 12-13.   E2 reaction mechanism.

BASE

Strong bases such as hydroxide (HO^-) and alkoxide (RO^-) anions are commonly used in E2 reactions.

SUBSTRATE

The order of reactivity of alkyl halides in *E2* reactions is $3° > 2° > 1°$. This is opposite of the order seen for bimolecular S_N2 reactions. The reaction rate in *substitution* reactions depends on the ability of the nucleophile to attack the carbon atom bonded to the leaving group. Increasing the number of alkyl groups bonded to this carbon atom hinders the approach by the nucleophile. In *elimination* reactions, the base can pull off a proton on the "outer edges" of the molecule. Even strong, bulky bases, such as *tert*-butoxide anion $[(CH_3)_3CO^-]$, can approach and remove these unhindered protons.

ZAITSEV'S RULE

Many molecules have more than one type of β hydrogen. As a result, β-elimination reactions can give different alkene products, as shown for 2-bromobutane, Structure 12-14a in Fig. 12-14. Alkene stability increases with increasing alkyl substitution of the alkene. Tetraalkyl-substituted alkenes are more stable than trialkyl-substituted alkenes, which are more stable than dialkyl-substituted alkenes, and so on. (For a review, see the section entitled *Stability of Alkenes* in Chapter 6.) The major product of the dehydrohalogenation of 2-bromobutane is 2-butene, Structure 12-14b. 2-Butene is a dialkyl-substituted alkene, while the minor elimination product, 1-butene, Structure 12-14c, is a monoalkyl-substituted alkene.

According to the Hammond postulate, the stability of the transition state is related to the stability of the reactant or product. One would expect that the

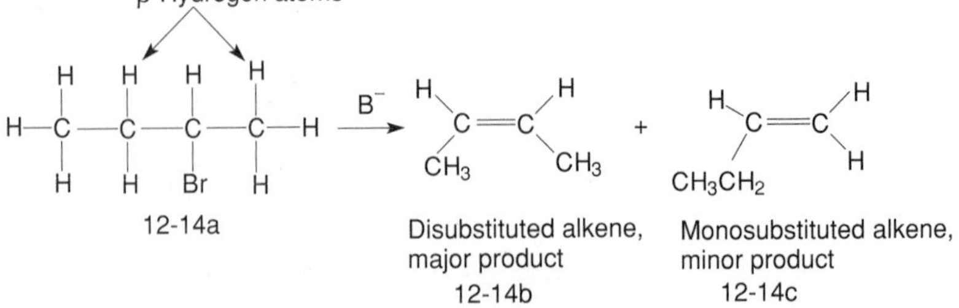

Fig. 12-14. Zaitsev's rule.

transition state leading to the most highly substituted (most stable) alkene would be lower in energy than is the transition state leading to the least substituted product. This is observed in elimination reactions, and the most highly substituted alkene is usually the major product. *Zaitsev's rule* states that elimination reactions tend to give the most highly substituted alkene as the major product.

QUESTION 12-15
Which is more stable, propene $(CH_3CH=CH_2)$ or 2-methylpropene $(CH_3CH(CH_3)=CH_2)$?

ANSWER 12-15
2-Methylpropene is more stable as it is a disubstituted alkene and propene is a monosubstituted alkene.

Substituted alkenes

The greater reactivity of 3° alkyl halides toward dehydrohalogenation is explained by considering the stability of the transition state leading to the alkene product. Figure 12-15 has examples of a 3° alkyl iodide giving a trisubstituted alkene and a 2° alkyl iodide giving a disubstituted alkene. The greater stability of the trisubstituted alkene product will be reflected in the greater stability of the transition state leading to that product. The lower activation energy for dehydrohalogenation of a 3° alkyl halide explains why it reacts faster than a 2° or 1° alkyl halide.

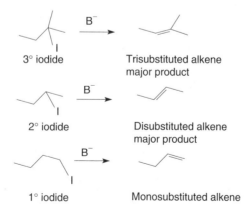

Fig. 12-15.   Reactivity of alkyl iodides in the E2 elimination reaction.

LEAVING GROUPS

The order of leaving group ability in elimination reactions is the same as that observed in substitution reactions (see Table 12-2).

STEREOCHEMISTRY

The dehydrohalogenation of some haloalkanes can result in cis or trans alkenes. An example of the dehydrohalogenation of 2-bromobutane is shown in Fig. 12-16. In this reaction, the α-, and β-carbon atoms rehybridize from sp^3 in the alkyl halide to sp^2 in the resulting alkene. In order for the p orbitals of the α- and β-carbon atoms to overlap and form a π bond, they need to be in the same plane. Alkene formation is most energetically favorable, if the H and Br leaving groups are in the same plane. The H and Br atoms are in the same plane when in eclipsed or anti positions as shown in Fig. 12-16. The conformation where the H and Br atoms are in eclipsed positions is higher in energy (due to steric interactions) than if the *groups are anti to each other*. If the groups are anti to each other, they are still in the same plane. The major product of the reaction is the cis alkene that results from dehydrohalogenation when the H and X groups are anti in the starting alkyl halide.

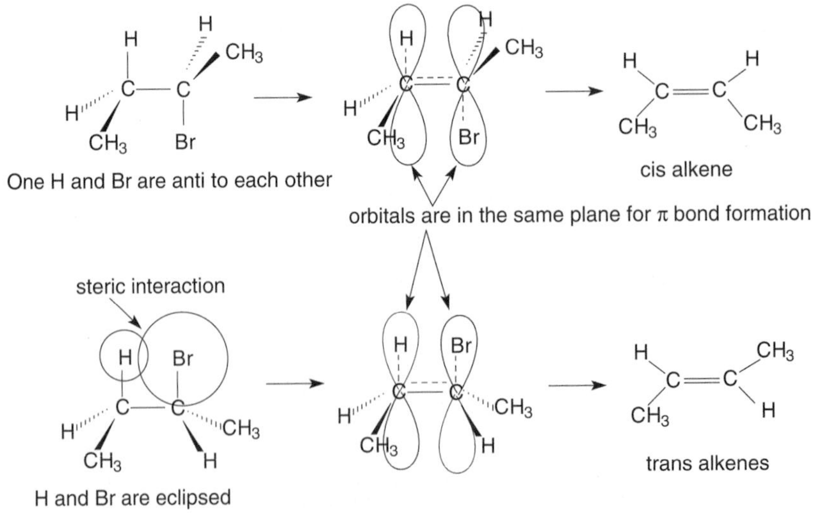

Fig. 12-16. Stereochemistry of alkene formation.

QUESTION 12-16

Do dehydrohalogenation reactions always give the more stable trans isomer rather than the less stable cis isomer?

ANSWER 12-16

No, the leaving groups have to be anti to each other but the alkene may be cis or trans, depending on the configuration of the starting material.

DEHYDROHALOGENATION OF CYCLIC COMPOUNDS

The same stereochemical requirements for acyclic alkyl halides are observed for the *dehydrohalogenation of halocylcohexanes.* The lowest energy dehydrohalogenation reaction occurs when the leaving groups (H and X) are in anti (or trans) conformations. The halogen atom exists in the axial or equatorial positions when the ring flips, as shown in Fig. 12-17. There are two possible dehydrohalogenation products shown for *cis*-2-bromo-1-methylcyclohexane. The major product is the most highly substituted alkene, as predicted by Zaitsev's rule.

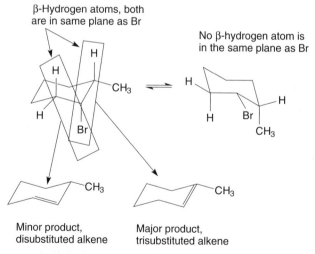

Fig. 12-17. E2 elimination in cyclohexyl halides.

First-Order Elimination (E1) Reactions

First-order elimination, E1, reactions compete with E2 and S_N1 reactions. The reaction that takes place depends upon three variables: the substrate, the leaving group, and the solvent. The reaction is independent of the base.

REACTION KINETICS

The reaction rate is first order in substrate and independent of the concentration of base:

$$\text{reaction rate} = k[\text{substrate}]$$

SUBSTRATE

The initial, rate-determining step is heterolytic cleavage of the carbon–halogen bond to give a carbocation and a halide anion. The stability of carbocations is $3° > 2° > 1°$ and, therefore, 3° alkyl halides are more reactive than 2° alkyl halides. Primary alkyl halides do not undergo E1 reactions since primary carbocations seldom, if ever, exist in solvent-based reactions.

LEAVING GROUPS

The best leaving groups have the same order of reactivity as observed for substitution reactions (see Table 12-2).

SOLVENTS

The rate of reaction in E1 reactions is dependent on the stability of the transition state leading to the carbocation reactive intermediate. Polar protic solvents are the solvents of choice since they stabilize the carbocation and the transition state leading to the carbocation to a greater extent than do polar aprotic solvents. Stabilizing the transition state lowers the activation energy, E_a, and results in a greater rate of reaction.

STEREOCHEMISTRY

A carbocation may have more than one type of β-hydrogen atom that can be removed to give an alkene. Formation of the carbocation is the rate-determining step. Deprotonation of the carbocation is not the rate-determining step for the

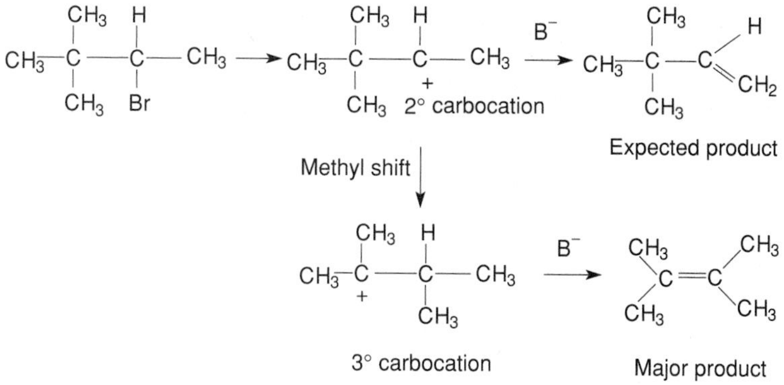

Fig. 12-18.   Elimination with carbocation rearrangement.

E1 reaction, but an activation energy is required to form the transition state leading from the carbocation to the alkene. The most highly substituted (most stable) alkene will have the lowest transition state energy leading to that alkene (the Hammond postulate). Alkene formation via E1 reactions gives the most highly substituted alkene as the major product (Zaitsev's rule).

QUESTION 12-17
Do E1 reactions undergo Walden inversion?

ANSWER 12-17
No, since E1 reactions form an alkene no inversion of configuration is involved in the reaction.

CARBOCATION REARRANGEMENTS

Whenever a carbocation is involved as a reactive intermediate in a reaction, one always has to consider if a rearrangement reaction (hydride or alkyl shift) leads to a more stable carbocation. If a rearrangement occurs, the alkene formed may be different from the one predicted from the starting reagent, as shown in Fig. 12-18.

Summary of E1 and E2 Reactions

Primary alkyl halides undergo only E2 reactions. Secondary and tertiary alkyl halides can undergo E1 and E2 reactions. High concentrations of strong base and polar aprotic solvents favor E2 reactions, while weak bases and polar protic solvents favor E1 reactions.

Competition between Substitution and Elimination Reactions

Four possible reaction mechanisms (S_N1, S_N2, E1, and E2) can result from the reaction of nucleophiles or bases with halogenated substrates. Predicting which reaction will take place may seem like a daunting task at this point. Reactions even proceed by multiple, competing pathways. Here are some guidelines that can be used to predict the major reaction pathway.

- *Primary alkyl halides* undergo only S_N2 reactions with good nucleophiles. A good nucleophile that is also a strong bulky base, such as *tert*-butoxide, tends to give E2 reactions.
- *Tertiary alkyl halides* undergo S_N1, E1, and E2 reactions. E1 reactions are favored with weak bases (e.g., H_2O) in polar protic solvents. E2 reactions are favored with strong bases (OH^-) and polar aprotic solvents. S_N1 reactions occur with good nucleophiles in polar protic solvents.
- *Secondary alkyl halides* are more difficult to predict since they can undergo all four mechanisms. Conditions favoring S_N1/E1 reactions are weak nucleophiles/bases and polar protic solvents. Conditions that favor S_N2 and E2 reactions are strong nucleophiles/bases and polar aprotic solvents.

Quiz

1. Nucleophiles have
   (a) only bonding electron pairs
   (b) nonbonding electron pairs
   (c) no nonbonding electron pairs
   (d) a positive charge

2. Nucleophiles have _____ nonbonding electron pairs.
   (a) one
   (b) two
   (c) three
   (d) all of the above

3. Electrophiles are
   (a) electron rich
   (b) electron phobic
   (c) electron poor
   (d) electron hating

4. Good leaving groups are
    (a) strong bases
    (b) weak bases
    (c) strong acids
    (d) always hydroxide anions

5. Second-order reactions require _____ species in the rate-determining step.
    (a) zero
    (b) one
    (c) two
    (d) three

6. The strongest nucleophile is
    (a) H_3O^+
    (b) H_2O
    (c) OH^-

7. Good substrates in S_N2 reactions are
    (a) primary alkyl halides
    (b) secondary alkyl halides
    (c) tertiary alkyl halides
    (d) quaternary alkyl halides

8. An example of a polar aprotic solvent is
    (a) H_2O
    (b) NH_3
    (c) H_2S
    (d) $CH_3C(O)CH_3.$

9. An example of a polar protic solvent is
    (a) H_2O
    (b) NH_3
    (c) CH_3CH_2OH
    (d) all of the above

10. Polar protic solvents _____ the potential energy of nucleophiles.
    (a) increase
    (b) decrease
    (c) do not affect

11. S_N1 reactions are
    (a) nonmolecular
    (b) unimolecular

(c) bimolecular

(d) tetramolecular

12. An S_N1 reaction shows _____ -order kinetics.
    (a) zero
    (b) first
    (c) second
    (d) third

13. _____ alkyl halides react fastest in S_N1 reactions.
    (a) Primary
    (b) Secondary
    (c) Tertiary
    (d) Quaternary

14. A carbocation intermediate results (ideally) in _____ configuration.
    (a) inversion of
    (b) racemization of
    (c) retention of
    (d) no change in

15. Polar protic solvents favor_____ reactions.
    (a) S_N1
    (b) S_N2
    (c) no preference for either

16. Elimination reactions of alkyl halides result in
    (a) alkyl halides (with inversion)
    (b) alkyl alcohols
    (c) alkenes
    (d) alkynes

17. Zaitsev's rule states the _____ is formed most readily.
    (a) least substituted alkane
    (b) most substituted alkane
    (c) least substituted alkene
    (d) most substituted alkene

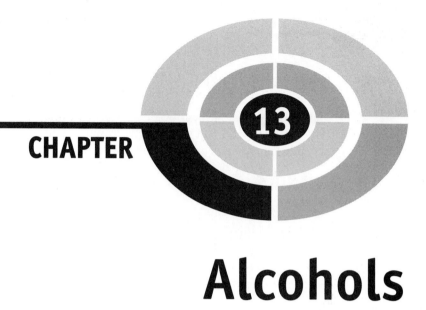

CHAPTER 13

Alcohols

Introduction

ETHANOL

Alcohols are a family of compounds containing a hydroxyl (—OH) group bonded to an sp^3 hybridized carbon atom. The most widely known alcohol is ethanol (ethyl alcohol, CH_3CH_2OH). It is the alcohol in alcoholic beverages, is used as a solvent, and is added to gasoline to aid combustion efficiency. Ethanol is also known as grain alcohol as it is a fermentation product of grains. Ethanol used in industrial (nonbeverage) applications is made by the hydration of (addition of water to) ethylene.

OTHER ALCOHOLS

Two other common alcohols are methanol (methyl alcohol or wood alcohol, CH_3OH) and 2-propanol (isopropanol, isopropyl alcohol or rubbing alcohol, $(CH_3)_2CHOH$). Methanol, ethanol, and 2-propanol each contains one hydroxyl (alcohol) group. Ethylene glycol, a diol, contains two hydroxyl groups ($HOCH_2CH_2OH$). It is used as an engine coolant in automobile radiators. Glycerol (glycerine) is a triol and contains three hydroxyl groups ($HOCH_2CH(OH)CH_2OH$). It is hydrophilic and has a slippery feel. It is used as a lubricant and as a humectant in personal care products to keep skin moist.

PHENOLS

Hydroxyl groups can also be bonded to sp^2 hybridized carbon atoms. Phenol is a compound that has a hydroxyl group bonded to one carbon atom in a benzene ring. Each carbon atom in a benzene ring is sp^2 hybridized. Although phenol is an alcohol, its properties are quite different from "normal" alcohols. Phenol and similar ring compounds are called *aromatic* or *aryl* alcohols. Their properties are discussed in Chapter 17. Hydroxyl groups bonded to sp^2 hybridized carbon atoms in *nonaromatic* compounds are called *enols*. Enols are in equilibrium with their ketone (or aldehyde) tautomers and exist mainly in the carboxyl form. Enols are discussed in the section entitled *Tautomerization* in Chapter 9.

Properties of Alcohols

Alcohols are classified as primary (1°), secondary (2°), or tertiary (3°) based on the carbon atom to which the hydroxyl group is bonded. The hydroxyl group of a 3°, 2°, or 1° alcohol is bonded to a 3°, 2°, or 1° carbon atom, respectively. The carbon atom bonded to a hydroxyl group is known as the *carbinol* carbon. Examples of these alcohols are shown in Fig. 13-1.

QUESTION 13-1
Can you draw the structure of a quaternary alcohol?

ANSWER 13-1
No, a quaternary carbon atom has four alkyl groups bonded to it and cannot accommodate a fifth bond to a hydroxyl group.

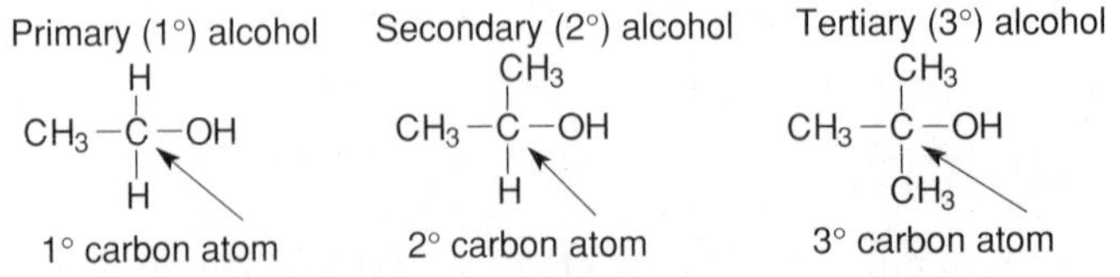

Fig. 13-1. Classification of alcohols.

The melting and boiling points of alcohols increase with increasing molecular weight. Alcohols up to about six carbon atoms are quite soluble in water. Their hydrophilic nature, and water solubility, decreases as the hydrophobic alkyl group gets larger in an homologous series of alcohols. Alcohols have much higher boiling points than do alkanes of similar molecular weight. The boiling point of methanol (CH_3OH; $MW = 32$ g/mol) is 65 °C, while the boiling point of ethane (C_2H_6; $MW = 30$ g/mol) is –89 °C. The higher boiling point of methanol is attributed to hydrogen bonding between the alcohol molecules. Energy is required to disrupt this strong intermolecular interaction for vaporization to occur.

QUESTION 13-2
Which would you expect to have a higher boiling point, ethanol (CH_3CH_2OH) or ethanethiol (CH_3CH_2SH)?

ANSWER 13-2
Ethanol has a higher boiling point since hydrogen bonding occurs between ethanol molecules while hydrogen bonding does not occur between ethanethiol molecules.

Nomenclature

When the hydroxyl group is the only functional group in a molecule, the IUPAC naming methodology is similar to that used for alkanes.

1. Select the longest continuous carbon atom chain containing the carbinol group(s).
2. Number the chain, giving the hydroxyl (alcohol) substituent(s) the lowest number possible.
3. Name the longest chain as an alkane, but drop the terminal -e and add -ol. Ethan*e* would become ethan*ol*.
4. If more than one hydroxyl group is present, the name becomes -diol, -triol, etc. For polyols, the terminal -e in the parent name is not dropped from the alkane name. Ethane with two hydroxyl groups would become ethanediol. Since the hydroxyl groups could be on different carbon atoms, one needs to specify where the hydroxyl groups are attached. The name would be 1,2-ethanediol if the hydroxyl groups are on adjacent (vicinal) carbon atoms. Updated nomenclature rules suggest that the position of substituent attachment should precede the functional group name, e.g., ethane-1,2-diol rather than 1,2-ethanediol.

COMMON NAMES

Common names are accepted for the most widely used alcohols. In the common name, the name of the *alkyl* group precedes the word *alcohol*. Methanol is called methyl alcohol. Ethanol is called ethyl alcohol. Diols are called glycols. Glycols usually have the hydroxyl groups on adjacent carbon atoms as in ethylene glycol, $HOCH_2CH_2OH$. 1,2,3-Propanetriol has the common names glycerine (glycerin) and glycerol.

HIERARCHY IN NOMENCLATURE

Nomenclature rules given in this and previous chapters are for compounds containing only one functional group. Many organic compounds have more than one functional group. One has to determine which functional group is used in the parent name and which groups are named as substituents. A hierarchy for naming functional groups is listed in Table 13-1. The highest priority functional group is considered the parent and lower priority groups are named as substituents. The substituents are named alphabetically, not on a hierarchical basis. Alcohols have a higher priority than alkanes and halides, so CH_2BrCH_2OH is called 2-bromoethanol and not 2-hydroxy-1-bromoethane.

QUESTION 13-3
Give the formal name of $CH_3CH(Br)C{\equiv}CCH_2OH$. A group/atom in parentheses means it is bonded to the preceding carbon atom.

ANSWER 13-3
4-Bromo-2-pentyn-1-ol or 4-bromopent-2-yn-1-ol. An alcohol has a higher priority than an alkyne or a halide.

Table 13-1. Nomenclature hierarchy for the parent name.

Carboxylic acid (highest priority)	Amine
Carboxylic ester	Alkene
Amide	Alkyne
Nitrile	Alkane
Aldehyde	Ether
Ketone	Halide (lowest priority)
Alcohol	

Acidity and Basicity of Alcohols

ACIDITY

Alcohols can act as acids or bases, depending upon the presence of other reagents. Alcohols have an acidity similar to that of water ($pK_a = 15.7$). Methanol has a pK_a value of 15.5 and ethanol a pK_a value of 15.9. Smaller pK_a values indicate stronger acids. (See the section entitled *Acids and Bases* in Chapter 3 for a discussion of pK_a.) The acidity of alcohols decreases in the order primary > secondary > tertiary. The pK_a values for ethanol, isopropyl alcohol, and *tert*-butyl alcohol are 15.9, 17.1, and 18.0, respectively.

Inductive effect

An inductive effect refers to electrons being withdrawn or donated through a sigma bond. Electron-withdrawing groups attached to a carbinol carbon atom increase the acidity of the alcohol by withdrawing electrons from and stabilizing the alkoxide anion (see Fig. 13-2). The alkoxide oxygen atom is shown with a full negative charge in Structure 13-2b. Reducing the "concentration" of a negative charge on an atom by dispersing the charge (electron density) throughout the molecule increases the stability of that species. This inductive effect is shown in Structure 13-2c. Increasing the stability of the alkoxide anion causes the equilibrium to shift, favoring the formation of more alkoxide ion. This shift also increases the H^+ (acid) concentration.

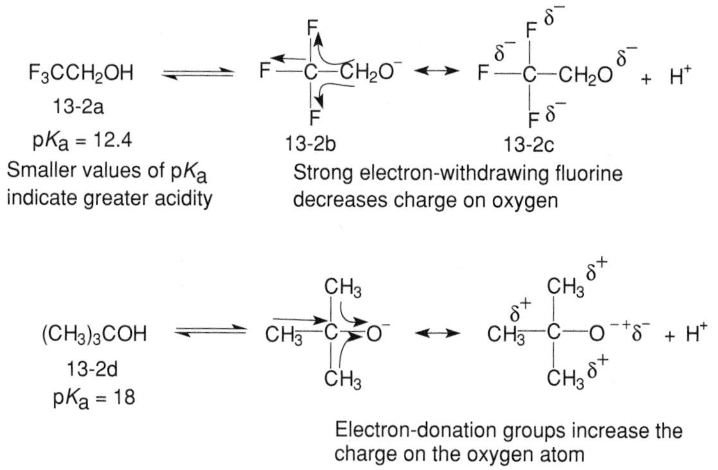

F_3CCH_2OH

13-2a

$pK_a = 12.4$

Smaller values of pK_a indicate greater acidity

13-2b   13-2c

Strong electron-withdrawing fluorine decreases charge on oxygen

$(CH_3)_3COH$

13-2d

$pK_a = 18$

Electron-donation groups increase the charge on the oxygen atom

Fig. 13-2.   Acidity of alcohols.

Groups that donate electrons, such as alkyl groups, destabilize the alkoxide anion by increasing the electron density on the oxygen atom. Increasing the electron density on the oxygen atom decreases the alkoxide's stability and its (and the H^+) equilibration concentration. Examples of electron-withdrawing and electron-donating groups are shown in Fig. 13-2. The electron-withdrawing fluorine atoms make 2,2,2-trifluoroethanol ($pK_a = 12.4$) (Structure 13-2a), a stronger acid than ethanol ($pK_a = 15.9$). The electron-donating methyl groups make 2-methyl-2-propanol (*tert*-butyl alcohol, $pK_a = 18$) (Structure 13-2d), a weaker acid than ethanol.

QUESTION 13-4
Which is a stronger acid, F_3CCH_2OH or Cl_3CCH_2OH?

ANSWER 13-4
F_3CCH_2OH; since F is more electronegative than Cl, the fluorine atoms stabilize the resulting alkoxide anion to a greater extent than do the chlorine atoms. The equilibrium for trifluoroethanol is shifted further to the right, increasing the alkoxide anion and proton concentrations.

BASICITY

Alcohols act as bases in the presence of strong acids. The two nonbonding electron pairs on the oxygen atom in an alcohol are nucleophilic and can bond with a proton forming a protonated alcohol, ROH_2^+. Protonated alcohols are very strong acids with pK_a values of about -2. Protonated alcohols are involved as catalysts and intermediates in reactions that are discussed in other chapters. A protonated hydroxyl group is also a good leaving group in substitution and elimination reactions. (See the section entitled *Properties of Leaving Groups* in Chapter 12 for an explanation of protonated leaving groups.)

QUESTION 13-5
Which is a stronger acid, CH_3OH or $CH_3OH_2^+$?

ANSWER 13-5
The protonated alcohol, $CH_3OH_2^+$, is a much stronger acid. The conjugate base (CH_3OH) of the protonated alcohol is a weaker base (more stable) than is the conjugate base $(CH_3O)^-$ of an alcohol (survival of the weakest).

Strong bases such as the amide anion, $^-NH_2$, can be used to remove a proton from an alcohol to form the corresponding alkoxide anion. The hydroxide anion is not a strong enough base for the complete conversion of alcohol to alkoxide anion. To determine if a base is strong enough to remove (essentially completely) the proton from an alcohol, compare the pK_a values of the alcohol and conjugate

$$CH_3CH_2OH + {}^-NH_2 \rightleftharpoons CH_3CH_2O^- + NH_3$$
$$pK_a = 15.9 \qquad\qquad pK_a = 36$$
Much weaker acid

$$CH_3CH_2OH + {}^-OH \rightleftharpoons CH_3CH_2O^- + H_2O$$
$$pK_a = 15.9 \qquad\qquad pK_a = 15.7$$
Slightly weaker acid

Fig. 13-3. Acid-base reactions of alcohols.

acid of the base and use the "survival of the weakest" criteria to predict which way the equilibrium will be shifted. Two acid-base equilibrium reactions are given in Fig. 13-3. Ammonia ($pK_a = 36$) is a much weaker acid than ethanol ($pK_a = 15.9$) and, therefore, the reaction is strongly shifted toward the weaker acid. Water ($pK_a = 15.7$) has about the same pK_a value as ethanol ($pK_a = 15.9$) and the equilibrium for this reaction is not shifted strongly in either direction. Alkoxide anions are used as reagents and catalysts in organic synthesis.

Reaction with Active Metals

Alcohols react with active metals such as sodium and potassium, forming an alkoxide anion (RO^-) and hydrogen gas. These are oxidation–reduction reactions, where the alcohol is acting as an acid and oxidizing agent.

$$2\ ROH + 2\ Na \rightarrow 2\ RO^- + 2\ Na^+ + H_2$$

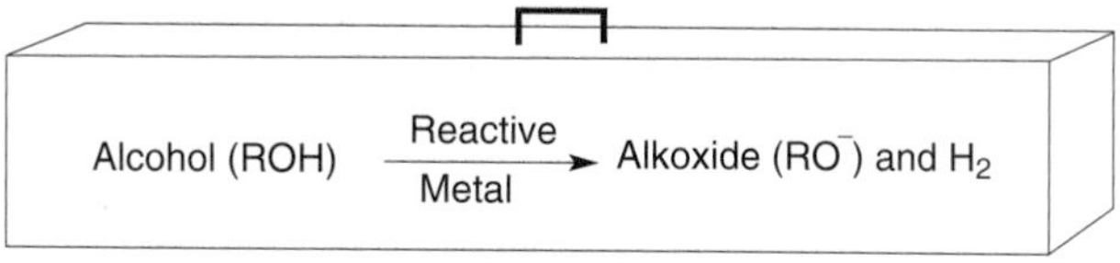

Alcohol (ROH) $\xrightarrow[\text{Metal}]{\text{Reactive}}$ Alkoxide (RO^-) and H_2

Toolbox 13-1.

Preparation of Alcohols

ADDITION AND SUBSTITUTION REACTIONS

Alkenes are excellent starting materials for the preparation of alcohols. Several methods of preparing monoalcohols and diols from alkenes by addition reactions are described in the section entitled *Reaction of Alkenes* in Chapter 8. The preparation of alcohols by nucleophilic substitution reactions of alkyl halides is discussed in Chapter 12 and also below.

OXIDATION–REDUCTION REACTIONS

Inorganic compounds undergo oxidation when an atom loses an electron and undergo reduction when an atom gains an electron. In organic reactions, electrons are more often shared (but not necessarily equally) between atoms rather than being transferred from one atom to another atom.

In organic compounds a carbon atom undergoes oxidation when it becomes bonded to an atom that is more electronegative than that carbon atom. The electron density of the carbon atom is decreased to some extent but an electron is not completely "lost." A compound undergoes oxidation when an electron-withdrawing N, O, or halogen atom is added (bonded) to a carbon atom in that compound. A compound undergoes reduction when a N, O, or halogen is removed. Bonding of a hydrogen atom to a carbon atom is a reduction reaction and removal of a hydrogen atom from a carbon atom is an oxidation reaction. Some oxidation and reduction reactions are shown in Fig. 13-4.

QUESTION 13-6
Are the following reactions oxidation or reduction: (a) formation of 1,2-diiodoethane from ethene (ethylene), (b) formation of ethane from ethene, and (c) formation of ethanol from ethene?

ANSWER 13-6
(a) Oxidation, (b) reduction, and (c) neither since addition of OH is oxidation and addition of H is reduction.

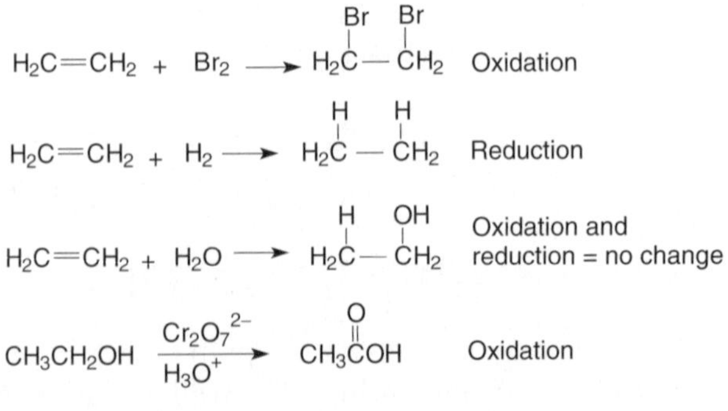

Fig. 13-4.   Organic oxidation and reduction reactions.

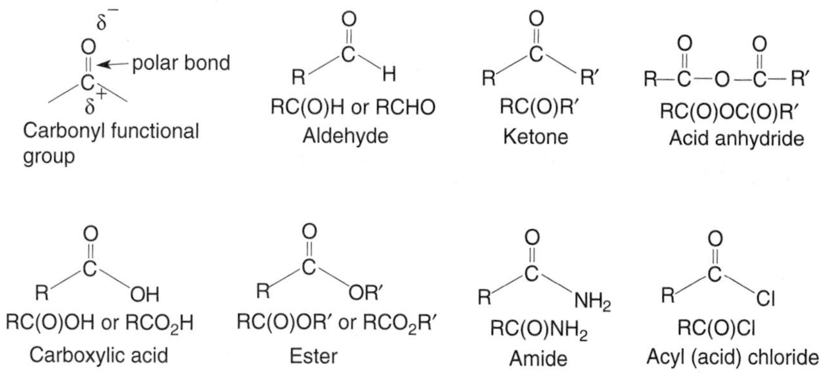

Fig. 13-5.   Carbonyl compounds.

REDUCTION OF CARBONYL COMPOUNDS

Types of carbonyl compounds

A carbonyl group consists of a carbon atom doubly bonded to an oxygen atom ($C=O$). The carbonyl group appears in aldehydes, ketones, carboxylic acids, esters, amides, acyl (acid) chlorides, and acid anhydrides. Structures of these compounds are shown in Fig. 13-5. The carbonyl double bond is a polar bond. The carbon atom has a partial positive charge and the oxygen atom has a partial negative charge, as shown in the first structure in Fig. 13-5. The carbon atom is electrophilic and the oxygen atom is nucleophilic. The carbon atom reacts with bases or nucleophiles and the oxygen atom reacts with acids or electrophiles.

Representation of a carbonyl group

In text form, the oxygen atom in a carbonyl group is often shown in parentheses following the carbon atom to which it is bonded. Ketones are shown as $RC(O)R$ and aldehydes are written as $RC(O)H$, or more commonly, RCHO. If you are uncertain what compound is represented by a formula, remember each carbon atom (that has no net charge) must have four bonds. The first carbon atom in $H_3CC(O)H$ (or H_3CCHO) has three bonds to hydrogen atoms and one bond to the second carbon atom (four bonds). The second carbon atom is bonded to the first carbon atom (one bond), to one hydrogen atom (one bond), and to an oxygen atom (with two bonds) for a total of four bonds to that carbon atom.

Reactions with LiAlH$_4$ and NaBH$_4$

Aldehydes and ketones. Two reagents commonly used to reduce aldehydes and ketones to alcohols are sodium borohydride ($NaBH_4$) and lithium aluminum

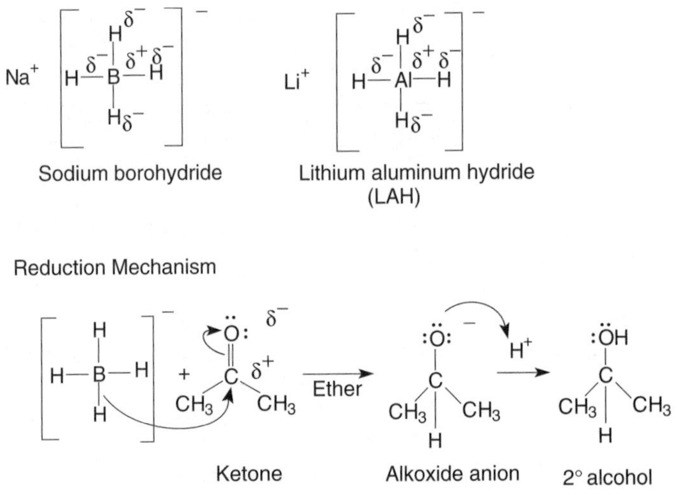

Sodium borohydride

Lithium aluminum hydride (LAH)

Reduction Mechanism

Ketone   Alkoxide anion   2° alcohol

Fig. 13-6.   Reduction of the carbonyl function.

hydride (LiAlH$_4$ or LAH). Structures of these reagents are shown in Fig. 13-6. Since hydrogen is more electronegative than B or Al, the hydrogen atom has a partial negative charge in these compounds. NaBH$_4$ and LAH react with carbonyl groups by transferring a hydride anion to the carbon atom of the carbonyl group. This is a nucleophilic attack by the hydride anion on the electrophilic carbon atom.

Aldehydes are reduced to primary alcohols and ketones are reduced to secondary alcohols. The mechanism for the reduction of acetone, a ketone, is shown in Fig. 13-6. The initial product is an alkoxide anion. In a second step, addition of aqueous acid converts the alkoxide anion to an alcohol. If the aqueous acid were added during the first step, it would react with NaBH$_4$ (or LAH) before the carbonyl group has had a chance to undergo reduction.

QUESTION 13-7
What product results from the reaction of acetaldehyde (CH$_3$CHO) with sodium borohydride followed by treatment with aqueous acid?

ANSWER 13-7
Ethanol, CH$_3$CH$_2$OH.

Solvent requirements. LAH is very reactive and reacts vigorously with molecules that have acidic (reactive) hydrogen atoms such as water and alcohols. LAH reduction reactions must therefore be carried out in inert (unreactive) solvents, usually an ether such as diethyl ether (CH$_3$CH$_2$OCH$_2$CH$_3$) or the cyclic ether tetrahydrofuran (THF). Sodium borohydride is much less reactive than LAH. It reacts more slowly with water and alcohols than with the carbonyl group at room temperature. Thus, water and alcohols can be used as solvents for reduction reactions with $NaBH_4$.

Aldehyde $\xrightarrow[\text{or 1. LAH   2. } H_3O^+]{\text{1. } NaBH_4 \text{ 2. } H_3O^+}$ Primary alcohol

Ketone $\xrightarrow[\text{or 1. LAH   2. } H_3O^+]{\text{1. } NaBH_4 \text{ 2. } H_3O^+}$ Secondary alcohol

Toolbox 13-2.

Reduction of acids, esters, and acid halides Sodium borohydride is too sluggish to react with the carbonyl group of acids, esters, and acyl halides, but LAH reacts with these compounds to give initially an aldehyde which is subsequently reduced to a primary alcohol. The mechanism for the reaction of an ester with LAH is shown in Fig. 13-7. Note an alcohol is also produced from the alkoxide (RO^-) by product.

QUESTION 13-8
What are the products of the reaction of methyl propionate, $CH_3CH_2C(O)OCH_3$, with LAH followed by treatment with acid?

ANSWER 13-8
1-Propanol from the propionate group and methanol from the methyl ester group.

Ester or Acid or Acid chloride $\xrightarrow[\text{2. } H_3O^+]{\text{1. LAH}}$ Primary alcohol (and alcohol from the ester group)

Toolbox 13-3.

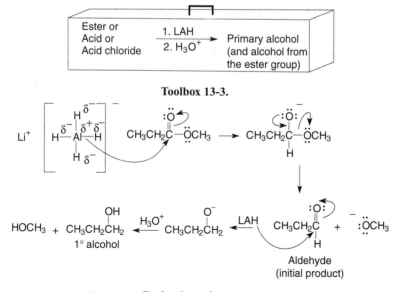

Fig. 13-7.   Reduction of an ester with LAH.

(a) $CH_3CH=CHCH_2\overset{\overset{\displaystyle O}{\|}}{C}H$ $\xrightarrow[\text{2. } H_3O^+]{\text{1. NaBH}_4}$ $CH_3CH=CHCH_2CH_2OH$

(b) $CH_3CH=CHCH_2\overset{\overset{\displaystyle O}{\|}}{C}H$ $\xrightarrow[\text{Ni (Raney)}]{H_2}$ $CH_3CH_2CH_2CH_2CH_2OH$

Fig. 13-8. Reduction of an unsaturated aldehyde.

Unsaturated carbonyl compounds

Neither LAH nor NaBH$_4$ reduces alkene or alkyne functional groups. Unsaturated carbonyl compounds can be converted into unsaturated alcohols by both reagents. This reaction is shown in Reaction(a) in Fig. 13-8. *A key point: LAH and NaBH$_4$ reduce carbonyl groups but not alkene or alkyne groups.*

Hydrogenation

Hydrogen gas, in the presence of a special catalyst (Raney Ni), reduces aldehydes to 1° alcohols and ketones to 2° alcohols. This reagent also reduces alkenes and alkynes to alkanes. This reaction is shown in Reaction(b) in Fig. 13-9.

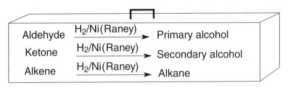

Toolbox 13-4.

Organometallic Compounds

Organometallic compounds contain an organic (organo) segment and a metal atom (Na, K, Li, Mg, etc.). The degree of polarity of the organo–metal bond is solvent dependent. The organometallic bond is more highly polarized in polar solvents, because polar solvents stabilize charged species to a greater extent than do nonpolar solvents. Polar ethers are good solvents for organometallic compounds. The nonbonding electron pairs on the oxygen atom of the ether coordinate with the metal atom and solubilize organometallic compounds. Although organo–metal bonds are covalent, their reactivity is more characteristic of ionic bonds.

ORGANOLITHIUM (RLi) COMPOUNDS

Lithium reacts with 1°, 2°, or 3° alkyl halides, alkenyl (vinyl) halides, and aryl (aromatic) halides to form organolithium compounds, RLi . The halide can be

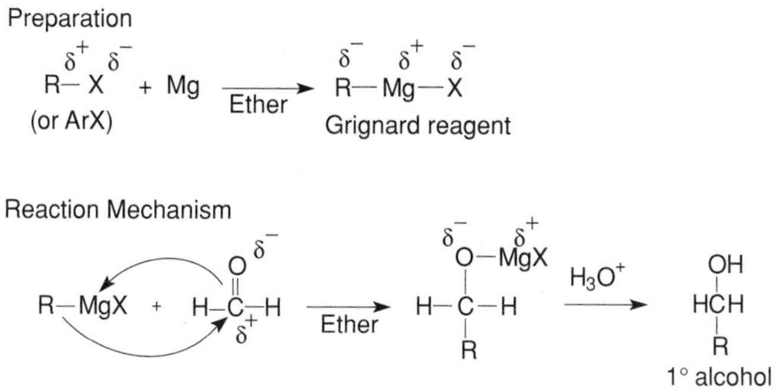

Preparation

Reaction Mechanism

Fig. 13-9. Grignard reagent preparation and mechanism of the reaction with carbonyl compounds.

Cl, Br, or I. The carbon atom bonded to the halide in the alkyl halide has a partial positive charge, since the halide is more electronegative than the carbon atom. That same carbon atom in an organometallic product has a partial negative charge.

GRIGNARD REAGENTS (RMgX)

Another very versatile organometallic reagent is the Grignard reagent. This reagent is made by reacting magnesium metal with organic halides as shown in Fig.13.9. The organic group can be a $1°$, $2°$, or $3°$ alkyl, an alkenyl (vinyl), or an aryl (aromatic) group halide. The halide can be Cl, Br, or I. Grignard reagents can be easily prepared in inert ether solvents at room temperature. They are less reactive than organolithium or organosodium compounds.

The first equation in Fig. 13-9, the preparation of an organomagnesium Grignard reagent, shows how the partial charge on a carbon atom changes from δ^+ to δ^-. *A key point: the carbon atom in the alkyl (or organo) group bonded to the metal atom in an organometallic compound is electron rich and acts as a nucleophile, while the carbon atom bonded to the halide in an alkyl halide is electron poor and acts as an electrophile.*

Preparation of Alcohols Using Organometallic Reagents

Grignard reagents react with carbonyl compounds to give alcohols. The mechanism for this reaction is shown in Fig. 13-9. The initial product is the magnesium salt of the alkoxide anion. In a second step, a proton source (H_3O^+) is added to protonate the alkoxide giving the alcohol. Analogous reactions and similar resulting products are obtained with organolithium or organosodium compounds.

GRIGNARD REACTIONS WITH CARBONYL COMPOUNDS

Figure 13-10 shows a series of reactions of a Grignard reagent with various carbonyl compounds. These reactions must be carried out in inert solvents (usually ethers) since solvents with active (acidic) protons, like water and alcohols, would react with (and destroy) the Grignard reagent before it could react with the carbonyl compound. Note that different alcohols (1°, 2°, or 3°) are formed depending on the starting carbonyl compound. The reaction of 1 equiv. of a Grignard reagent with an ester, acyl chloride, or carboxylate salt forms a ketone. The ketone reacts with a second equivalent of Grignard reagent to give a 3° alkoxide anion. Protonaton of the alkoxide anion gives a 3° alcohol. The reaction of a Grignard reagent with an ester also gives a second alcohol from the ester function as shown in Fig. 13-10, Reaction(e).

Molecules with active (acidic) hydrogen atoms (protons) react with Grignard reagents by transferring a proton to the R group in RMgX, giving RH. Molecules with active hydrogen atoms include $-C{\equiv}CH, -NH_2, -OH,$ and $-SH$ functional groups.

QUESTION 13-9
What product results from the reaction of acetone, $CH_3C(O)CH_3$, with methyl-magnesium bromide and subsequent treatment with H_3O^+?

ANSWER 13-9
tert-Butyl alcohol.

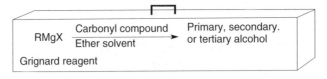

Toolbox 13-5.

QUESTION 13-10
What products result from the reaction of methyl acetate, $CH_3C(O)OCH_3$, with 2 equiv. of ethylmagnesium chloride followed by treatment with acid?

ANSWER 13-10
3-Methyl-3-pentanol (or 3-methylpentan-3-ol) and methanol (from the methyl ester group.)

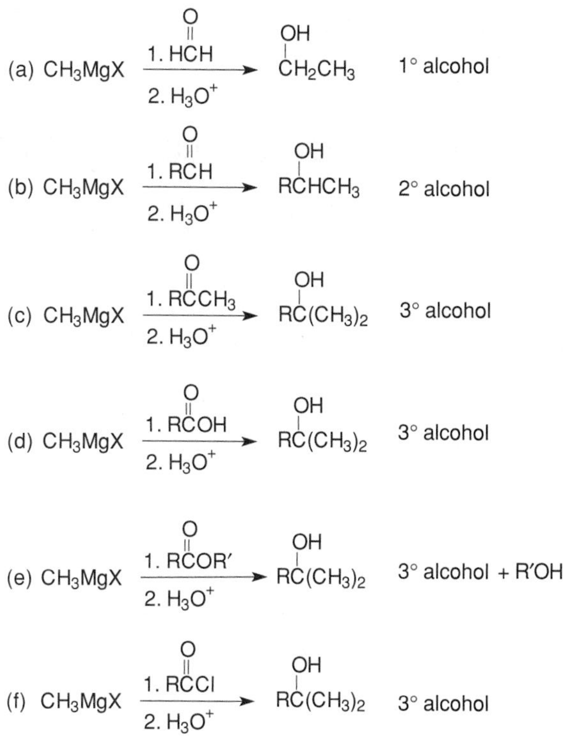

(a) CH_3MgX $\xrightarrow[\text{2. } H_3O^+]{\text{1. } HCH}$ $\underset{CH_2CH_3}{\overset{OH}{|}}$ 1° alcohol

(b) CH_3MgX $\xrightarrow[\text{2. } H_3O^+]{\text{1. } RCH}$ $\underset{RCHCH_3}{\overset{OH}{|}}$ 2° alcohol

(c) CH_3MgX $\xrightarrow[\text{2. } H_3O^+]{\text{1. } RCCH_3}$ $\underset{RC(CH_3)_2}{\overset{OH}{|}}$ 3° alcohol

(d) CH_3MgX $\xrightarrow[\text{2. } H_3O^+]{\text{1. } RCOH}$ $\underset{RC(CH_3)_2}{\overset{OH}{|}}$ 3° alcohol

(e) CH_3MgX $\xrightarrow[\text{2. } H_3O^+]{\text{1. } RCOR'}$ $\underset{RC(CH_3)_2}{\overset{OH}{|}}$ 3° alcohol + R'OH

(f) CH_3MgX $\xrightarrow[\text{2. } H_3O^+]{\text{1. } RCCl}$ $\underset{RC(CH_3)_2}{\overset{OH}{|}}$ 3° alcohol

Fig. 13-10. Grignard reactions leading to alcohols.

GRIGNARD REACTIONS WITH NONCARBONYL GROUPS

In addition to reacting with carbonyl compounds, Grignard reagents react with other *polar multiple bonds* including S=O, C=N, N=O, and C≡N. When Grignard reactions are carried out, one has to consider all the functional groups present in the molecule. The desired reaction with a functional group may occur in one part of the molecule and a undesired reaction with a different functional group could occur in another part of the molecule.

Reactions of Alcohols

OXIDATION OF ALCOHOLS

Alcohols are very useful starting materials for making other functional groups. There are a large variety of alcohols that are readily available and inexpensive.

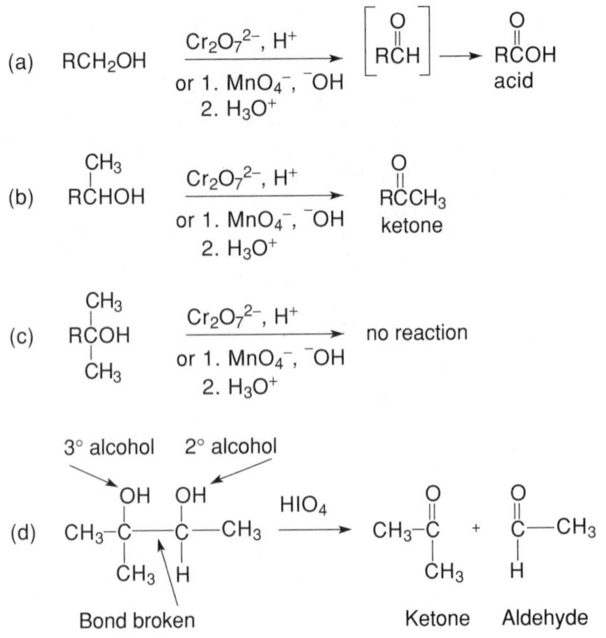

(a) RCH_2OH $\xrightarrow[\begin{array}{c}\text{or 1. } MnO_4^-,\ ^-OH \\ \text{2. } H_3O^+\end{array}]{Cr_2O_7^{2-},\ H^+}$ $\left[\ \overset{\overset{\displaystyle O}{\parallel}}{R}CH\ \right] \longrightarrow \underset{\text{acid}}{\overset{\overset{\displaystyle O}{\parallel}}{R}COH}$

(b) $\overset{\overset{\displaystyle CH_3}{|}}{R}CHOH$ $\xrightarrow[\begin{array}{c}\text{or 1. } MnO_4^-,\ ^-OH \\ \text{2. } H_3O^+\end{array}]{Cr_2O_7^{2-},\ H^+}$ $\underset{\text{ketone}}{\overset{\overset{\displaystyle O}{\parallel}}{R}CCH_3}$

(c) $\overset{\overset{\displaystyle CH_3}{|}}{\underset{\underset{\displaystyle CH_3}{|}}{R}}COH$ $\xrightarrow[\text{or 1. } MnO_4^-,\ ^-OH]{Cr_2O_7^{2-},\ H^+}$ no reaction

(d)

3° alcohol   2° alcohol

$CH_3-\overset{\overset{\displaystyle OH}{|}}{\underset{\underset{\displaystyle CH_3}{|}}{C}}-\overset{\overset{\displaystyle OH}{|}}{\underset{\underset{\displaystyle H}{|}}{C}}-CH_3 \xrightarrow{HIO_4} \underset{\underset{\displaystyle CH_3}{}}{CH_3-\overset{\overset{\displaystyle O}{\parallel}}{C}} + \underset{\underset{\displaystyle H}{}}{\overset{\overset{\displaystyle O}{\parallel}}{C}-CH_3}$

Bond broken   Ketone   Aldehyde

Fig. 13-11.   Oxidation of alcohols.

Alcohols can be oxidized to aldehydes, ketones, and carboxylic acids. Figure 13-11 shows examples of these oxidation reactions.

Primary alcohols are oxidized to carboxylic acids in acidic solutions of dichromate $(Cr_2O_7^{2-})$ or basic solutions of permanganate (MnO_4^-). The 1° alcohol is first oxidized to an aldehyde, which is rapidly oxidized to a carboxylic acid.

Secondary alcohols are oxidized to ketones in acidic solutions of dichromate or basic solutions of permanganate.

Tertiary alcohols do not react with dichromate or permanganate. Oxidation of 3° alcohols would require cleavage of a carbon–carbon bond. Carbon–carbon bond cleavage does not occur under conditions used to oxidize 1° and 2° alcohols.

Primary alcohols can be oxidized to aldehydes (and not to acids) by reaction with *pyridinium chlorochromate (PCC)* in anhydrous solvents.

Periodic acid (HIO_4) reacts with 1,2-diols (vicinal diols) to give aldehydes or ketones. If an alcohol is tertiary, it is oxidized to a ketone. If the alcohol is secondary or primary, it is oxidized to an aldehyde. Under these reaction conditions a carbon–carbon bond is broken.

Primary alcohol $\xrightarrow[\text{or 1. MnO}_4^-, \ \bar{}\text{OH} \ \ 2. \ \text{H}_3\text{O}^+]{\text{Cr}_2\text{O}_7^{2-}, \ \text{H}_3\text{O}^+}$ Acid

Secondary alcohol $\xrightarrow[\text{or 1. MnO}_4^-, \ \bar{}\text{OH} \ \ 2. \ \text{H}_3\text{O}^+]{\text{Cr}_2\text{O}_7^{2-}, \ \text{H}_3\text{O}^+}$ Ketone

Tertiary alcohol $\xrightarrow[\text{or 1. MnO}_4^-, \ \bar{}\text{OH} \ \ 2. \ \text{H}_3\text{O}^+]{\text{Cr}_2\text{O}_7^{2-}, \ \text{H}_3\text{O}^+}$ No reaction

Toolbox 13-6.

Diol (primary and secondary) $\xrightarrow{\text{HIO}_4}$ Aldehyde

Diol (tertiary) $\xrightarrow{\text{HIO}_4}$ Ketone

Toolbox 13-7.

Conversion of Alcohols to Alkyl Halides

HYDROGEN HALIDES

Alcohols can be converted to the corresponding halides by reaction with concentrated HCl, HBr, or HI. Tertiary alcohols react quite readily with the hydrogen halides at room temperature or with slight heating. These reactions, shown in Fig. 13-12, proceed via a carbocation (S_N1 reaction).

Secondary alcohols and primary alcohols also react with hydrogen halides, but higher reaction temperatures are usually required. Secondary alcohols can undergo rearrangement reactions, if the reaction procedes by an S_N1 mechanism (a carbocation is an intermediate). High reaction temperatures can result in undesirable changes in other parts of the molecule. To avoid these difficulties, other halogenating reagents are used.

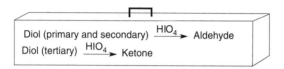

Better leaving group than $\bar{}\text{OH}$

Fig. 13-12.   Acid-catalyzed alkyl halide formation.

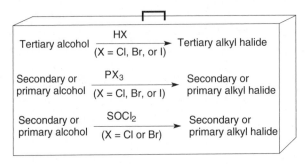

Fig. 13-13. Conversion of alcohols to haloalkanes (alkyl halides).

PHOSPHOROUS TRIHALIDE AND THIONYL HALIDE

Primary and secondary alcohols react readily, under mild conditions, with phosphorous trihalide (Cl, Br, or I) to form the corresponding alkyl halide. Thionyl chloride ($SOCl_2$) and thionyl bromide react with 1° and 2° alcohols to form the corresponding alkyl halides. Tertiary alcohols react sluggishly with these reagents for steric reasons. The reaction of 3° alcohols with hydrogen halides is often a better synthetic method for making 3° halides.

Hydroxyl groups are poor leaving groups. Phosphorous trihalide and thionyl halides convert the alcohol group into a better leaving group. The halide (X^-) byproduct of this reaction acts as a nucleophile displacing the newly formed leaving group. These reactions are shown in Fig. 13-13 and occur via an S_N2 mechanism, giving the Walden inversion products.

Tertiary alcohol	$\xrightarrow[\text{(X = Cl, Br, or I)}]{\text{HX}}$	Tertiary alkyl halide
Secondary or primary alcohol	$\xrightarrow[\text{(X = Cl, Br, or I)}]{PX_3}$	Secondary or primary alkyl halide
Secondary or primary alcohol	$\xrightarrow[\text{(X = Cl or Br)}]{SOCl_2}$	Secondary or primary alkyl halide

Toolbox 13-8.

Fig. 13-14.   Dehydration of alcohols.

Dehydration Reactions

Water undergoes acid-catalyzed addition to alkenes to give alcohols. This is a reversible reaction. The reaction of alcohols with concentrated sulfuric acid results in dehydration of the alcohol to give an alkene as shown in Fig. 13-14. If the dehydration reaction involves a carbocation intermediate, rearrangement reactions may occur. *A key point: if a carbocation is involved as an intermediate in a reaction, the possibility of a rearrangement reaction must always be considered.* A rearrangement in a dehydration reaction is shown in Fig. 13-14.

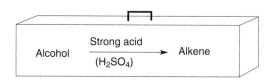

Toolbox 13-9.

Quiz

1. Alcohols have _____ boiling points than do alkanes of similar molecular weight.
    (a) lower
    (b) higher
    (c) similar

2. The formal (IUPAC) names for alcohols end in
   (a) -al
   (b) -ol
   (c) -one
   (d) -ane

3. Glycols have ———— hydroxyl group(s).
   (a) one
   (b) two
   (c) three
   (d) four

4. Alcohols can act as acids.
   (a) True
   (b) False

5. Alcohols can act at bases.
   (a) True
   (b) False

6. Alcohols react with active metals (Na, K, etc.) to produce ————
   gas.
   (a) oxygen
   (b) hydrogen
   (c) nitrogen
   (d) helium

7. The formation of ethylene glycol ($HOCH_2CH_2OH$) from ethylene
   ($H_2C=CH_2$) is an example of
   (a) oxidation
   (b) reduction
   (c) dehydration
   (d) hydrogenation

8. An example of a compound that contains a carbonyl group is
   (a) $H_2C=O$
   (b) CH_3OH
   (c) $CH_2=CH_2$
   (d) CH_3OCH_3

9. One mole of $NaBH_4$ reacts with 1 mole of acetone, $CH_3C(O)CH_3$, to
   give, after treatment with aqueous acid,–
   (a) an acid

(b) a ketone

(c) an alcohol

(d) hydrogen

10. LAH reacts with glyoxal, $HC(O)C(O)H$, to give
    (a) CH_3CO_2H
    (b) $HOCH_2CH_2OH$
    (c) $H_2C=CH_2$
    (d) $HOCH_2CO_2H$

11. LAH reacts with $H_2C=CHC(O)H$ to give
    (a) $CH_3CH_2C(O)H$
    (b) $CH_3CH_2CH_2OH$
    (c) $H_2C=CHCO_2H$
    (d) $H_2C=CHCH_2OH$

12. Hydrogen (H_2), in the presence of Raney Ni catalyst, reacts with $H_2C=CHCHO$ to give
    (a) CH_3CH_2CHO
    (b) $CH_3CH_2CH_2OH$
    (c) $H_2C=CHCO_2H$
    (d) $H_2C=CHCH_2OH$

13. CH_3CHO reacts first with ethyllithium (CH_3CH_2Li) and then with dilute acid to give
    (a) ethane
    (b) butan-2-ol
    (c) butane
    (d) no reaction

14. Lithium reacts with *n*-butyl chloride to give
    (a) *n*-butane
    (b) *n*-butyllithium
    (c) chlorine
    (d) hydrogen

15. Formaldehyde (HCHO) reacts first with CH_3MgCl and then with dilute acid to give a
    (a) primary alcohol
    (b) secondary alcohol
    (c) tertiary alcohol
    (d) quaternary alcohol

16. The oxidation of CH_3CH_2OH with acidic dichromate $(Cr_2O_7^{2-})$ gives
    (a) a ketone
    (b) ethyl alcohol
    (c) an aldehyde
    (d) an acid

17. CH_3CH_2OH reacts with pyridinium chlorochromate (PCC) to give
    (a) a ketone
    (b) an alcohol
    (c) an aldehyde
    (d) an acid

18. PBr_3 reacts with ethanol to give
    (a) tribromoethane
    (b) bromoethanol
    (c) bromoethane
    (d) no reaction

19. Thionyl chloride, ClS(O)Cl, reacts with isopropyl alcohol, $(CH_3)_2CHOH$, to give
    (a) 2-chloropropane
    (b) 1-chloropropane
    (c) 2-propanol
    (d) acetone (a ketone)

20. Isopropyl alcohol, $(CH_3)_2CHOH$, reacts with hot concentrated H_2SO_4 to give
    (a) propane
    (b) 2-propanol
    (c) propene
    (d) propyne

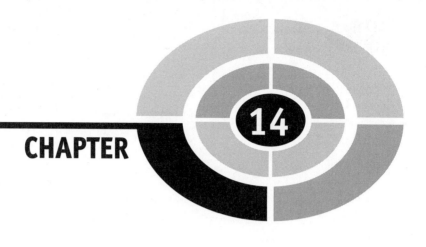

Ethers

Introduction

Ethers (ROR) can be considered derivatives of water (HOH) or alcohols (ROH) by replacing an H with an R group. The R groups in ethers can be alkyl, aryl, or alkenyl. The R groups can be the same or different. If the R groups are the same, the ethers are called symmetrical ethers. If the R groups are different, the ethers are called unsymmetrical ethers. Ethers can be cyclic or acyclic. Rings that contain an atom other than carbon are called *heterocyclic compounds*. Oxygen is the heteroatom in cyclic ethers. Cyclic ethers with five or more atoms in the ring (including the oxygen atom) have chemical reactivity properties similar to those of acyclic ethers. Cyclic ethers containing three atoms are highly strained and more chemically reactive than acyclic ethers.

Nomenclature

Formal (IUPAC) names for acyclic ethers use the largest R group as the parent and the RO segment is named as an alkoxy substituent. Common names

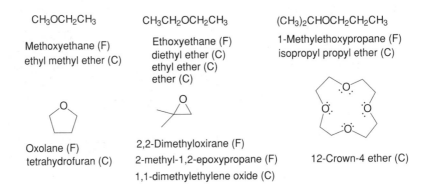

$CH_3OCH_2CH_3$

Methoxyethane (F)
ethyl methyl ether (C)

$CH_3CH_2OCH_2CH_3$

Ethoxyethane (F)
diethyl ether (C)
ethyl ether (C)
ether (C)

$(CH_3)_2CHOCH_2CH_2CH_3$

1-Methylethoxypropane (F)
isopropyl propyl ether (C)

Oxolane (F)
tetrahydrofuran (C)

2,2-Dimethyloxirane (F)
2-methyl-1,2-epoxypropane (F)
1,1-dimethylethylene oxide (C)

12-Crown-4 ether (C)

Fig. 14-1.   Names of ethers: formal (F) and common (C).

list each R group alphabetically followed by the word *ether*. The formal name for $CH_3CH_2OCH_2CH_3$ is ethoxyethane and the common name is diethyl ether. Common names for symmetrical ethers often just name one R group. For example, diethyl ether is often called ethyl ether. Since diethyl ether is likely the most widely known ether, it is often just called ether (like alcohol often refers to ethanol). Examples of naming ethers are given in Fig. 14-1.

QUESTION 14-1
Give the formal and common names for $CH_3OCH_2CH_2CH_3$.

ANSWER 14-1
Methoxypropane and methyl *n*-propyl ether respectively.

Cyclic ethers have formal and common names. The formal name for a saturated five-membered cyclic ether is *oxolane*. The common name for this material is *tetrahydrofuran* (THF, a common solvent). A special class of cyclic ethers containing 10 to 20 atoms, including several oxygen atoms, is called *crown ethers*. Certain metal ions are "crowned by," or fit into the center of, the ether ring. A bonding interaction exists between the electrophilic metal ion and the nonbonding electrons on the oxygen atoms. Examples of these ethers are shown in Fig. 14-1.

Properties

Most ethers are quite unreactive toward strong bases, weak acids, and nucleophiles. Hence they are often used as inert solvents. Three- and four-membered cyclic ethers are, however, more reactive because of ring strain. Three-membered rings (epoxides) readily undergo acid- or base-catalyzed ring opening.

Solubility in Water, g/100 mL		Boiling Point, °C	
$CH_3CH_2OCH_2CH_3$	8.0	$CH_3CH_2OCH_2CH_3$ (MW = 74)	35
$CH_3CH_2CH_2CH_2OH$	7.4	$CH_3CH_2CH_2CH_2CH_3$ (MW = 72)	36
		$CH_3CH_2CH_2CH_2OH$ (MW = 74)	118

Fig. 14-2.　Physical properties of ethers.

The oxygen atom in ethers is sp^3 hybridized, and acyclic ethers have a bent structure (around the oxygen atom) with a bond angle of about 107°. Ethers are polar molecules because of this bent (nonlinear) structure. Ethers have no hydrogen atom available to form hydrogen bonds with other molecules. Hydrogen bonding requires an H—O bond, which ethers do not have. However, other molecules (e.g., water and alcohols) are capable of forming hydrogen bonds to ethers. Ethers and alcohols of comparable molecular weights have similar solubilities in water because water can form hydrogen bonds to ethers and alcohols. An example of comparable water solubility is given in Fig. 14-2.

QUESTION 14-2
Would you expect diethyl ether to be soluble in ethyl alcohol?

ANSWER 14-2
Yes, because the alcohol can hydrogen bond to the ether.

Since ether molecules cannot form hydrogen bonds with other ether molecules, they have boiling points similar to those of alkanes (that also do not form intermolecular hydrogen bonds) of comparable molecular weight, as shown in Fig. 14-2. Ethers are quite volatile because of their low boiling points. They are also very flammable. Their boiling points are much lower than those of alcohols of comparable molecular weight, since alcohols can form intermolecular hydrogen bonds, as shown in Fig. 14.2.

Ethers tend to solvate metal ions because the oxygen atom can share its nonbonding electron pairs with electrophilic metal ions. Crown ethers are an example of this interaction. Ethers are used as solvents in Grignard reactions because they solubilize Grignard reagents (RMgX) by forming complexes with the electron-poor magnesium atom.

Preparation of Ethers

Two methods of preparing ethers have been discussed in Chapter 8. One method is the acid-catalyzed addition of an alcohol to an alkene to give an ether (see the section entitled *Alcohol as a Solvent* in Chapter 8). The intermediate in this

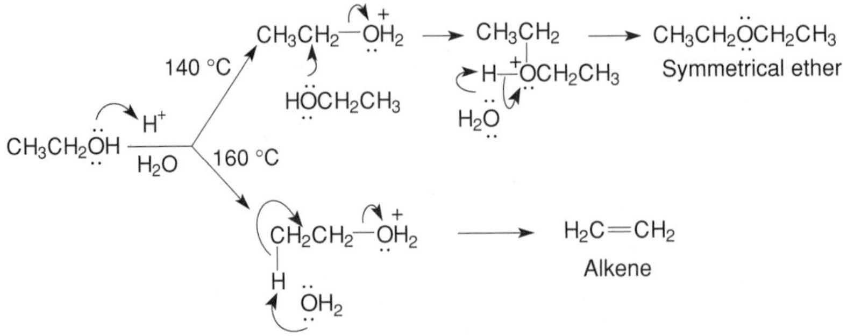

Fig. 14-3. Acid-catalyzed ether and alkene formation.

reaction is a carbocation that can undergo rearrangement reactions (hydride or methyl shifts). The second method is the oxymercuration of an alkene and subsequent demercuration (see the section entitled Oxymercuration-demercuration in Chapter 8). One advantage of this second method is that no carbocation is formed and no rearrangement reactions are observed.

ACID-CATALYZED DEHYDRATION REACTIONS

Ethers can be made from primary alcohols by an acid-catalyzed dehydration reaction. A single alcohol will give a symmetrical ether. If two different alcohols are used in a reaction, a mixture of three ether products is obtained. Reactions conducted at about 140 °C favor ether formation. Higher reaction temperatures favor alkene formation. These reactions are shown in Fig. 14-3.

QUESTION 14-3
What ethers are formed by the acid-catalyzed dehydration of a mixture of methanol and ethanol?

ANSWER 14-3
Dimethyl ether, diethyl ether, and ethyl methyl ether.

WILLIAMSON ETHER SYNTHESIS

The Williamson ether synthesis involves the reaction of a 1°, 2°, or 3° alkoxide anion (RO^-) with a methyl halide or a primary alkyl halide. This is an S_N2 reaction. Secondary and tertiary alkyl halides react with alkoxide anions to give elimination products (alkene formation) rather than substitution (ether) products. Examples of ether and alkene formation are shown in Fig. 14-4. *A key point: methyl halide and primary alkyl halides can be reacted with primary, secondary, or tertiary alkoxide anions to make ethers; but secondary and tertiary alkyl halides react with alkoxides gives alkenes.*

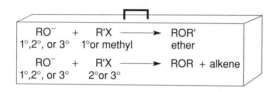

Fig. 14-4. Williamson ether synthesis and its limitations.

QUESTION 14-4

(a) What product is formed by the reaction of *tert*-butoxide and methyl iodide?

(b) What products are formed by the reaction of *tert*-butyl iodide and methoxide?

RO⁻   +   R'X ⟶ ROR'
1°,2°, or 3°   1°or methyl   ether

RO⁻   +   R'X ⟶ ROR + alkene
1°,2°, or 3°   2°or 3°

Toolbox 14-1.

ANSWER 14-4

(a) *tert*-butyl methyl ether. (b) Isobutylene (2-methylpropene) and methanol.

Reactions of Ethers

Most ethers are unreactive toward bases, weak acids, and nucleophiles. An exception is the reaction of concentrated aqueous solutions of HBr or HI (but not that after HCl) that breaks (cleaves) the carbon–oxygen bond in ethers. The reaction mechanism is shown in Fig. 14-5. The first step in this reaction is protonation of the ether oxygen. The protonated oxygen atom withdraws electrons from the two carbon atoms bonded to each side of the protonated oxygen atom. These carbon atoms now have a partial positive charge and are susceptible to nucleophilic attack. Both Br⁻ and I⁻ are good nucleophiles (Cl⁻ is not a strong enough nucleophile for this reaction) and attack the least hindered carbon atom (on either side of the ether oxygen atom). The reaction mechanism is S_N2 if the alkyl groups are primary or secondary.

Concentrated HBr or HI cleaves ethers in an S_N1 manner if a tertiary, allyl, or benzyl group is bonded to the oxygen atom. The O–C bond breaks, giving an alcohol and an alkyl carbocation. The carbocation reacts with a halide anion to give an alkyl halide.

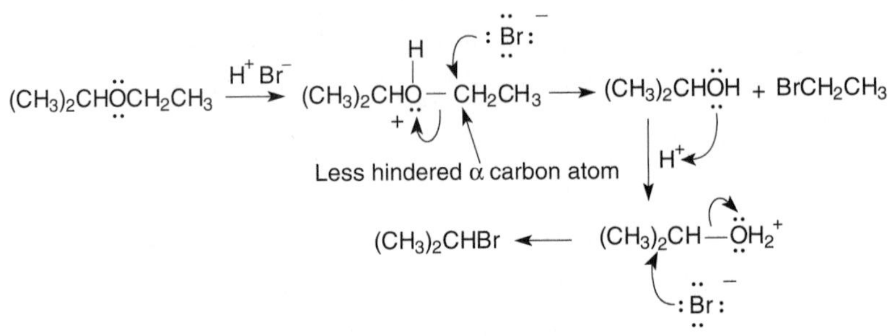

$$(CH_3)_2CH\ddot{O}CH_2CH_3 \xrightarrow{H^+ Br^-} (CH_3)_2CH\overset{+}{\ddot{O}} - CH_2CH_3 \longrightarrow (CH_3)_2CH\ddot{O}H + BrCH_2CH_3$$

Less hindered α carbon atom

$$(CH_3)_2CHBr \longleftarrow (CH_3)_2CH - \ddot{O}H_2^+$$

Fig. 14-5. Acid-catalyzed cleavage of an ether.

The alcohol produced in the ether cleavage reaction can undergo further reaction with HBr or HI to give the corresponding alkyl halide. This reaction (shown in Fig. 14-5) is discussed in detail in the section entitled *Preparation of Halides* in Chapter 7.

ETHERS AS PROTECTING GROUPS

Many molecules contain more than one functional group. If a molecule contains an alcohol group that would react or interfere with a planned reaction in another part of the molecule, the alcohol group can be "protected" by converting it temporarily into an unreactive ether. The hydrogen atom is removed from the alcohol group and replaced with another group forming an ether. This new group is called a *protecting group*. There are two common protecting groups: the *tert*-butyl group and the trimethylsilyl group. After the alcohol is protected and the desired reaction is carried out in another part of the molecule, the protecting group is removed (cleaved) from the oxygen atom by an acid-catalyzed hydrolysis reaction and the original alcohol group is regenerated. Trimethylsilyl and *tert*-butyl ethers are examples of ethers that are readily cleaved under acidic conditions. Protecting group formation is shown in Fig. 14-6.

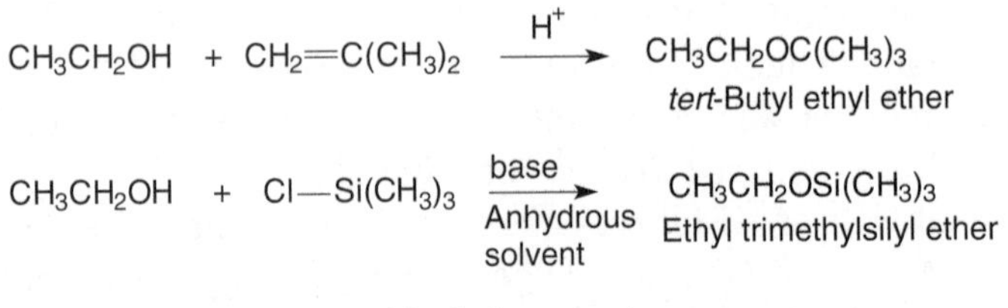

$$CH_3CH_2OH + CH_2=C(CH_3)_2 \xrightarrow{H^+} CH_3CH_2OC(CH_3)_3$$
tert-Butyl ethyl ether

$$CH_3CH_2OH + Cl-Si(CH_3)_3 \xrightarrow[\substack{Anhydrous \\ solvent}]{base} CH_3CH_2OSi(CH_3)_3$$
Ethyl trimethylsilyl ether

Fig. 14-6. Alcohol-protecting groups.

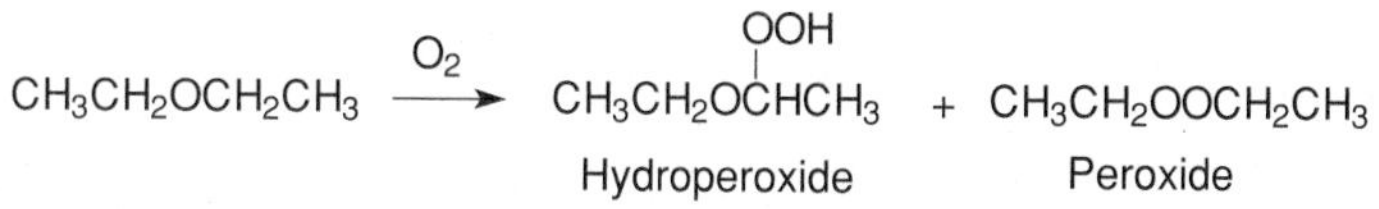

$$CH_3CH_2OCH_2CH_3 \xrightarrow{O_2} \overset{\overset{\displaystyle OOH}{\displaystyle |}}{CH_3CH_2OCHCH_3} + CH_3CH_2OOCH_2CH_3$$

Hydroperoxide   Peroxide

Fig. 14-7. Oxidation of ethers.

OXIDATION REACTIONS

Ethers undergo a generally undesirable oxidation reaction with oxygen at ambient conditions. This reaction is shown in Fig. 14-7. Most of the liquid ethers sold commercially contain stabilizers that minimize this oxidation reaction. Ethers can absorb sufficient oxygen when exposed to air for this reaction to occur. The peroxides and hydroperoxides that form are potentially explosive and care must be taken with containers that have been opened. Distilling (heating) ethers containing peroxides or letting the ether evaporate from a container leaving a peroxide residue can be dangerous. When using ethers, they should be checked for the presence of peroxides and the peroxides should be destroyed if detected.

Three-Membered Ether Rings

NOMENCLATURE

The formal (IUPAC) names for three-membered ether rings are oxacyclopropane, oxirane, and 1,2-epoxyethane. The common name is ethylene oxide. Derivatives of ethylene oxide are called epoxides.

PROPERTIES

Each atom in the ethylene oxide ring is sp^3-hybridized. Internal ring bond angles of 60° result in high angle strain. Because of this strain, carbon–oxygen bond cleavage occurs readily resulting in opening of the three-membered ring. A common product containing the ethylene oxide ring is an epoxy glue. This glue (adhesive) comes in two parts (tubes): one tube contains the epoxide (cyclic ether) compound that has an epoxide group at each end of the molecule. The other tube contains molecules (the hardener) that have an amine group at each end of the molecule. When mixed, the amine reacts with and opens the three-membered ring, resulting in one very large (polymer) molecule. This is analogous to a room full of people who "react" by randomly holding hands. You now have a single long "people molecule".

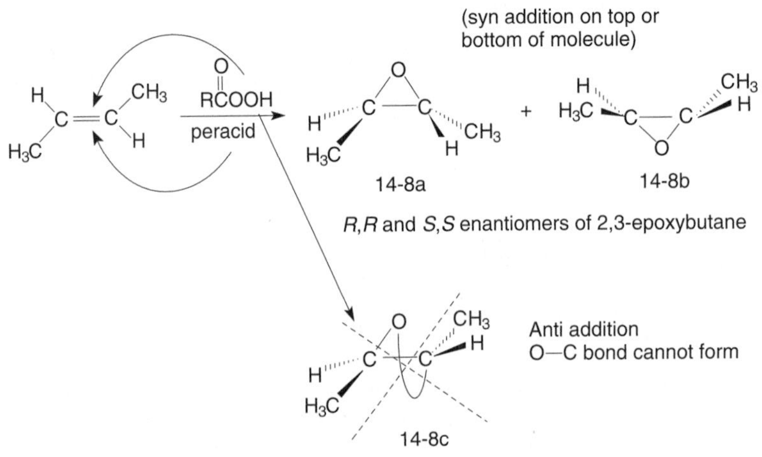

Fig. 14-8. Epoxide formation with peracids.

PREPARATION OF EPOXIDES

Epoxides are readily prepared by reaction of a peroxyacid with an alkene. A peroxyacid has one more oxygen atom (RCO_3H) than a typical carboxylic acid (RCO_2H). The reaction of a peracid with an alkene is a stereospecific reaction where the oxygen atom adds in a syn (same side) manner. The epoxidation of *trans*-2-butene, shown in Fig. 14-8, results in enantiomers 14-8a and 14-8b.

Anti (opposite) addition of the oxygen atom does not occur, since one of the bonds in the ring would have to "stretch" from the "top" side of the molecule to the "bottom" side, as shown in Structure 14-8c. This would require a very long and a very weak bond and there would be little (if any) orbital overlap and bond formation.

QUESTION 14-5
What is the stereochemistry of the product(s) from the epoxidation of *trans*-2-pentene?

ANSWER 14-5
Two enantiomers are formed, the $2R$, $3R$ and $2S$, $3S$ isomers of 2,3-epoxypentane.

Epoxides can also be prepared by reacting an alkene with an aqueous bromine or chlorine solution to form a halohydrin, an α-halo alcohol. The halohydrin is treated with base to form an alkoxide anion. The alkoxide anion displaces the halide in an internal S_N2 reaction to give an epoxide. This reaction is shown in Fig. 14-9.

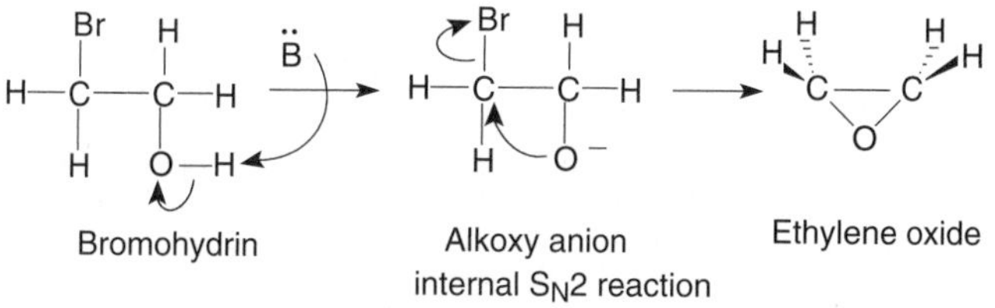

Fig. 14-9. Epoxide formation from halohydrins.

REACTIONS OF EPOXIDES WITH ACIDS

Epoxides react with water in aqueous acid solutions to form diols. The initial reaction is protonation of the epoxy oxygen atom. Nucleophiles (water in this case) attack the *less hindered* carbon atom of the protonated ring when the ring carbon atoms are primary or secondary. If one of the ring carbon atoms is tertiary, the nucleophile attacks the *more sterically hindered* tertiary carbon atom. This is rationalized by examining three resonance structures of the protonated intermediate, shown in Fig. 14-10. One resonance form has a positive charge on the primary carbon atom (Structure 14-10a). Another resonance form (Structure 10-14b) has a positive charge on a tertiary carbon atom. The resonance form with the 3° carbocation is expected to contribute more to the resonance hybrid than does the resonance form with the 1° carbocation. The nucleophile preferentially attacks this hindered tertiary carbon atom because it contributes more to the structure of the resonance hybrid. (A similar explanation is given for attack on bromonium ions, discussed in the section entitled *Regioselectivity* in Chapter 8.)

QUESTION 14-6
What product results from the acid-catalyzed ring-opening hydrolysis of 2-methyl-1,2-epoxypropane?

ANSWER 14-6
2-Methylpropan-1,2-diol.

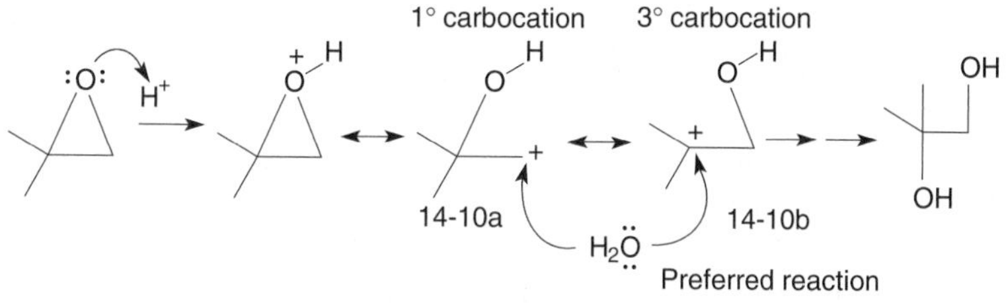

Fig. 14-10.   Acid-catalyzed ring opening of epoxides.

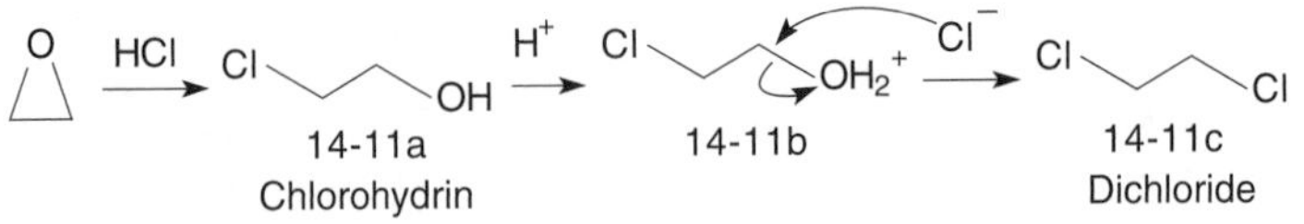

Fig. 14-11. Addition of a hydrogen halide to ethylene oxide.

HALOHYDRINS AND DIHALIDES

Hydrogen halides (HCl, HBr, or HI) react with epoxides in inert solvents to give (α-haloalcohols). This reaction is shown in Fig. 14-11. The hydroxy group in the halohydrin (Structure 14-11a) subsequently undergoes protonation under strong acid conditions, giving Structure 14-11b. The protonated alcohol group withdraws electrons from and activates the carbinol carbon atom toward nucleophilic attack. A second halide anion attacks this carbon atom displacing a molecule of water and forming a dihalide (Structure 14-11c).

REACTION OF EPOXIDES WITH NUCLEOPHILES

Epoxides react with nucleophiles such as OH^-, RO^-, $RC\equiv C^-$, R^-Li^+, and $:NH_3$ (and substituted amines). Reactions with these reagents, shown in Fig. 14-12, give after protonation: diols, hydroxy ethers, chain-extended alcohols, and amino alcohols. *A key point: nucleophiles react with the less-substituted (less sterically hindered) carbon atom in the ring-opening reactions of epoxides under basic conditions.*

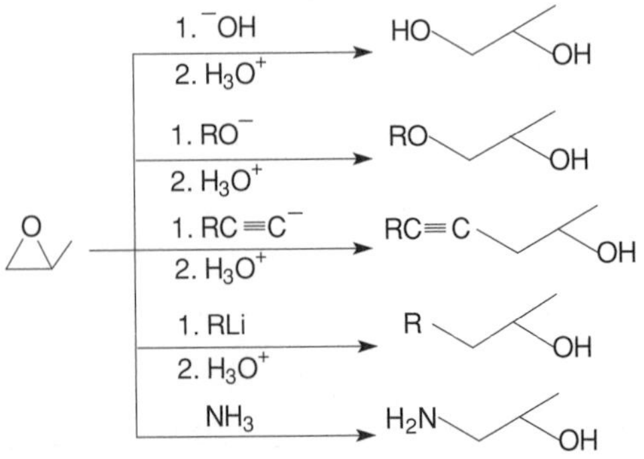

Fig. 14-12. Reactions of propylene oxide with nucleophiles.

Fig. 14-13. Polymerization of ethylene oxide.

QUESTION 14-7
What product results from the reaction of sodium methoxide with 2-methyl-1,2-epoxypropane?

ANSWER 14-7
2-Hydroxy-2-methylpropyl methyl ether.

Chain reactions

Hydroxide ions react with epoxides to give, initially, an alkoxide anion. The alkoxide anion reacts with another epoxide molecule to generate another, higher molecular weight alkoxide anion. This reaction is repeated many times to form a very high molecular weight molecule (a polyether) as shown in Fig. 14-13. When ethylene oxide is used in this reaction, the large molecules are water soluble and are used as thickeners in food, personal care and pharmaceutical products. A similar chain mechanism is involved when epoxy adhesives (epoxide and an amine) are mixed to give a high molecular weight polymer.

Quiz

1. Acyclic ethers are cleaved by
    (a) weak acids
    (b) strong bases
    (c) strong acids
    (d) nucleophiles

2. Ethers are used as solvents because they
    (a) are reactive
    (b) are nonpolar
    (c) are generally inert
    (d) are highly flammable

3. The oxygen atom in ethers is ———— hybridized.
    (a) sp^3
    (b) sp^2

(c) sp

(d) non-

4. Ethers can hydrogen-bond with
   (a) ethers
   (b) alcohols
   (c) alkanes
   (d) alkenes

5. The acid-catalyzed dehydration of ethanol gives
   (a) an unsymmetrical ether
   (b) a symmetrical ether
   (c) an alkane
   (d) an alcohol

6. The ether formed by the reaction of ethoxide and methyl chloride is
   (a) dimethyl ether
   (b) diethyl ether
   (c) ethyl methyl ether
   (d) trimethyl ether

7. The reaction of concentrated HI with *tert*-butyl methyl ether gives
   (a) *tert*-butyl iodide and methanol
   (b) *tert*-butanol and methyl iodide
   (c) dimethyl ether
   (d) di-*tert*-butyl ether

8. Hydroperoxides are
   (a) stable
   (b) unstable
   (c) nitrogen compounds
   (d) oxides of the element hydroper (Hp)

9. The reaction of HI with 1,2-epoxypropane gives
   (a) 2-iodo-1-propanol
   (b) 1-iodo-2-propanol
   (c) 1,2-dihydroxypropane
   (d) 2-iodo-1,2-epoxypropane

10. Sodium methoxide reacts with 2-methyl-1,2-epoxypropane to give, after reaction with dilute acid, a
    (a) primary alcohol
    (b) secondary alcohol
    (c) tertiary alcohol
    (d) ketone

11. The reaction of the Grignard reagent methylmagnesium bromide with ethylene oxide gives, after addition of aqueous acid, a
    (a) primary alcohol
    (b) secondary alcohol
    (c) tertiary alcohol
    (d) ketone

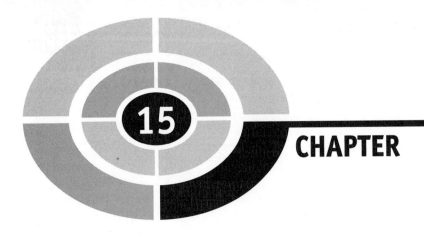

Sulfur Compounds

The element sulfur (S) is directly below oxygen in Group VIB in the periodic table (see Appendix A) and many sulfur compounds can be considered analogs of oxygen-containing alcohols (ROH) and ethers (ROR). Sulfur compounds are perhaps best known for their undesirable odors in rotten eggs, skunk spray, and additives that give odorless natural gas, which is burned in home furnaces, a detectable odor.

Nomenclature

As with other families of compounds, sulfur compounds have formal (IUPAC) and common names. Sulfur compounds with the general formula RSH are formally known as alkanethiols or sulfanyl compounds. The term thiol is used if the —SH group is the parent group. R retains the terminal letter, for example, the letter e in ethanethiol (not ethanthiol). The R group may be an alkyl, aryl, or alkenyl group. If an —SH group is present as a substituent (another functional group is the parent) the prefix sulfanyl is used.

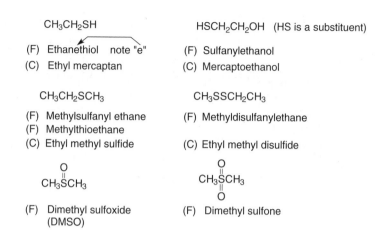

Fig. 15-1. Formal (F) and common (C) names for sulfur compounds.

The common name for an —SH group is mercaptan. If the —SH group is the parent group, an alkyl RSH compound is called an alkyl mercaptan. If the —SH group is a substituent, it is named as a mercapto group. Examples for naming these compounds are given in Fig. 15-1.

QUESTION 15-1
What are the formal and common names of CH_3SH?

ANSWER 15-1
Methanethiol and methyl mercaptan.

Compounds with the generic formula RSR are formally called sulfanyl alkanes or alkylthioalkanes. The R group may be alkyl, aryl, or alkenyl. The largest R group is the parent and the RS segment is the alkylsulfanyl or alkylthio group. The common name for RSR compounds is dialkyl sulfide. The R groups are listed alphabetically in sulfides. Examples are given in Fig. 15-1.

Compounds with the structure RSSR′ are disulfides, RS(O)R′ are sulfoxides, and RS(O)(O)R′ are sulfones. Examples of these structures are also shown in Fig. 15-1.

Properties

BOILING POINTS

Since the sulfur atom is larger and more polarizable than an oxygen atom, one might predict sulfur compounds would have higher boiling points than their oxygen analogs. This trend is seen for sulfides and ethers. The boiling point of diethyl sulfide is 91 °C and the boiling point of diethyl ether is 34 °C. The reverse trend is seen for alcohols and thiols. The boiling point of ethanethiol is 37 °C while the boiling point of ethanol is 78 °C. Thiols have intermolecular dipolar attractions but do not have intermolecular hydrogen bonding, as do alcohols. Thus, alcohols have much higher boiling points than do thiols of comparable molecular weight.

QUESTION 15-2
Which has a higher boiling point, CH_3SH or CH_3OH?

ANSWER 15-2
CH_3OH, since it can form intermolecular hydrogen bonds.

ACIDITY

Alkanethiols are weaker acids ($pK_a \sim 10$) than carboxylic acids ($pK_a \sim 5$) but stronger acids than water ($pK_a = 15.7$). (Smaller pK_a values indicate stronger acids.) Aqueous hydroxide solutions are sufficiently basic to form the alkanethiolate anion (RS^-) from alkanethiols. Thiols are stronger acids than alcohols. The pK_a value of ethanethiol is 8.5, while that of ethanol is 15.9.

The reason for the increased acidity of thiols is that the alkanethiolate anion (RS^-) can accommodate a negative charge much better than can an alkoxide anion. The larger diameter of the sulfur atom can disperse electrons throughout a larger area than can the oxygen atom with its smaller diameter. The stability of a species is increased if the charge on that species (the electron density) can be delocalized (spread throughout) in an atom or a molecule. The more stable the anion, the weaker it is as a conjugate base. The weaker the conjugate base, the stronger the corresponding acid. These equilibrium reactions are shown in Fig. 15-2.

$$RSH + H_2O \rightleftarrows RS^- + H_3O^+$$
$$pK_a \cong 10 \qquad\qquad pK_a = -1.7$$

$$RSH + {}^-OH \rightleftarrows RS^- + H_2O$$
$$pK_a \cong 10 \qquad\qquad pK_a = 15.7$$

Fig. 15-2.   Acidity of thiols.

$$H_2S + OH^- \longrightarrow H_2O + HS^- \text{ (Hydrosulfide anion)}$$

$$HS^- + RX \longrightarrow X^- + RSH \text{ (Thiol)}$$

$$RSH + OH^- \text{ (or } HS^-) \longrightarrow RS^- + H_2O \text{ (Thiolate anion)}$$

$$RS^- + RX \longrightarrow X^- + RSR \text{ (Sulfide)}$$

Alkene
E2 reaction

Fig. 15-3. Formation of thiols and sulfides.

QUESTION 15-3

Which is a stronger acid, $CH_3CH_2CH_2CH_2SH$ or $CH_3CH_2SCH_2CH_3$?

ANSWER 15-3

$CH_3CH_2CH_2CH_2SH$. $CH_3CH_2SCH_2CH_3$ has no acidic proton.

Reactions of Sulphur Compounds

PREPARATION OF THIOLS

Thiols are prepared by the reaction of the hydrosulfide anion (HS^-) with an alkyl halide. Sulfur is a good nucleophile because the electrons in the large sulfur atom are readily polarizable. Primary alkyl halides undergo an S_N2 reaction with the hydrosulfide anion to give a alkanethiol. An excess of hydrosulfide ion must be used since the initially formed alkanethiol can form an alkanethiolate anion and undergo a further S_N2 reaction with the alkyl halide to give a symmetrical sulfide, as discussed below. These reactions are shown in Fig. 15-3.

Hydrosulfide anion + 1° Alkyl halide $\longrightarrow$ Alkanethiol
S_N2 reaction

Toolbox 15-1.

QUESTION 15-4

What product is obtained from the reaction of 1 equiv. of CH_3Br with 1 equiv. of HS^-?

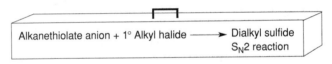

Fig. 15-4. Alkylating reagent.

ANSWER 15-4
Methanethiol.

PREPARATION OF SULFIDES

Alkyl sulfides are prepared by the reaction between an alkanethiolate anion and an alkyl halide. Primary alkyl halides react with alkanethiolate anions to give mainly sulfides. The reaction between tertiary alkyl halides and alkanethiolate anions primarily give elimination products (alkenes). These reactions are shown in Fig. 15-3.

Alkanethiolate anion + 1° Alkyl halide ⟶ Dialkyl sulfide
S_N2 reaction

Toolbox 15-2.

PREPARATION OF SULFONIUM SALTS

Sulfides are also good nucleophiles (the sulfur atoms have two nonbonding electron pairs) and react with primary alkyl halides to give sulfonium salts. This reaction is shown in Fig. 15-4. Sulfonium salts are themselves good alkylating reagents. Sulfonium salts play a major role as alkylating (methylating) reagents in biological systems.

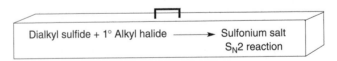

Dialkyl sulfide + 1° Alkyl halide ⟶ Sulfonium salt
S_N2 reaction

Toolbox 15-3.

OXIDATION REACTIONS

Sulfides are oxidized to sulfoxides by the controlled reaction with hydrogen peroxide/acetic acid or sodium metaperiodate ($NalO_4$). Dimethyl sulfoxide ($CH_3S(O)CH_3$, DMSO) is probably the best known sulfoxide. It is a very powerful solvent, but must be handled with care as it readily penetrates the skin (carrying along any material dissolved in the DMSO).

Sulfoxides can be further oxidized to sulfones using peroxide or metaperiodate. A well-known sulfonate anion is *p*-toluenesulfonate, which is a very good leaving group in S_N2 reactions.

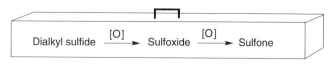

Toolbox 15-4.

QUESTION 15-5

Name the compound that results from the reaction of CH_3SCH_3 with excess H_2O_2.

ANSWER 15-5

Dimethyl sulfone.

Thiols form disulfides when treated with oxidizing reagents including oxygen, bromine, or iodine. Thiols readily undergo oxidation in (oxygen-containing) air and containers of thiols should be kept sealed to prevent undesirable oxidation reactions. Disulfides are very important in biological systems. The amino acid *cysteine* contains an —SH (thio) group. It can react with the —SH group in another cysteine molecule to form *cystine*, a molecule with a disulfide bond. These disulfide bonds are responsible for giving proteins (enzymes) the three-dimensional structure required for biological activity.

REDUCTION REACTIONS

Disulfides can be reduced to thiols by zinc/acid or ammonium thioglycolate $(HSCH_2CO_2NH_4)$ solutions. Hair consists of protein molecules that contain disulfide bonds. Hair "permanents" involve breaking cystine disulfide bonds between adjacent protein molecules, reorienting hair strands (and the protein molecules), and then reforming the disulfide bonds so the hair strands now stay in the new style.

Quiz

1. The common name for CH_3SCH_3 is
    (a) methane methane sulfide
    (b) dimethyl sulfide

(c) dimethyl sulfoxide
(d) dimethyl sulfone

2. The boiling point of $(CH_3)_2CHSH$ is———that of $(CH_3)_2CHOH$.
    (a) higher than
    (b) lower than
    (c) the same as

3. CH_3OH is a(n)———acid than/to CH_3SH.
    (a) stronger
    (b) weaker
    (c) identical

4. The hydrosulfide anion (HS^-) reacts with primary alkyl halides by an
    ———mechanism.
    (a) S_N1
    (b) S_N2
    (c) E1
    (d) E2

5. The R group in RSR is an———group.
    (a) alkyl
    (b) aryl
    (c) alkenyl
    (d) all of the above

6. DMSO is a
    (a) sulfide
    (b) disulfide
    (c) thiol
    (d) sulfoxide
    (e) sulfone

7. A disulfide bond is between———atoms.
    (a) two carbon
    (b) two oxygen
    (c) two sulfur
    (d) oxygen and sulfur

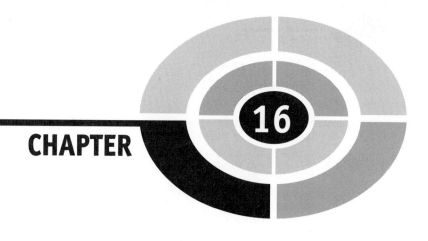

CHAPTER

16

Conjugated Systems

Introduction

Double bonds that are separated by *one* single (sigma) bond are *conjugated double bonds*. If two double bonds are separated by more than one sigma bond, the double bonds are *nonconjugated* or *isolated double bonds*. Double bonds can be between two carbon atoms, between carbon and oxygen atoms, or between carbon and other heteroatoms. Conjugated systems can consist of two or more (poly) alternating double–single bond sequences. Compounds containing conjugated double bonds have different chemical and physical properties than do similar compounds with nonconjugated double bonds. Double bonds that share a common carbon atom are called cumulated double bonds or *allenes*. Examples of these alkenes are shown in Fig. 16-1.

Stability of Dienes

The relative stability of alkenes can be determined by measuring the heat given off during a hydrogenation (the addition of hydrogen) reaction. If different

Fig. 16-1. Structures of mono-, di-, and polyenes.

alkenes are hydrogenated and all give the same final product, the amount of heat given off (per mole of alkene compound) in each reaction is a measure of the potential energy (relative stability) of the starting alkenes.

CONJUGATED DIENES

Four alkenes are shown in Fig. 16-2. Structures 16-2a and 16-2b are monoenes. Monoenes contain just one carbon–carbon double bond. These two molecules will be used to approximate the shape (and energy) of the double bonds in conjugated diene 16-2c. For comparison, a nonconjugated diene, 16-2d, is also shown. The amount of heat given off (252 kJ/mol) in the hydrogenation of nonconjugated diene 16-2d is about twice the heat of hydrogenation (125 kJ/mol) of monoene 16-2a. The heat of hydrogenation (115 kJ/mol) of trans dialkyl-substituted monoene 16-2b is slightly less than the heat of hydrogenation (125 kJ/mol) of monoalkyl-substituted monoene 16-2a. (The stability of alkenes as a function of degree of substitution is discussed in the section entitled *Stability of Alkenes* in Chapter 6.) Conjugated diene 16-2c is 15 kJ/mol more stable than the sum of the heats of hydrogenation of the two monoenes 16-2a and 16-2b, and

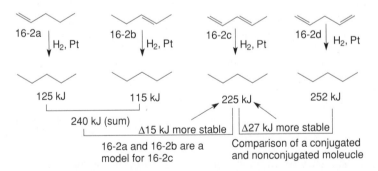

Fig. 16-2. Relative molar stability of alkenes.

27 kJ/mol more stable than nonconjugated diene 16-2d. Heats of hydrogenation are shown as positive numbers for this discussion while the values are usually shown as negative numbers since the potential energy of the products is less than that of the reactants. *A key point: molecules containing conjugated double bonds are more stable than molecules with nonconjugated double bonds of similar structure (composition and molecular weight).*

QUESTION 16-1
Which molecule is more thermodynamically stable, 2,4-hexadiene or 1,4-hexadiene?

ANSWER 16-1
Conjugated 2,4-hexadiene. 1,4-Hexadiene is nonconjugated.

RESONANCE ENERGY

The increased stability (decrease in potential energy) of conjugated dienes relative to nonconjugated dienes is called the *resonance energy* of the molecule. *Conjugation energy*, *delocalization energy*, and *stabilization energy* are synonymous terms with resonance energy. Two explanations for this increased stability are usually given: hybridization and molecular orbital stabilization.

Stabilization due to hybridization

Figure 16-3 shows the structure of a conjugated and nonconjugated diene. A double-stemmed arrow points to a single bond in each structure. In the nonconjugated diene (16-3a), the indicated single bond results from the overlap of sp^3 and sp^2 orbitals on adjacent carbon atoms. In the conjugated diene (16-3b), the indicated single bond results from the overlap of two sp^2 orbitals, one from each adjacent carbon atom. Hybridized orbitals on carbon atoms result from the combination (mixing) of 2s and 2p orbitals. The electrons in the 2s orbital are held closer and more tightly to the nucleus than are the electrons in a 2p orbital. There is more s character in an sp^2 orbital (33%, from one s orbital and two

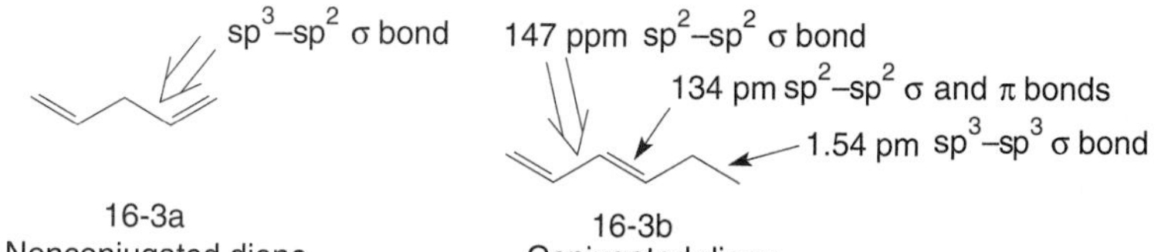

16-3a
Nonconjugated diene

16-3b
Conjugated diene

Fig. 16-3. Hybridization stability.

p orbitals) than in an sp^3 orbital (25%, from one s orbital and three p orbitals) and therefore electrons in sp^2 orbitals are held more closely (tightly) to the nucleus than are electrons in sp^3 orbitals. Consequently, a bond formed by the overlapping of two sp^2 orbitals is shorter and stronger than a bond formed by the overlapping of sp^2 and sp^3 orbitals. The bond formed by overlapping of two sp^3 orbitals is even longer and weaker than the bonds formed by overlapping sp^2–sp^2 and sp^2–sp^3 orbitals.

QUESTION 16-2
Which bond is stronger, one formed by overlapping sp^2–sp orbitals or sp^3–sp^2 orbitals?

ANSWER 16-2
There is more s character in the overlapping orbitals of an sp^2–sp bond, making it shorter and stronger than an sp^3–sp^2 bond.

Bond lengths. The lengths of carbon–carbon bonds in a conjugated diene are shown for Structure 16-3b in Fig. 16-3. The length of the single bond (147 pm) located between sp^2–sp^2 hybridized carbon atoms is shorter than a single bond (154 mp) between sp^3–sp^3 hybridized carbon atoms and longer than a typical double bond (134 pm). The single bond between the conjugated double bonds is shorter than a typical isolated single bond and has some degree of double bond character but not enough to prevent rotation about this single bond at room temperature.

QUESTION 16-3
Which bond is shorter, one formed by overlapping sp^2–sp orbitals or sp^3–sp^2 orbitals?

ANSWER 16-3
There is more s character in the sp^2–sp bond, making it shorter and stronger than an sp^3–sp^2 bond.

Molecular orbital stabilization

The second reason for the increased stability of conjugated systems is explained by the formation of bonding molecular orbitals (MOs). Bonding orbitals are described as being localized or delocalized. Localized atomic orbitals (AOs) include unhybridized s, p, and d orbitals and hybridized sp, sp^2, and sp^3 orbitals. Bonding electrons are shared between localized orbitals on adjacent atoms but the orbitals remain assigned to a specific atom. MOs encompass two or more atoms. The electrons in these orbitals can be delocalized over several atoms. The method of forming MOs by mathematically combining the wave functions for

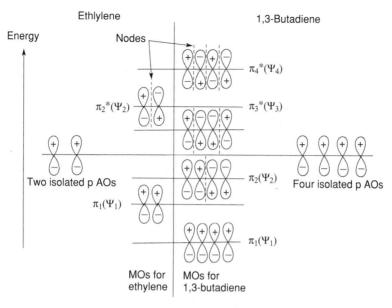

Fig. 16-4.   Relative stability of AOs and MOs for ethylene and 1,3-butadiene.

the AOs is called the linear combination of atomic orbitals (LCAO). Resonance energy will be discussed in terms of the π MOs and not the σ MOs.

Ethylene molecular orbitals.   The AO description of the pi (π) bond in ethylene involves the overlapping of two atomic p orbitals, one on each adjacent atom. The MO description mathematically combines the p AOs to form π MOs. Orbital conservation is observed in that two p AOs combine to form two π MOs. The orbitals (or the wave functions) combine in two ways, by constructive overlap (combining AO lobes of the same sign) and destructive overlap (combining AO lobes of opposite sign). The individual p AOs no longer exist in the MO model.

Figure 16-4 shows the energy levels of p AOs and the corresponding π MOs for ethylene. The addition (constructive overlap) of the two p AOs forms a *bonding* πMO (π_1) and subtraction (destructive overlap) of the two p AOs forms an *antibonding* πMO (π_2*). The asterisk (*) refers to antibonding orbitals. The bonding π_1 MO is lower in energy (more stable) than is the isolated p AO. The antibonding π_2* MO is less stable than an isolated p AO.

The bonding π_1 MO has a high electron density between the carbon atoms. High electron density contributes to bonding. The antibonding π_2* MO has one node (a region of zero electron density) between the two carbon atoms. If electrons occupied this MO, they would make little, if any, contribution to bonding. The two electrons in the π system of ethylene go into the lowest energy bonding MO (aufbau principle). *The MO model for ethylene predicts that two*

electrons in the bonding MO is a more stable arrangement than are two electrons in two p AOs (one in each p AO). This energy difference explains the resonance stabilization for the MO model and is shown in Fig. 16-4 for ethylene.

QUESTION 16-4
Does a bonding MO in ethylene result from the overlap of the lobes of p orbitals of the same or opposite sign?

ANSWER 16-4
Bonding MOs result from overlap of lobes of the same sign. Don't confuse the mathematical sign of an orbital with the charge of an electron, which is always negative.

1,3-Butadiene molecular orbitals. The AO description of 1,3-butadiene, a conjugated molecule, consists of four separate (isolated) p AOs. The four p AOs combine (LCAO) to form four π MOs. The p AOs no longer exist when the MOs are formed. Figure 16-4 shows an energy diagram for the four AOs and the four MOs. The lowest energy bonding π MO (π_1) results from constructive (same sign) overlap of all four p AOs. Electrons are shared by all four carbon atoms in this MO. The lowest bonding π MO (π_1) for 1,3-butadiene is lower in energy than the bonding π MO (π_1) of ethylene (see Fig. 16-4). The next higher energy π MO in 1,3-butadiene (π_2) has one node (destructive overlap) between the two internal carbon atoms and constructive overlap between two pairs of outer orbitals. Since this MO is also lower in energy than are the isolated p AOs, it is also called a bonding MO. The next higher energy π MO ($\pi_3{}^*$) for 1,3-butadiene has two nodes and bonding between only the central carbon atoms. The next, and highest, energy MO ($\pi_4{}^*$) has three nodes and no bonding between orbitals on any carbon atoms. The two highest energy MOs are higher in energy than are the isolated p AOs and are called antibonding *MOs*. The terms ψ_1, ψ_2, ψ_3, and ψ_4 are used in some textbooks to designate the four MOs orbitals: π_1, π_2, $\pi_3{}^*$, and $\pi_4{}^*$. The p AOs no longer exist when MOs are formed.

Electron configuration in 1,3-butadiene. There are four pi electrons in 1,3-butadiene. The *aufbau principle* is used to fill π MOs. Each MO can hold a maximum of two electrons. The first two electrons go into the lowest energy MO, π_1. The next two electrons go into the next higher MO, π_2. Both MOs are lower in energy than are the isolated p AOs. The MO model predicts that a conjugated 1,3-butadiene molecule containing four electrons will be more stable than if the four electrons occupied four isolated p AOs. The conjugated diene molecule is also lower in energy than is the sum of the energies of the nonconjugated double bonds in two ethylene molecules. This agrees with the experimental observation that molecules with two conjugated double bonds are

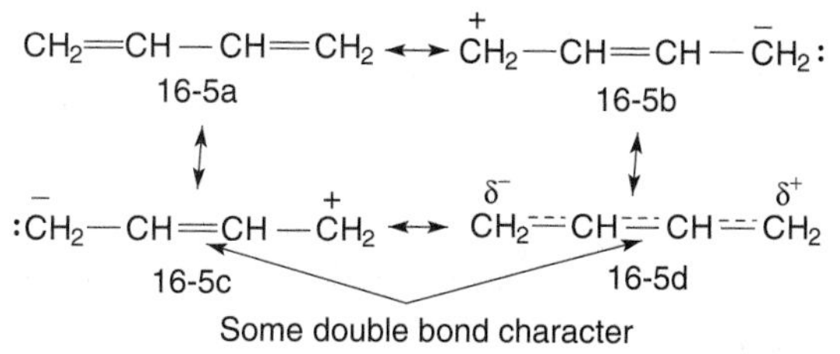

$$CH_2\!=\!CH\!-\!CH\!=\!CH_2 \longleftrightarrow \overset{+}{C}H_2\!-\!CH\!=\!CH\!-\!\overset{-}{C}H_2\!:$$

16-5a   16-5b

$$:\overset{-}{C}H_2\!-\!CH\!=\!CH\!-\!\overset{+}{C}H_2 \longleftrightarrow \overset{\delta^-}{CH_2}\!=\!=\!\!CH\!=\!\!CH\!=\!=\!\overset{\delta^+}{CH_2}$$

16-5c   16-5d

Some double bond character

Fig. 16-5. Lewis resonance structures of 1,3-butadiene.

more stable (lower in energy) than molecules with two isolated double bonds. This is called resonance stabilization.

The lowest MO, π_1, in 1,3-butadiene shows bonding between the central carbon atoms, indicating some π character between these carbon atoms. Since π bonds are shorter and stronger than single bonds, the MO description explains the additional stability (shorter, stronger carbon–carbon bond) observed for the C2–C3 bond in 1,3-butadiene.

QUESTION 16-5
Would 1,3-butadiene be more stable if it contained four or five π electrons?

ANSWER 16-5
A fifth electron would go into an antibonding orbital, making the molecule less stable than if it contained only four electrons in bonding orbitals.

Lewis structures

Lewis resonance structures can be drawn showing a double bond character for the central bond in 1,3-butadiene. Figure 16-5 shows three resonance structures for 1,3-butadiene. The structures showing charge separation (16-5b and 16-5c) are not as important as the structure without charge separation (16-5a), but they contribute to the partial double-bond character for the central bond in the hybrid structure (16-5d).

s-CIS AND s-TRANS ISOMERS

The central bond in 1,3-butadiene has some double bond character but it is not sufficient to prevent rotation around this bond at room temperature. The orientation around the central bond can have a cis-like structure and a trans-like structure, as shown in Fig. 16-6. The terms *s-cis* and *s-trans* are used to indicate the stereochemistry around a single (s) bond.

QUESTION 16-6
Can 1,3-cyclohexadiene exist in s-cis and s-trans conformations?

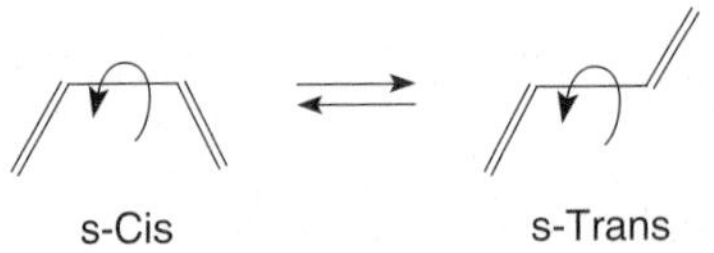

Fig. 16-6. s-Cis and s-trans structures.

ANSWER 16-6

No, the ring would only allow the s-cis conformation. Draw the ring structure to visualize the cis conformation.

Electrophilic Addition to Conjugated Dienes

The addition of HCl, HBr, or HI to isolated double bonds occurs by Markovnikov addition. (For a review, see the section entitled *Reactions of Alkenes* in Chapter 8.) The addition of HCl to 1,3-butadiene gives three products: 3-chloro-1-butene (16-7c), *cis*-1-chloro-2-butene (16-7d), and *trans*-1-chloro-2-butene (16-7e) shown in Fig. 16-7. These products result from an allylic carbocation intermediate.

ALLYLIC CARBOCATIONS

The term *allylic* means "*next to a double bond*." Figure 16-7 shows the electrophilic addition of a proton to an electron-rich double bond. Markovnikov

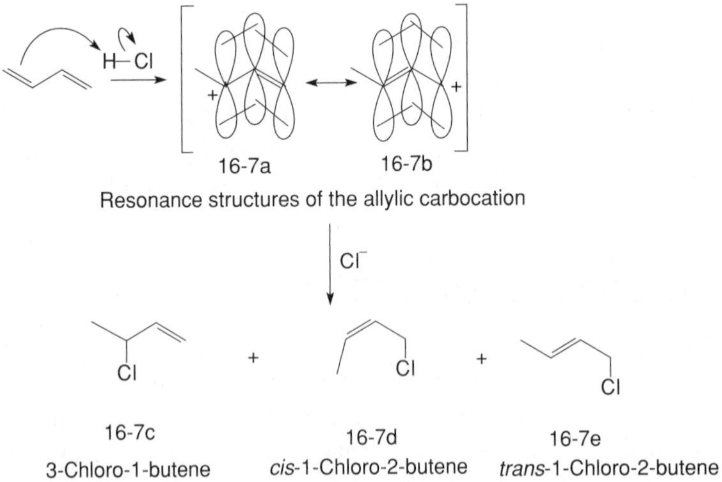

16-7a   16-7b

Resonance structures of the allylic carbocation

16-7c   16-7d   16-7e

3-Chloro-1-butene   *cis*-1-Chloro-2-butene   *trans*-1-Chloro-2-butene

Fig. 16-7. Electrophilic additions to 1,3-butadiene. The addition of the hydrogen atom is shown for clarity.

addition gives Structure 16-7a. An allylic carbocation consists of a positively charged carbon atom next to a double bond. The allylic carbon atom is sp^2 hybridized and contains an unhybridized p orbital. If this p orbital is in the same plane as the p orbitals in the double bond, these orbitals can overlap and share electrons. This overlap is shown in resonance structures 16-7a and 16-7b. Resonance structures 16-7a and 16-7b show a carbocation on two alternate carbon atoms. The chloride anion can attack either carbocation to form the three observed monochloro products, 16-7c, 16.7d, and 16.7e. Note that both *cis-* and *trans*-1-chloro-2-butene are formed. It may help if you draw the reaction mechanism to visualize how both products are formed.

Structure 16-7a is a secondary allylic carbocation ($CH_3-CH^+-CH=CH_2$) and Structure 16-7b is a primary allylic carbocation ($CH_3-CH=CH-CH_2{}^+$). Experimental studies have shown the order of stability of carbocations to be *secondary allylic ≈ tertiary alkyl > primary allylic ≈ secondary alkyl > primary > vinyl.* If the secondary allylic carbocation (Structure 16-7a) is more stable than the primary allylic carbocation (Structure 16-7b), the transition state for the secondary allylic carbocation reacting with Cl$^-$ should be lower in energy than the transition state for the primary allylic carbocation reacting with Cl$^-$ (Hammond postulate). One would then predict that 3-chloro-1-butene would be the major product, and it is—sometimes!

QUESTION 16-7
What products would result from the addition of HBr to 1,3-butadiene?

ANSWER 16-7
3-Bromo-1-butene and *cis-* and *trans*-1-bromo-2-butene.

PRODUCT COMPOSITION AS A FUNCTION OF TEMPERATURE

If the reaction of HCl with 1,3-butadiene is carried out at −40 °C, the major product results from 1,2-addition, Structure 16-7c. If the reaction is carried out at +40 °C, the major products result from 1,4-addition, Structures 16-7d and 16-7e. 1,2-addition means new bonds are formed to adjacent carbon atoms while 1,4-addition means bonds are formed to atoms separate by two carbon atoms. These results can be explained by analyzing the energy–reaction pathway diagram shown in Fig. 16-8.

Kinetic control

At low temperatures, molecules have less energy than at higher temperatures. Initial protonation of 1,3-butadiene gives the resonance-stabilized intermediate

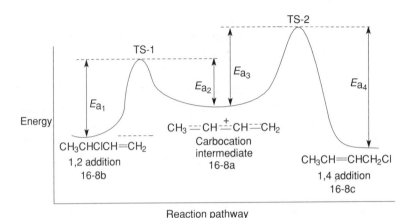

Fig. 16-8. Energy–reaction pathway for electrophilic addition to 1,3-butadiene.

16-8a. A given amount of energy, E_a, is required for 16-8a to react with Cl^-. This reaction can proceed in two ways: one gives the 1,2-addition product via transition state TS-1 and the second gives the 1,4-addition product via transition state TS-2. The activation energy, E_{a2}, needed to form transition state TS-1 leading to the 1,2-addition product (16-8b) is less than the activation energy, E_{a3}, required to form transition state TS-2 leading to the 1,4-addition product, 16-8c. TS-1 involves the reaction of the more stable 2° allylic carbocation with chloride anion and is lower in energy than TS-2 that involves the reaction of a primary allylic carbocation. With a limited amount of energy available (at low temperatures), the reaction takes the lowest energy pathway and the major product (16-8b) results from 1,2-addition. This is a reversible reaction but at low temperatures few molecules have sufficient energy (E_{a1}) required for the reverse reaction. At low temperatures, the 1,2-addition product, 16-8b, is the major product even though the 1,4-addition product, 16-8c, is more thermodynamically stable (it is lower in energy). *This is called kinetic control.*

Thermodynamic control

At higher reaction temperatures, there is sufficient kinetic energy for the carbocation intermediate (16-8a) to climb both energy hills (E_{a2} and E_{a3}) to form the 1,2- and 1,4-addition products. The activation energy for the 1,2-addition reaction is lower than that for the 1,4-addition product. Consequently, the 1,2-product is formed faster than the 1,4-addition product (kinetic control). *It is important to note that all the reactions are reversible, equilibrium reactions.* There is sufficient thermal energy at +40 °C for 3-chloro-1-butene (16-8b) to undergo a reverse reaction (dechlorination) and reform the allylic carbocation 16-8a.

The energy requirement (E_{a4}) for the 1,4-addition product to re-form 16-8a is higher than the energy requirement (E_{a1}) for the 1,2-addition product to reform 16-8a. It is more difficult (energy intensive) for the 1,4 addition to undergo the reverse reaction. Higher temperatures provide sufficient energy for both the 1,2- and 1,4-addition reactions to achieve equilibrium by way of the carbocation intermediate. *Under equilibrium conditions, the most thermodynamically stable (lowest energy) product is the major product.* This is called *thermodynamic control.* At higher temperatures, the 1,4-addition product (the most stable, highly substituted alkene) is the major product.

QUESTION 16-8
If a solution of *cis*-1-chloro-2-butene is held at 40 °C until equilibrium conditions are reached, what is/are the major product(s)?

ANSWER 16-8
cis- and *trans*-1-Chloro-2-butene, since they are thermodynamically more stable than 3-chloro-1-butene.

Allylic Cations, Radicals, and Anions

The MOs of an allylic system result from the linear combination of three p atomic orbitals (LCAO). An MO diagram for an allyl group is shown in Fig. 16-9. Three p AOs combine to give three MOs. The lowest MO (π_1) results

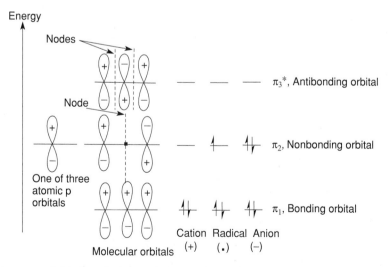

Fig. 16-9.   MO description and electronic configuration of an allylic system.

CHAPTER 16   Conjugated Systems

from constructive overlap of three p orbitals and contains no nodes (areas of zero electron density). It is called a *bonding MO*. The next higher MO (π_2) has one node and the only symmetrical arrangement exists if the node is in the center of the molecule. It has isolated orbitals on each terminal carbon atom and has the *same stability* as two isolated p AOs. This MO is called a *nonbonding MO* as it has the same energy as the isolated p AOs. The third MO (π_3^*) has two nodes between the orbitals on each carbon atom. It is higher in energy than the p AOs and is called an *antibonding MO*.

QUESTION 16-9

How many MOs result, when the three p AOs of an allyl anion are mathematically combined?

ANSWER 16-9

Three; the number of MOs formed equals the number of AOs used to form the MOs (orbital conservation).

ALLYLIC ELECTRONIC STRUCTURES

Figure 16-9 shows the electronic configurations for the allyl carbocation, radical, and anion. The allyl carbocation contains two pi electrons that go into the lowest energy π_1 MO (aufbau principle). The allyl radical has three pi electrons, two go into the lowest energy π_1 MO and the third goes into the next higher π_2 MO, the nonbonding MO. The anion has four pi electrons that go into the bonding and nonbonding π_1 and π_2 MOs.

The π_2 MO indicates the electrons in that orbital will have an equal probability of being on either terminal atom, since there is a node at the center carbon atom. The formal charge on either terminal atom will, therefore, be +1, 0, or −1 for the cation, the radical, and the anion respectively. Each terminal carbon atom will therefore, on average, have 1/2 positive charge, 1/2 radical, or 1/2 negative charge. *The MO description is consistent with the resonance description shown in Fig. 16-10, where each terminal carbon atom has an equal probability of being a carbocation, radical, or anion in an allyl species.*

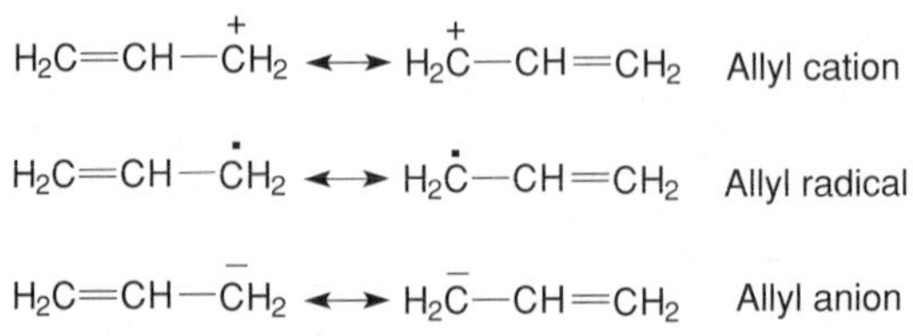

Fig. 16-10.   Resonance structures for the allylic cation, radical, and anion.

Fig. 16-11. Diels–Alder reaction.

Diels–Alder Reactions

Diels–Alder (D–A) reactions represent a convenient way of making six-membered rings. This reaction is also called a *(4+2) cycloaddition* reaction, because four pi electrons of a *diene* interact with two pi electrons of a *dienophile*. A D–A reaction is shown in Fig. 16-11. The reaction is not stepwise, as Fig. 16-11 may imply, but is a *concerted reaction* where all bond breaking and making occur at the same time (in one step). No radicals or ionic species are involved. The net result is two pi bonds are broken and two sigma bonds are formed. Since sigma bonds tend to be stronger than pi bonds, the reactions are thermodynamically favorable (energy is given off). The reaction is stereospecific (syn addition) and regiospecific.

The diene is an electron-rich species and the dienophile is electron-poor (by comparison). Electron-donating alkyl (inductive effect) and alkoxy (resonance effect) groups bonded to the diene increase the electron density of the diene and make it more reactive. Electron-withdrawing groups like carbonyl and nitrile (cyano) groups bonded to the dienophile decrease the electron density of the dienophile and make it a stronger (and more reactive) dienophile. Various substituted dienes and dienophiles are shown in Fig. 16-12.

s-CIS AND s-TRANS ORIENTATIONS

The diene is capable of undergoing D–A reactions if it has the correct orientation. Several diene molecules are shown in Fig. 16-13. 1,3-Butadiene exists as an equilibrium between the s-cis and s-trans conformations. The s refers to orientation

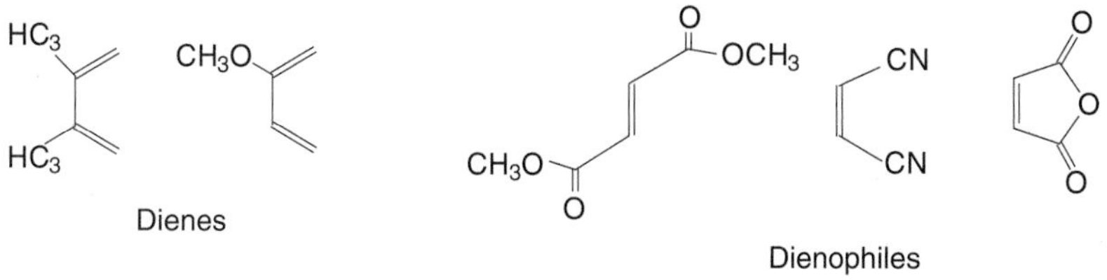

Fig. 16-12. Dienes and dienophiles.

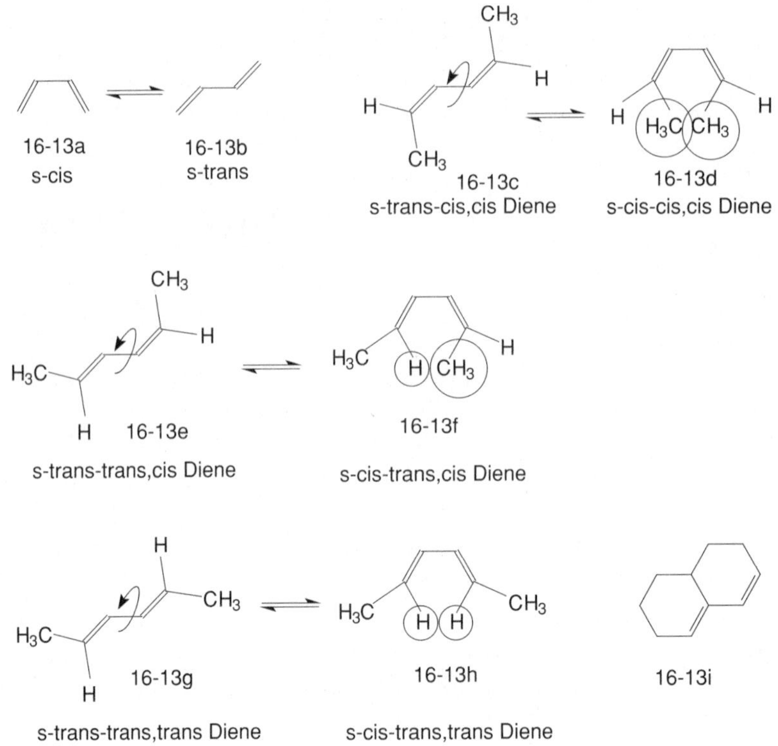

Fig. 16-13. s-Cis and s-trans conformations.

about a single sigma bond. Structure 16-13a is the s-cis conformation and Structure 16-13b is the s-trans conformation for 1,3-butadiene. The s-trans conformation is more thermodynamically stable than the s-cis conformation and the molecule exists primarily in the s-trans conformation. However, only the s-cis structure has the appropriate orientation for the orbitals on the terminal carbon atoms to overlap with the orbitals of a dienophile in a D–A reaction. Overlapping orbital lobes of the same sign can result in bond formation. Although the central single bond in the diene has some pi character, rotation around this bond can occur readily at room temperature. The s-trans conformation rotates into the s-cis conformation and D–A reactions can occur.

DIENE STEREOCHEMISTRY

Not all dienes can achieve a planar s-cis conformation, and these dienes cannot undergo D–A reactions. 2,4-Hexadiene can exist in a cis,cis (Structure 16-13c), a trans,cis (Structure 16-13e), and a trans,trans (Structure 16-13g) configuration about the double bonds. (It is more correct to use E and Z nomenclature here, but structures are easier to visualize using cis and trans terms.) Rotation around the central single bond in Structure 16-13c results in an s-cis-cis,cis conformation (Structure 16-13d). The steric strain between the methyl groups (represented by circles in Structure 16-13d) is sufficiently large to prevent this conformation from achieving the necessary planar orientation to undergo D–A reactions. s-Trans-trans,cis Structure 16-13e undergoes rotation about the central single bond to form the s-cis-trans,cis conformation 16-13f. The H–CH$_3$ strain interaction is not great enough to prevent this compound from achieving a planar structure and undergoing D–A reactions. The s-trans-trans,trans compound (Structure16-13g) undergoes rotation to give s-cis structure 16-13h, which readily undergoes D–A reactions. Dienes like 16-13i are locked into an s-trans conformation and cannot undergo D–A reactions.

QUESTION 16-10

Is s-*trans-trans,trans*-2,4-octadiene capable of acting as a diene in a D–A reaction?

ANSWER 16-10

Yes, rotation can occur to give the planar s-cis orientation since the two hydrogen atoms can exist in the same plane. This is similar to the explanation for the reaction of compound 16–13e.

SYN STEREOCHEMISTRY

The stereochemistry of the D–A cyclohexene product is dependent on the stereochemistry of the diene and the dienophile. The s-cis conformation of *trans,cis*-2,4-hexadiene (16-14a) reacts with a dienophile, *cis*-2-butenedioic acid (maleic acid, 16-14b), to give *trans*-3,6-dimethylcyclohexene-*cis*-4,5-dicarboxylic acid (16-14c), as shown in Fig. 16-14. The s-cis conformation of *trans,trans*-2,4-hexadiene (16-14d) reacts with 16-14b to give *cis*-3,6-dimethylcyclohexene-*cis*-4,5-dicarboxylic acid (16-14e). The acid groups that are cis in maleic acid remain cis in the cyclohexene product. *The stereochemistry in D–A reactions results from syn addition of the dienophile to the diene.*

QUESTION 16-11

Does *trans,cis*-2,4-hexadiene react with ethylene to give *cis-* or *trans-* 3,6-dimethylcyclohexene? Look at the structures of the dienes in Fig. 16-14 for suggestions to answer this question.

ANSWER 16-11

trans-3,6-Dimethylcyclohexene. The trans, cis diene gives the trans dimethyl product while the trans, trans diene gives the cis dimethyl product.

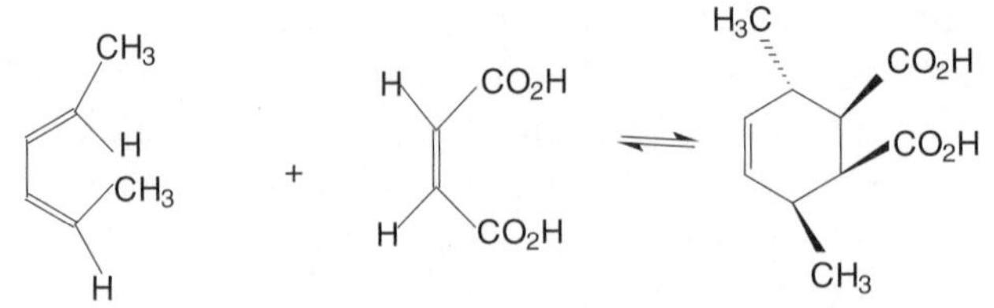

Fig. 16-14. Stereochemistry of D–A reactions.

CONSERVATION OF ORBITAL SYMMETRY IN D–A REACTIONS

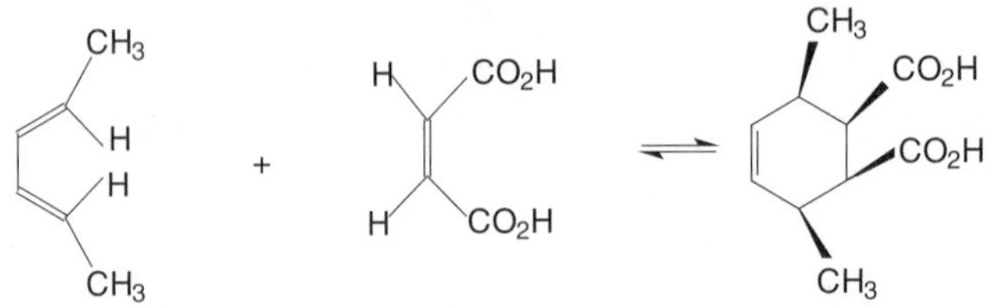

The stereochemistry of syn addition observed in D–A reactions can be explained by considering the MOs involved in the cycloaddition reaction. The electron-rich diene donates electrons to the electron-poor dienophile. The electrons from the diene will come from the highest energy MO that contains (is occupied by) pi electrons. The *h*ighest *o*ccupied *m*olecular *o*rbital is called the HOMO orbital. The dienophile will accept the electrons in its lowest energy orbital that does not contain (is unoccupied by) electrons. This is its *l*owest *u*noccupied *m*olecular *o*rbital, or its LUMO orbital. A D–A cycloaddition reaction will

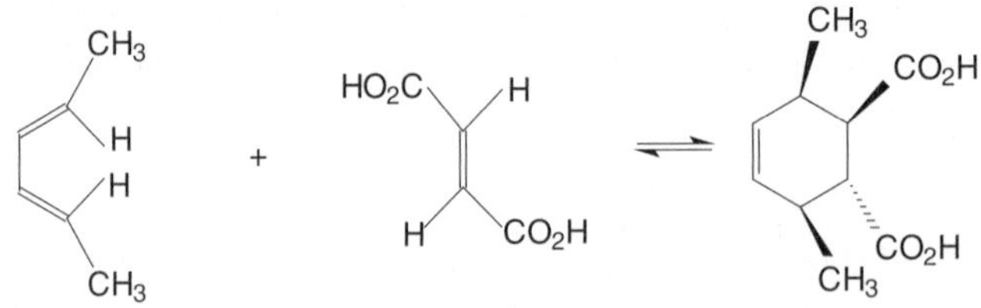

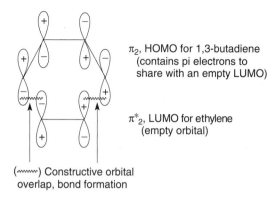

π_2, HOMO for 1,3-butadiene
(contains pi electrons to
share with an empty LUMO)

π^*_2, LUMO for ethylene
(empty orbital)

(〰〰) Constructive orbital
overlap, bond formation

Fig. 16-15.   MO description of D–A reactions.

occur, if overlapping (bond-forming) orbitals of the diene and dienophile have the same sign ($+$ or $-$).

Figure 16-15 shows the HOMO for 1,3-butadiene, π_2. (See Fig. 16-4 for a review of the π MOs for butadiene and ethylene.) The LUMO for ethylene is also shown in Fig. 16-15. Only the sign of the terminal (frontier) orbitals of the HOMO and LUMO need be considered since they overlap to form the new sigma bonds. When these orbitals approach each other in a syn manner, constructive overlap occurs and bond formation takes place. The stereochemistry of the starting materials is maintained in the products. The cis acid groups in maleic acid remain cis in the resulting cyclohexene product. The trans acid groups in fumaric acid (16-14f) remain trans in the resulting cyclohexene product, as shown in Fig. 16-14. A more detailed discussion of the stereochemistry of these reactions is found in most organic textbooks.

QUESTION 16-12
Which MO is higher in energy in butadiene, the HOMO or LUMO?

ANSWER 16-12
The LUMO.

Endo and exo addition reactions

The positions of atoms or groups bonded to bicyclic (two-ring) compounds are identified as *endo* and *exo* positions. Adamantane, shown in Fig. 16-16, shows two hydrogen atoms, H_a and H_b. The bicyclic molecule has two bridgehead carbon atoms and two bridges, a one-carbon bridge and a two-carbon bridge. The exo and endo positions of the hydrogen atoms are named relative to the

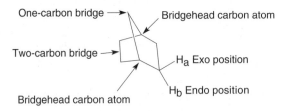

Fig. 16-16. Endo and exo positions in adamantane.

largest bridge (or the largest ring). The endo (meaning inside) hydrogen, H_b, fits into the "pocket" of the six-membered ring (that contains the two-carbon bridge). The exo (outside) hydrogen, H_a, is anti to the largest bridge. The exo hydrogen fits into the "pocket" of the five-membered ring (that contains the one-carbon bridge).

The endo rule

Dienophiles containing pi electrons that are *not* involved in the D–A cycloaddition reaction can interact with a diene according to the *endo rule*. This interaction is shown in Fig. 16-17. The pi electrons in the nonreacting carbonyl $(C = O)$ double bond of the dienophile overlap with the p orbitals on C-2 and C-3 of a 1,3-diene. This interaction positions the dienophile "under" the diene resulting in formation of the endo product.

QUESTION 16-13
In a bicyclic compound, is the endo position closest to the face of the largest ring or the smallest ring?

ANSWER 16-13
The endo group is closest to the face of the largest ring.

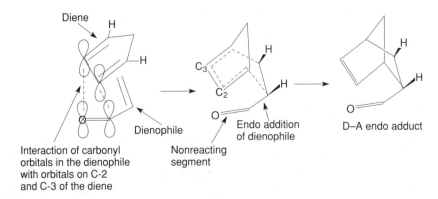

Fig. 16-17. The endo rule.

Quiz

1. Conjugated dienes contain
    (a) only two double bonds
    (b) two double bonds connected to the same carbon atom
    (c) alternating single and double bonds
    (d) two double bonds separated by two single bonds

2. Conjugated double bonds _____ nonconjugated double bonds.
    (a) have greater stability than
    (b) have less stability than
    (c) have no relationship to the stability of
    (d) have equal stability to

3. Resonance energy results from
    (a) sharing electrons
    (b) overlapping p orbitals
    (c) conjugation
    (d) all of the above

4. sp^2 orbitals form stronger bonds than sp^3 orbitals because the orbitals
    (a) hold electrons further from the nucleus
    (b) have more p character
    (c) have more s character
    (d) are not hybridized

5. Bonding MOs are _____ the AOs they are made from.
    (a) more stable than
    (b) less stable than
    (c) equal in stablity to

6. MOs are classified as
    (a) nonbonding
    (b) antibonding
    (c) bonding
    (d) all of the above

7. Nonbonding MOs _____ antibonding MOs.
    (a) are equal in stability to
    (b) are less stable than
    (c) are more stable than
    (d) have no energy relationship with

8. The lowest energy MO has ——— nodes.
   (a) zero
   (b) one
   (c) two
   (d) three

9. The s-cis structure refers to
   (a) configuration around a double bond
   (b) conformation around a single bond
   (c) configuration around an alkene bond
   (d) configuration around an alkyne bond

10. The addition of 1 mol of HBr to 1 mol of 1,3-butadiene gives ——— constitutional isomers.
    (a) zero
    (b) one
    (c) two
    (d) four

11. A primary allylic carbocation is less stable than a ——— carbocation.
    (a) methyl
    (b) vinyl
    (c) secondary alkyl
    (d) tertiary alkyl

12. Kinetic control refers to
    (a) formation of the most stable product
    (b) the stability of reactants relative to products
    (c) how fast a product forms
    (d) the amount of potential energy an atom has

13. Thermodynamic control refers to
    (a) how fast a product forms
    (b) the amount of kinetic energy an atom has
    (c) formation of the most stable (lowest energy) product
    (d) control of the movement of heat

14. An allyl radical has
    (a) an unpaired electron localized on one atom
    (b) an unpaired electron delocalized on two atoms
    (c) a positive charge localized on one atom
    (d) a negative charge delocalized on two atoms

15.   A Diels–Alder reaction is called a (4+2) cycloaddition reaction because
(a) four bonds react with two bonds
(b) four molecules react with two molecules
(c) four pi electrons interact with two pi electrons
(d) four pi electrons interact with two sigma electrons

16.   The ———— requires an s-cis conformation to undergo a Diels–Alder reaction.
(a) diene
(b) dienophile
(c) monoene
(d) allyl carbocation

17.   The Diels–Alder reaction is
(a) isotopic
(b) stereophonic
(c) stereospecific
(d) isothermal

18.   HOMO refers to
(a) an unemployed person
(b) high-order molecular orbital
(c) highest occupied molecular orbital
(d) a homosapien

19.   LUMO refers to
(a) luxurious sedan driven by a chauffeur
(b) lowest unoccupied molecular orbital
(c) a rock band
(d) lowest unimolecular orbital

20.   An exo position refers to
(a) annex a position
(b) being outside
(c) being inside
(d) a position that no longer exists

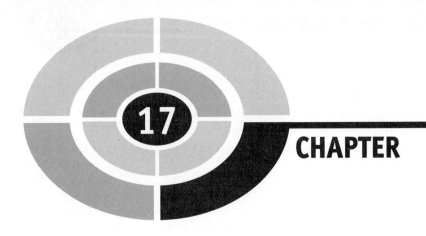

CHAPTER

Aromatic Compounds

Introduction

Aromatic compounds are conjugated, unsaturated cyclic compounds that have a much greater stability than their acyclic conjugated analogs. They also undergo different chemical reactions than their alkene analogs. The best-known aromatic compound is benzene. Benzene, C_6H_6, has a low hydrogen-to-carbon ratio (1:1) in comparison to monoalkenes (2:1). Some of the first benzenoid (benzene-like) compounds were isolated from natural products. These compounds had pleasant aromas and were called aromatic compounds. Not all aromatic compounds have pleasant aromas and several are very toxic. Aromatic compounds are called *arenes*. Arenes are called *aryl (Ar-) groups* when present as substituents.

Reactivity of Aromatic Compounds

Aromatic compounds are cyclic compounds with alternating (conjugated) double and single carbon–carbon bonds. They do not undergo many of the reactions characteristic of alkenes. Alkenes react readily with bromine to form dibromides and are oxidized with permanganate ion. (For a review of alkene reactivity, see the section entitled *Reaction of Alkenes* in Chapter 8.) Benzene does not react with bromine (without a catalyst) or with permanganate ion under mild conditions.

Nomenclature

Many benzene derivatives have common names as well as formal names. Figure 17-1 shows the structures and names of benzene and substituted benzene compounds. It is not necessary to state the location of the substituent in a mono-substituted benzene. The single substituent is always on carbon atom number one (C–1). Positions of the substituents must be stated in polysubstituted (more than one substituent) benzenes.

The positions of the substituents in disubstituted benzenes can be on carbon atoms C–1 and C–2, C–1 and C–3, C–1 and C–4. Examples are shown in Fig. 17-2. The 2, 3, and 4 positions are also called the *ortho* (*o*), *meta* (*m*), and *para* (*p*) positions relative to C–1. If there are three or more substituents, numbers must be used to indicate points of attachment to the benzene ring. The substituents are named in alphabetical order, not numerical order.

Numbers are assigned to the substituents to give the lowest number combination. Figure 17-2 gives an example of a trisubstituted benzene, assigning

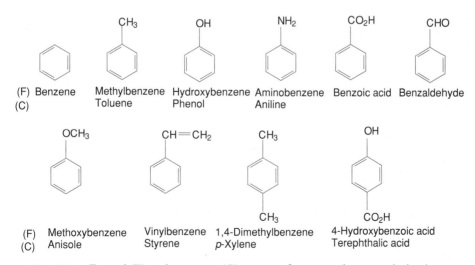

Fig. 17-1. Formal (F) and common (C) names of common benzene derivatives.

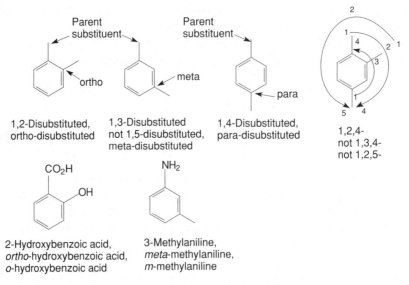

Fig. 17-2.   Disubstituted benzenes.

substituents as 1,2,4-, 1,3,4-, and 1,2,5-. The 1,2,4- positions give the lowest number combination to the substituents.

QUESTION 17-1
Give an alternate name for 1,3-dimethylbenzene.

ANSWER 17-1
meta-Dimethylbenzene or *m*-dimethylbenzene.

   If common names are used, the substituent the common name is based on is always attached to carbon atom number one (C–1). The position of additional substituents is relative to C–1. For example, the –NH$_2$ group of aniline is on C–1 and derivatives of aniline (e.g., 3-methylaniline, *meta*-methylaniline, or *m*-methylaniline) will be assigned positions relative to the amine (nitrogen) group. The structure of *m*-methylaniline is shown in Fig. 17-2.

QUESTION 17-2
On what numbered carbon atom is the hydroxyl group attached in *p*-hydroxybenzoic acid?

ANSWER 17-2
On carbon atom number 4 (C–4).

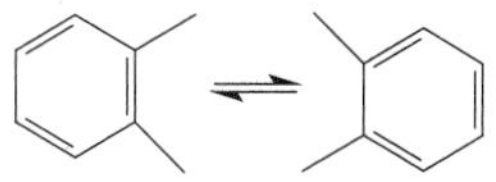

17-3a   17-3b

Fig. 17-3. "Nonequivalent" disubstituted benzene molecules.

Kekulé Structures

Kekulé (a famous chemist who visualized the cyclic structure of benzene as a snake biting its tail) proposed two structures for benzene, 17-3a and 17-3b shown in Fig. 17-3. The two molecules were suggested to be in *equilibrium* with each other. Both structures have alternating single and double bonds. Each molecule represents a different 1,2-disubstituted compound. The two substituents could be separated by a double bond (Structure 17-3a) or a single bond (Structure 17-3b). However, only one 1,2-disubstituted benzene has ever been observed for a given molecule. This led to the conclusion that the two *Kekulé structures* are resonance forms, shown in Fig. 17-4.

The actual molecule is a hybrid of the two resonance structures and not two structures that are in equilibrium. Experimental studies have shown that the length of all carbon–carbon bonds in benzene is 139.7 pm. This length is intermediate between the length of a single bond (148 pm) and a double bond (135 pm) in a model compound, 1,3-butadiene, $H_2C=CH-CH=CH_2$.

QUESTION 17-3
Why are the Kekulé equilibrium structures not valid for (vicinal) 1,2-dimethylbenzene?

ANSWER 17-3
One structure would be a 1,2-dimethylalkane and the other a 1,2-dimethylalkene. They would be expected to have different properties, but only one 1,2-dimethylbenzene has ever been obtained.

The π electrons in benzene are sometimes drawn as a ring (Structure 17-4a). It is more common, however, for benzene to be represented by one of the

17-4a   17-4b   17-4c

Fig. 17-4. Resonance forms of benzene.

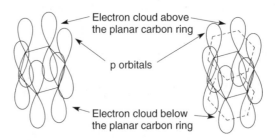

Fig. 17-5. Overlapping p orbitals in benzene.

resonance structures, 17-4b or 17-4c. This text will show the double bonds, as it is easier this way to demonstrate π electron movement in chemical reactions.

PI ELECTRON CLOUDS

Each carbon atom in the benzene ring is sp^2 hybridized and contains one unhybridized p orbital. The *ring is planar* (flat) and the p *orbitals are perpendicular* to the ring. The p orbitals are parallel to each other and are close enough to overlap, forming a continuous p-orbital electron cloud above and below the plane of carbon atoms, as shown in Fig. 17-5. The p orbitals do overlap, although this is not shown in these structures. The dashed lines indicate orbital overlap. Resonance, delocalization of (sharing) electrons over several atoms, stabilizes molecules and is especially important in explaining the stability of benzene. (Resonance is discussed in the section entitled *Conjugated Systems* in Chapter 16.)

QUESTION 17-4
Why aren't nonplanar conjugated rings aromatic?

ANSWER 17-4
If adjacent p orbitals are not in the same plane and parallel to each other, they will not overlap sufficiently to give a continuous electron cloud.

Stability of Benzene

The stability of benzene is demonstrated by comparing it to cyclohexene, 1,3-cyclohexadiene, 1,4-cyclohexadiene, and theoretical "cyclohexatriene." *Hydrogenation* (addition of hydrogen) of each of these unsaturated cyclohexenes gives the same product, cyclohexane. The amount of heat given off in each hydrogenation reaction is a measure of the potential energy in each of the starting materials. The heats of hydrogenation are shown in Fig. 17-6. Heats of hydrogenation are shown as positive numbers for this discussion while the values are

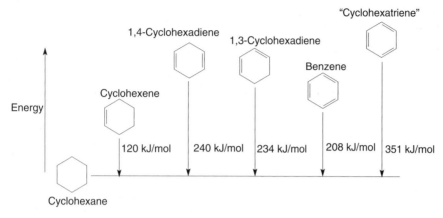

Fig. 17-6.   Heats of hydrogenation of cyclohexenes.

usually shown as negative numbers since the potential energy of the products is less than that of the reactants.

The *heat of hydrogenation* of cylcohexene is 120 kJ/mol. The heat of hydrogenation of nonconjugated 1,4-cyclohexadiene is 240 kJ/mol, twice the value for that of cyclohexene, as expected. The heat of hydrogenation of conjugated 1,3-cyclohexadiene is 234 kJ/mol. Conjugation makes this molecule a little more stable (6 kJ/mol) than the nonconjugated cyclic diene. The heat of hydrogenation of benzene is 208 kJ/mol. The heat of hydrogenation of three nonconjugated double bonds would be 360 kJ/mol. Benzene is 152 kJ/mol more stable than three isolated double bonds. It is better to compare benzene to a conjugated system. If we take the heat of hydrogenation as 1.5 times that of conjugated 1,3-cyclohexadiene (the equivalent of three conjugated double bonds), the heat of hydrogenation would be 351 kJ/mol. Benzene is still 143 kJ/mol more stable than the theoretical conjugated cyclohexatriene. This comparison quantitatively shows the unexpected stability of benzene relative to other 6-membered cycloalkenes.

When benzene is hydrogenated to give conjugated 1,3-cyclohexadiene (one double bond is hydrogenated) the resulting diene is higher in energy by 26 kJ/mol. Usually hydrogenation of a compound makes the resulting compound more stable, as in the other examples above, not less stable.

QUESTION 17-5
Why is it thermodynamically unfavorable to form 5,6-dibromo-1,3- cyclohexadiene from benzene?

ANSWER 17-5
The dibromo compound is no longer aromatic and is higher in energy than is the benzene starting material.

The Resonance Model

Resonance (delocalization of π electrons) is expected to occur for molecules with conjugated double or triple bonds. Benzene is an example of a conjugated system containing three double bonds. Molecules can show conjugation if three or more adjacent p orbitals overlap. The p orbitals must be in the same plane for maximum overlapping. Benzene is a planar molecule with p orbitals on all adjacent carbon atoms. All p orbitals are perpendicular to the ring carbon atoms and are parallel to each other. This allows all the p orbitals to overlap.

QUESTION 17-6
Would you expect 1,4-cyclohexadiene to be aromatic?

ANSWER 17-6
No, two carbon atoms in the ring are sp^3 hybridized and contain no p orbital so there is not a continuous ring of p orbitals.

ANNULENES

Resonance theory predicts that planar ring compounds with alternating double and single bonds should have an unexpectedly large stability, as is observed for benzene. Rings with alternating double and single bonds are called *annulenes*. Ring size is indicated by the number in a bracket, [ ], preceding the word annulene. Examples are shown in Fig. 17-7. 1,3-Cyclobutadiene (17-7a) is called [4]-annulene. Benzene (17-7b) is called [6]-annulene. 1,3,5,7-Cyclooctatetraene (17-7c) is called [8]-annulene. The smaller annulene rings are usually called by systemic or common names.

1,3-Cyclobutadiene (17-7a) is a cyclic molecule with p orbitals on each adjacent carbon atom. 1,3,5,7-Cyclooctatetraene (17-7c) is also a cyclic molecule with a p orbital on all adjacent carbon atoms. Increasing the ring size by two atoms gives 1,3,5,7,9-cyclodecapentaene, with a p orbital on each carbon atom in the ring. Resonance theory suggests that these three molecules should have

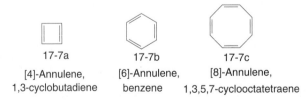

Fig. 17-7.   Annulenes.

the same unexpected stability observed for benzene. However, 1,3-Cyclobutadiene is a very reactive molecule and is too unstable to be isolated. The double bonds in 1,3,5,7-cyclooctatetraene undergo reactions more typical of alkenes rather than benzene. 1,3,5,7,9-Cyclodecapentaene also undergoes reactions more characteristic of alkenes. Is benzene the only molecule with this unusual stability? Some larger rings do show increased stability, or aromatic character, and are discussed below.

Molecular Orbital Description of Aromaticity

BENZENE

Since p orbitals are involved in π bonding, we will consider only the molecular orbitals (MOs) formed by the mathematical combination of the wave functions for p atomic orbitals (AOs). Six p AOs in benzene combine to form six MOs. These orbitals are shown in Fig. 17-8. The lowest energy MO (π_1) results from the constructive (all lobes have the same sign) overlapping of all six p orbitals. This MO has no *nodes* (areas of zero electron density). The next two higher energy MOs (π_2, π_3) are degenerate (of equal energy) and each has one node. There are two ways of symmetrically combining p orbitals that result in one node and hence there are two degenerate MOs. As in the case of the allyl radical, a node exists at the position of an atom. It may seem strange to see a node centered at a carbon atom, but remember, this is a mathematical description of orbital electron density. The next higher energy MOs (π_4*, π_5*) consist of two more degenerate orbitals and each has two nodes. The highest energy level MO (π_6*) has three nodes.

QUESTION 17-7
How does the number of nodes in MOs change with increasing MO energy levels? (Consider the example of benzene.)

ANSWER 17-7
Each successive energy level has one additional node.

Orbital energy levels

Molecular orbital energy levels for benzene, including the six π electrons, are shown in Fig. 17-8. Three of the MOs (π_1, π_2, and π_3) are bonding MOs and three are antibonding MOs (π_4*, π_5*, and π_6*). The bonding MOs are lower in energy than are the (six degenerate) p AOs and the antibonding MOs are higher in energy than are the p AOs. Adding the six π electrons to the MOs using the

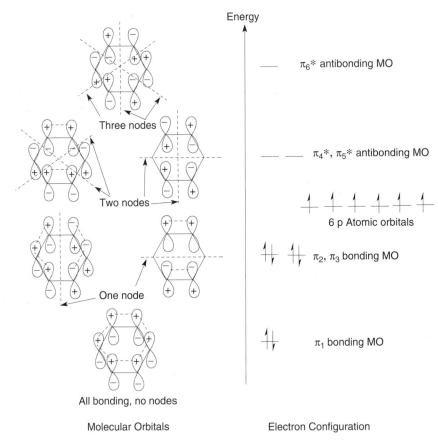

Energy

π₆* antibonding MO

π₄*, π₅* antibonding MO

6 p Atomic orbitals

π₂, π₃ bonding MO

π₁ bonding MO

Three nodes

Two nodes

One node

All bonding, no nodes

Molecular Orbitals   Electron Configuration

Fig. 17-8. Molecular orbital description of benzene.

aufbau principle gives an electronic configuration with all the electrons paired and in bonding MOs (as shown in Fig. 17-8). The MO model predicts benzene will be more stable (lower in energy) than the resonance model where electrons are shared in the six p AOs.

1,3-CYCLOBUTADIENE

The four p AOs of 1,3-cyclobutadiene combine to form four MOs as shown in Fig. 17-9. The lowest energy MO is a bonding MO (π_1). The next two higher energy MOs (π_2 and π_3) are degenerate MOs and are nonbonding MOs. The highest energy MO (π_4*) is antibonding. The two degenerate MOs have the same energy as the p AOs. Since these MOs do not contribute to bonding or antibonding, they are called nonbonding MOs. Adding the four π electrons to

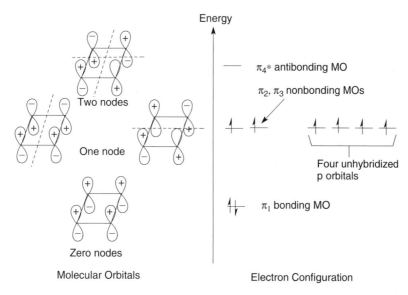

Fig. 17-9. Molecular orbital description of 1,3-cyclobutadiene.

the MOs by the aufbau principle (shown in Fig. 17-9) results in one pair of electrons in the lowest energy MO and one unpaired electron in each of the degenerate nonbonding orbitals. Molecules containing nonpaired electrons are very unstable and very reactive. This is consistent with the observation that 1,3-cyclobutadiene is unstable and very reactive. Molecules with one (or more) unpaired electrons are called (free) *radicals*.

1,3,5,7-CYCLOOCTATETRAENE

If 1,3,5,7-cyclooctatetraene were a planar molecule and all the double bonds had a cis configuration, each internal carbon–carbon bond angle would be 135°. The ideal angle for sp²–hybridized carbon orbitals is 120° A lower energy conformation results when the ring distorts to form a boat shape. Figure 17-10 shows the all-cis planar structure (17-10a) and the boat structure (17-10b). The boat shape, the preferred (most stable) structure, does not allow for p orbital overlap on all adjacent carbon atoms.

If 1,3,5,7-cyclooctatetraene were to assume a trans structure (17-10c) around two of the double bonds, all bond angles would be closer to 120°. However, Structure 17-10c cannot assume a planar structure, since the electron clouds of the two internal hydrogen atoms (shown circled in Structure 17-l0c) would potentially occupy much of the same space. The ring distorts from planarity to relieve the steric interaction between these two hydrogen atoms. The ring

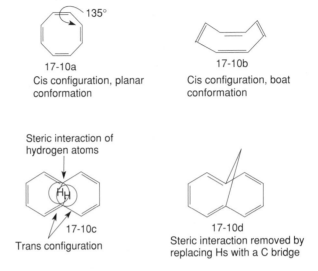

Fig. 17-10. 1,3,5,7-Cyclooctatetraene and a bridged analog.

distortion reduces ring planarity and therefore p orbital overlap and aromatic character. An analog of Structure 17-10c, Structure 17-10d, was prepared in which the two hydrogen atoms (that cause the steric strain interaction) were removed and replaced by a carbon atom bridge. The bridge carbon atom is above the plane of the molecule. The resulting conjugated pentaene system can now assume a planar conformation, the p orbitals overlap, and this molecule shows aromatic character (unexpected stability).

FROST CIRCLES

Relative MO energy levels of cyclic conjugated polyenes can be predicted by using *Frost circles (also called the polygon rule)*. The vertices in the polygon represent the carbon atoms in the cyclic polyene. A polygon is inscribed in a circle such that one of the vertices is at the bottom of the circle. The height of each of the vertices indicates the relative energy levels of the MOs. A horizontal line drawn through the center of the circle corresponds to the energy of the p AOs used to make the MOs. Examples using the Frost circle method to show the relative energy of MOs of conjugated four-, six-, and eight-membered rings are shown in Fig. 17-11.

Electrons are added to each MO using the aufbau principle. The resulting electron configuration is used to predict aromatic or antiaromatic character of the molecule. The electronic structure of 1,3-cyclobutadiene (17-11a) contains one pair of bonding electrons and two unpaired electrons, indicating an unstable,

Antibonding MO
Nonbonding MO
Bonding MO

17-11a
1,3-Cyclobutadiene

} Antibonding MOs
Nonbonding MOs
} Bonding MOs

17-11b
Benzene

17-11c
1,3,5,7-Cyclooctatetraene

Fig. 17-11.   Frost Circles.

reactive molecule. Benzene (17-11b) has six π electrons, all in bonding MOs, indicating a more stable molecule than the molecule with only AOs. 1,3,5,7-Cyclooctatetraene (17-11c) has an unpaired electron in each of two nonbonding orbitals. The unpaired electrons suggest a reactive molecule that does not have the stability of an aromatic molecule, as is observed for 1,3,5,7-cyclooctatetraene.

Properties of Aromatic, Nonaromatic, and Antiaromatic Compounds

Cyclic conjugated molecules (with alternating single and double bonds) that are *more stable* than their acyclic (noncyclic) counterparts are called aromatic compounds. An example is aromatic benzene and nonaromatic acyclic 1,3,5-hexatriene. Cyclic conjugated molecules that are *less stable* than their acyclic counterparts are called antiaromatic compounds. Cyclic compounds that do not have p orbitals on every ring atom cannot be aromatic or antiaromatic and are called nonaromatic. Acyclic compounds are also nonaromatic. Examples of representative aromatic, nonaromatic, and antiaromatic molecules are shown in Fig. 17-12.

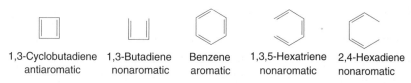

| 1,3-Cyclobutadiene | 1,3-Butadiene | Benzene | 1,3,5-Hexatriene | 2,4-Hexadiene |
| antiaromatic | nonaromatic | aromatic | nonaromatic | nonaromatic |

Fig. 17-12.   Aromatic, antiaromatic, and nonaromatic molecules.

QUESTION 17-8
Conjugated 1,3-hexadiene is more stable than nonconjugated 1,4-hexadiene. Is 1,3-hexadiene aromatic?

ANSWER 17-8
No, it is not cyclic. It would not be aromatic if the molecule were cyclic (a six-membered ring) since not all carbon atoms would contain a p orbital.

An analysis of the properties of the conjugated molecules discussed above, and many other molecules, leads to the conclusion that the following properties are required to achieve aromaticity.

(a) The molecule must be cyclic.
(b) Each atom in the ring has to have at least one p orbital.
(c) The molecule has to be planar or nearly so that p orbitals overlap.
(d) The p orbitals overlap to form a continuous electron cloud of π electrons above and below the planar ring atoms.
(e) The π electron cloud must contain an odd number of *pairs* of electrons.

Hückel's Rule

Hückel's rule states that planar monocyclic compounds containing one p orbital on each ring atom show the properties of aromatic compounds if the ring contains $(4n + 2)\ \pi$ electrons. The value of $n = 0$ or an integer, 1, 2, 3, etc. Rings that contain $4n$ pi electrons will not be aromatic. 1,3-Cyclobutadiene contains $4\ \pi$ electrons where $n = 1$ and is antiaromatic. 1,3,5,7-Cyclooctatetraene contains $8\ \pi$ electrons and is also predicted to be antiaromatic with $4n$ electrons where $n = 2$. Benzene has six π electrons and is predicted to be aromatic with $(4n + 2)$ electrons where $n = 1$.

QUESTION 17-9
Would you predict 1,3,5,7,9,11,13-cyclotetradecapentaene to be aromatic?

ANSWER 17-9
Hückel's rule would predict aromaticity, if the ring were planar. There are $14\ \pi$ electrons and $n = 3$.

Although Hückel's rule was developed for monocyclic molecules, it also predicts aromaticity in polycyclic molecules. It can also be used to predict aromaticity in *heterocyclic* (containing a ring atom other than carbon) molecules and in ionic molecules.

Heterocyclic Compounds

PYRIDINE

Three aromatic nitrogen-containing compounds are shown in Fig. 17-13. Each ring atom in *pyridine* (17-13a), including the nitrogen atom, is sp^2 hybridized and has one p orbital perpendicular to the planar ring. Pyridine has six $(4n + 2, n = 1)$ π electrons. Pyridine is aromatic and has a *resonance energy* of 113 kJ/mol. Like benzene, it undergoes substitution reactions rather than addition reactions. It has a pair of nonbonding electrons in a nitrogen sp^2 orbital. These electrons are not part of the π ring system and are available for sharing with electrophiles such as a proton. Pyridine has the properties of a Lewis base.

PYRROLE

Pyrrole (17-13b) also shows aromatic character with a resonance energy of 92 kJ/mol. It might appear that the ring nitrogen atom is sp^3 hybridized as it has four electron domains. It is bonded to two carbon atoms and a hydrogen atom, and contains a nonbonding electron pair. If the nitrogen atom was sp^2 hybridized, the nonbonding electrons would be in the nonhybridized p orbital. The p orbital on the nitrogen atom forms a continuous five-membered ring of p orbitals with the four p orbitals on the four carbon atoms. The ring contains six π electrons $(4n + 2, n = 1)$ and has aromatic character. The nonbonding electrons are part of the π ring system and not available for bonding to an electrophile. The molecule does not act as a Lewis base.

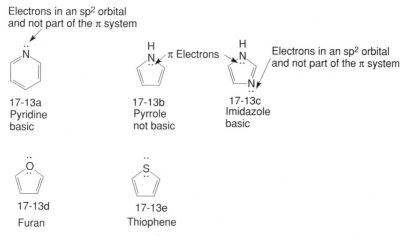

17-13a
Pyridine
basic

17-13b
Pyrrole
not basic

17-13c
Imidazole
basic

17-13d
Furan

17-13e
Thiophene

Fig. 17-13. Aromatic heterocycles.

IMIDAZOLE

Imidazole (17-13c) contains two nitrogen atoms in a five-membered ring. Both nitrogen atoms contain a pair of nonbonding electrons. Both nitrogen atoms are sp^2 hybridized. The electron pair on the nitrogen atom (that is bonded to the hydrogen atom) is in a p orbital and is part of the aromatic ring system. These electrons are not available for bonding with an electrophile and this nitrogen atom is not basic (as was the case in pyrrole).

The other nitrogen atom in imidazole (that is not bonded to a hydrogen atom) contains a nonbonding electron pair. This electron pair is in an sp^2 orbital (not a p orbital), and like pyridine, is available for sharing with an electrophile. Imidazole contains one basic and one nonbasic nitrogen atom. The π system in imidazole contains six π electrons and the molecule shows aromatic character.

FURAN

Furan (17-13d) shows aromaticity with 76 kJ/mol of resonance energy. Structure 17-3d shows the oxygen atom with two pairs of nonbonding electrons. If the oxygen atom is sp^2 hybridized, one of the electron pairs is in a p orbital and part of the π ring system. The π ring system now contains six ($4n + 2$, $n = 1$) π electrons. The second nonbonding electron pair is in an sp^2 orbital. If the oxygen atom were sp^3 hybridized, none of its nonbonding electrons would be part of the π ring system. It is more energetically favorable for the oxygen atom to be sp^2 hybridized, fulfilling the requirements for aromatic character.

THIOPHENE

Thiophene (17-13e) shows 121 kJ/mol of resonance energy. Structure 17-13e shows the sulfur atom with two pairs of nonbonding electrons. Even though sulfur's valence electrons are in principle energy level 3, an unhybridized 3p orbital still overlaps sufficiently with the 2p orbitals on the ring carbon atoms to form a six ($4n + 2$, $n = 1$) π electron ring system that shows aromatic character.

Aromatic Ions

CYCLOPROPENYL CATION

Several compounds show aromatic character when an electron pair is removed or an electron pair becomes available to form a conjugated π system. Cyclopropene is nonaromatic as it is not a conjugated molecule with a p orbital on each carbon atom. If 3-chlorocyclopropene (Structure 17-14a) reacts with antimony trichloride, the chloride anion (and its bonding electrons) is removed from the cyclopropene ring, giving a *cyclopropenyl carbocation* (Structure 17-14b). The carbon atom that was bonded to chlorine rehybridizes from sp^3 to sp^2, forming a planar ring with a p orbital on each ring atom. The π ring system contains two $(4n + 2, n = 0)$ π electrons and shows aromatic character.

QUESTION 17-10
Would you predict the cyclopropenyl anion to be aromatic?

ANSWER 17-10
No, it would have four $(4n, n = 1)$ π electrons.

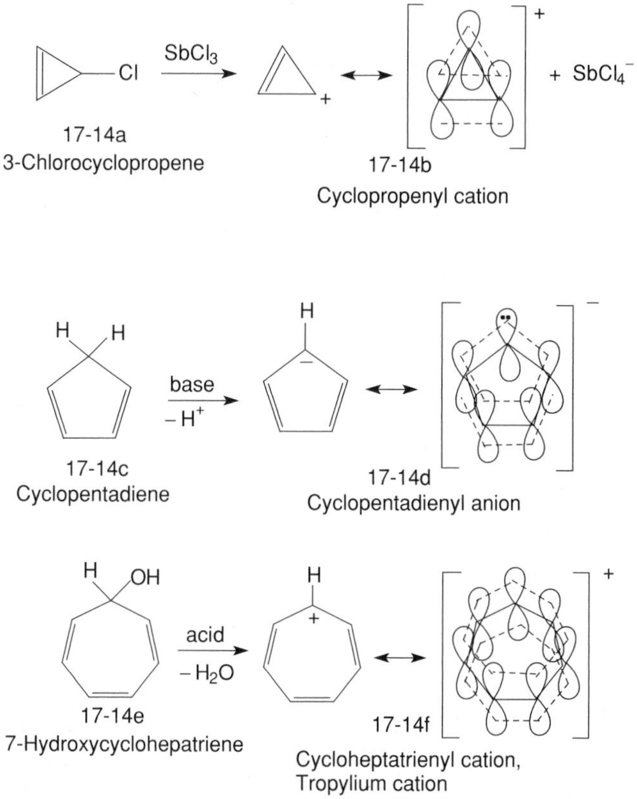

17-14a
3-Chlorocyclopropene

17-14b
Cyclopropenyl cation

17-14c
Cyclopentadiene

17-14d
Cyclopentadienyl anion

17-14e
7-Hydroxycyclohepatriene

17-14f
Cycloheptatrienyl cation,
Tropylium cation

Fig. 17-14.   Aromatic ions.

CYLCOPENTADIENYL ANION

Cyclopentadiene (17-14c) is a nonaromatic compound. When treated with a strong base, a proton is removed giving the *cyclopentadienyl anion* (17-14d). The carbon atom that lost the proton rehybridizes from sp^3 to sp^2 and now each carbon atom in the ring has a p orbital. The π ring system is planar, contains six $(4n + 2, n = 1)$ π electrons, and shows aromatic character. Cyclopentadiene is much more acidic ($pK_a = 16$) than a typical allylic proton because of the stability of the resulting cyclopentadienyl anion, the conjugate base.

QUESTION 17-11
Would you predict the cyclopentadienyl cation to be aromatic?

ANSWER 17-11
No, it would have four $(4n, n = 1)$ π electrons.

CYCLOHEPTATRIENYL CATION

Cycloheptatriene is nonaromatic. It does not contain a conjugated π ring system. Reaction of 7-hydroxycycloheptatriene (17-14e) with acid results in the loss of H_2O and formation of the *cycloheptatrienyl cation (tropylium ion, 17-14f)*. The carbon atom previously bonded to the alcohol group rehybridizes from sp^3 to sp^2 and the seven-membered ring now contains a conjugated π ring system with six $(4n + 2, n = 1)$ π electrons.

Polycyclic Aromatic Compounds

Although Hückel's rule was developed for monocyclic molecules, planar polycyclic molecules containing $(4n + 2)$ electrons also show aromatic character. Structures shown in Fig. 17-15 contain 10 $(n = 2)$ and 14 $(n = 3)$ π

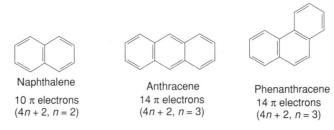

Naphthalene
10 π electrons
$(4n + 2, n = 2)$

Anthracene
14 π electrons
$(4n + 2, n = 3)$

Phenanthracene
14 π electrons
$(4n + 2, n = 3)$

Fig. 17-15. Polycyclic aromatic compounds.

electrons. The molecules are planar, contain a p orbital on each ring atom, and have resonance energies of 252, 351, and 381 kJ/mol, respectively.

Quiz

1. Aromatic compounds are
   (a) alkanes
   (b) linear
   (c) nonconjugated and cyclic
   (d) conjugated and cyclic

2. Ortho refers to groups that are ———— to each other.
   (a) 1,1-
   (b) 1,2-
   (c) 1,3-
   (d) 1,4-

3. The common name for hydroxybenzene is
   (a) aniline
   (b) benzoic acid
   (c) phenol
   (d) fenol

4. All bond lengths in benzene
   (a) are equal
   (b) alternate in length
   (c) are all different lengths
   (d) cannot be measured

5. Benzene is more stable than
   (a) cyclohexene
   (b) cyclohexane
   (c) 1,3-cyclohexadiene
   (d) all of the above

6. Annulenes are
   (a) linear compounds
   (b) conjugated cyclic compounds
   (c) conjugated linear compounds
   (d) annual events

7. The MO description of benzene has ———— different energy levels.
   (a) two

(b) three
(c) four
(d) six

8. 1,3-Cyclobutadiene is
    (a) aromatic
    (b) nonaromatic
    (c) antiaromatic
    (d) nonconjugated

9. 1,3,5,7-Octatetraene is
    (a) a cation
    (b) an anion
    (c) a diradical
    (d) a 10-membered ring

10. A Frost circle is used to predict
    (a) AO energy levels
    (b) condensation temperatures
    (c) MO energy levels
    (d) freezing points

11. Aromatic six-membered rings are ———— nonaromatic six-membered ring.
    (a) more stable than
    (b) less stable than
    (c) equal in stability to

12. Each ring atom in an aromatic compound contains
    (a) two hydrogen atoms
    (b) a p orbital
    (c) an sp^3 orbital
    (d) at least three p orbitals

13. Hückel's rule predicts aromaticity for a conjugated ring with ————
    π electrons.
    (a) one
    (b) three
    (c) four
    (d) six

14. Pyridine is aromatic because its nonbonding electrons
    (a) are included in the π electron system
    (b) are not included in the π electron system

(c) are unpaired

(d) are removed from the molecule

15. The cyclopropenyl cation has ———— π electrons.
    (a) one
    (b) two
    (c) three
    (d) four

16. The cyclopentadienyl anion has ———— π electrons.
    (a) two
    (b) four
    (c) six
    (d) eight

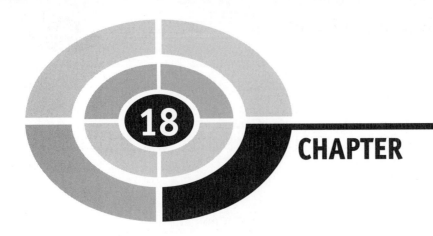

Reactions of Benzene and Other Aromatic Compounds

Introduction

Benzene and other aromatic compounds are conjugated cyclic polyenes. Their chemical reactivity and physical properties differ significantly from linear conjugated and nonconjugated mono- and polyenes. Ethylene and conjugated 1,3-butadiene react readily with bromine or chlorine at room temperature to yield halogen addition products. Benzene reacts with bromine or chlorine, in the presence of a catalyst, to yield a substitution product. *A key point: benzene undergoes substitution reactions, rather than addition reactions, with many reagents.* Substitution reactions include: halogenation, sulfonation, nitration, alkylation, and acylation. These reactions are common to various aromatic compounds. This discussion will be limited to benzene as a model aromatic compound.

Electrophilic Aromatic Substitution

Benzene has a cloud of π electrons above and below the plane of the cyclic carbon atom ring as shown in Structure 18-1a in Fig. 18-1. The π electrons in benzene are attracted to and react with electrophiles (electron-seeking species). Addition of an electrophile to a carbon atom in the benzene ring results in the rehybridization of that carbon atom from sp^2 to sp^3. This species is called a *sigma complex* as the electrophile is bonded to the ring with a sigma bond. (This species is also called an *arenium ion*. The term arenium implies an aromatic ion but this species is not aromatic.)

A carbon atom adjacent to the sp^3 hybridized ring carbon atom now has a formal positive charge (see Structure 18-1b). This carbocation is a resonance-stabilized *allylic carbocation*, shown as Structures 18-1b, 18-1c, and 18-1d. The addition of an electrophile is an endothermic reaction since the stability of the aromatic ring is lost. Loss of the hydrogen atom bonded to the sp^3 hybridized carbon atom (or reversible loss of the electrophile and hence no reaction) regenerates the stable aromatic ring (Structure 18-1e). The net result is that a hydrogen atom on the benzene ring is replaced by an electrophile and thus the name *electrophilic aromatic substitution reaction*.

If benzene were to react with bromine to form the nonaromatic addition product, 5,6-dibromo-1,3-cyclohexadiene, shown in Fig. 18-2, this product would be

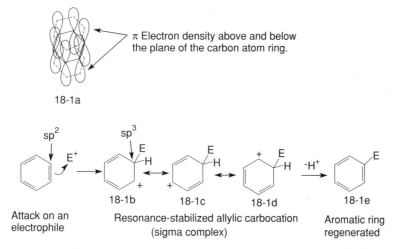

π Electron density above and below the plane of the carbon atom ring.

18-1a

sp^2 sp^3

18-1b   18-1c   18-1d   18-1e

Attack on an electrophile   Resonance-stabilized allylic carbocation (sigma complex)   Aromatic ring regenerated

Fig. 18-1.   Electrophilic aromatic substitution reaction.

Fig. 18-2. Addition of bromine to benzene.

higher in energy than the starting material. This would be a thermodynamically unfavorable endothermic reaction. It is for thermodynamic (energy) reasons that benzene undergoes substitution reactions, where it retains aromaticity (aromatic character), rather than addition reactions where aromaticity is lost.

QUESTION 18-1
Why is benzene unreactive toward addition reactions with chlorine?

ANSWER 18-1
Addition of chlorine to form dichlorocyclohexadiene would be an endothermic reaction and the stability of the aromatic ring would be lost.

HALOGENATION OF BENZENE

Benzene undergoes electrophilic substitution reactions with bromine, chlorine, and iodine in the presence of a catalyst or oxidant. Fluorine is too reactive to use under these reaction conditions and fluorinated derivatives of benzene are made by other methods.

Bromination of benzene

Bromination of benzene is shown in Fig. 18-3. Bromination requires the use of a Lewis acid catalyst such as $FeBr_3$ or $AlBr_3$. The Lewis acid reacts with the halogen to *polarize the halogen–halogen bond*, making one of the bromine atoms more electropositive and a stronger electrophile. The electrons in the π orbitals of the benzene ring are attracted to the electrophilic bromine atom. *A key point: by convention, electrons are always shown moving (indicated by a curved arrow) toward the electrophile.* The electrophile is never shown moving toward electrons. (In reality the most mobile species probably approaches the least mobile species.) The transition state in this bromination reaction represents some degree of bromine–bromine bond breaking and carbon–bromine bond making. When the bromine atom becomes bonded to the carbon atom, the bromine–bromine bond has broken and the intermediate carbocation (Structure 18-3a) is formed. This is an endothermic reaction and the rate-controlling step in the overall reaction.

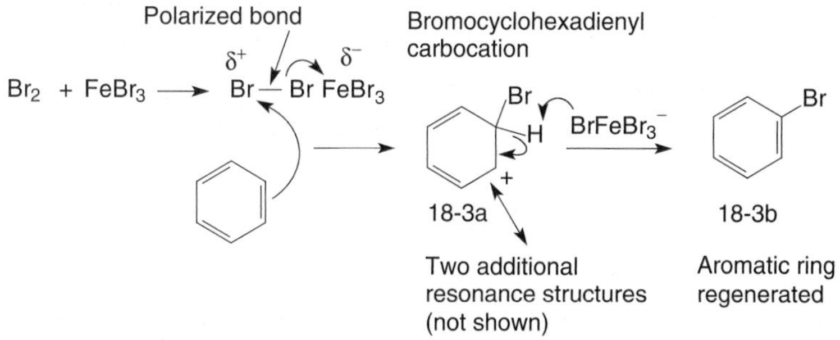

Fig. 18-3. Bromination of benzene.

$$\text{Halogen }(X_2) + \text{Benzene} \xrightarrow[\text{acid}]{\text{Lewis}} \text{Halobenzene}$$

Toolbox 18-1.

QUESTION 18-2
Why is a Lewis acid (electron-pair acceptor) required for the electrophilic substitution reaction of bromine with benzene?

ANSWER 18-2
The Lewis acid polarizes the bromine–bromine bond, making one bromine atom electron-poor and a better electrophile.

The bromocyclohexadienyl carbocation intermediate (18-3a) is an allylic carbocation that is resonance stabilized. However, it is still a very reactive carbocation. The tetrabromoferrate anion ($FeBr_4^-$) acts as a Lewis base to remove the proton from the same carbon atom to which the bromine atom is bonded in the cyclohexadiene ring. This is an exothermic reaction as the stable benzene ring (18-3b) is regenerated.

QUESTION 18-3
Why doesn't an electrophile replace a hydrogen atom by an S_N1 or S_N2 reaction?

ANSWER 18-3
The ring prevents backside S_N2 attack by the electrophile. An S_N1 reaction would involve a benzene carbocation that is very unstable (high in energy).

Chlorination and iodination of benzene

The reaction of chlorine with benzene involves a mechanism analogous to that of the bromination reaction. The Lewis acid catalyst used is commonly $FeCl_3$ or $AlCl_3$. Iodine reacts very slowly when iron and aluminum salt catalysts are

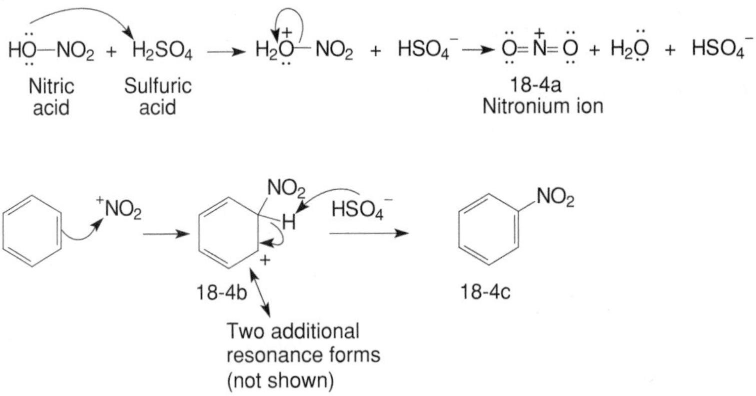

Fig. 18-4. Nitration of benzene.

used. Iodine is oxidized by nitric acid to give the iodonium cation, I^+ (or a highly polarized iodine molecule), which acts as the electrophile.

NITRATION OF BENZENE

Nitration of benzene involves a *nitronium cation* (Structure 18-4a in Fig. 18-4). This species is generated by the reaction between concentrated nitric acid and concentrated sulfuric acid. One Lewis structure for the nitronium ion (18-4a) has a formal positive charge on the nitrogen atom. Benzene reacts with this nitrogen atom, as shown in Fig. 18-4, by a mechanism analogous to that of halogenation reactions. The rate-controlling step is formation of the allylic nitrocyclohexadienyl carbocation (18-4b). Bisulfate anion (HSO_4^-) acts as a Lewis base removing a proton from the carbocation intermediate (18-4b) and regenerating the aromatic ring substituted with the nitro ($-NO_2$) group (Structure 18-4c).

SULFONATION OF BENZENE

Sulfonation of benzene is shown in Fig. 18-5. Fuming sulfuric acid (7% SO_3 dissolved in concentrated H_2SO_4) is used to sulfonate benzene. The electrophile is *either SO_3 or HSO_3^+* (depending upon the reaction conditions). Sulfur trioxide is a powerful electrophile because the sulfur atom is electron deficient, as shown in the resonance structures for SO_3 in Fig. 18-5. Π electrons in the benzene ring bond to SO_3 to form an allylic sulfocyclohexadienyl carbocation (Structure 18-5a). Since aromaticity of the benzene ring is lost, this is an endothermic reaction and the rate-controlling step. Bisulfate anion (HSO_4^-) acts as a base removing a proton from the sulfocyclohexadienyl carbocation (18-5a)

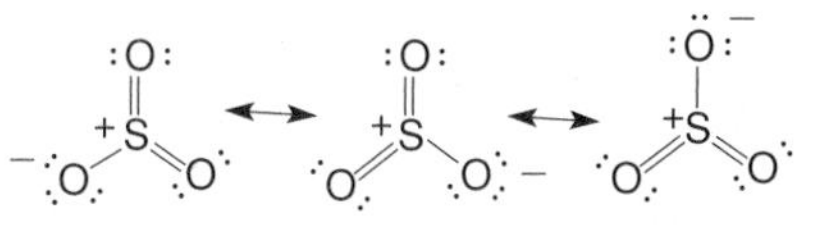

Resonance-stabilized electrophile (sulphur trioxide)

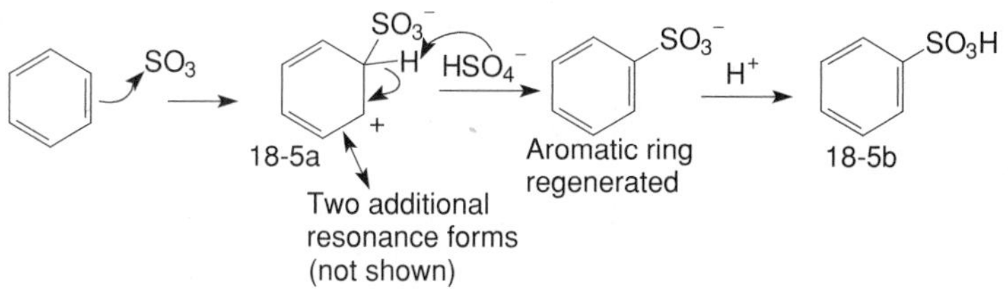

Fig. 18-5. Sulfonation of benzene.

and regenerating the aromatic benzene ring (Structure 18-5b). Since the reaction is carried out in strong acid, the product is benzenesulfonic acid (not the sulfonic acid salt).

| NO_2^+ + Benzene $\xrightarrow{HNO_3}$ Nitrobenzene |
| SO_3 + Benzene $\xrightarrow{H_2SO_4}$ Benzenesulfonic acid |

Toolbox 18-2.

QUESTION 18-4
Since SO_3 is a good electrophile, why isn't SO_2 a good nucleophile?

ANSWER 18-4
The three best resonance structures for SO_3 have a positive charge on the sulfur atom. The best resonance structure for SO_2 does not have a positive charge on the sulfur atom so SO_2 is a poor electrophile.

ALKYLATION OF BENZENE

A *Friedel–Crafts alkylation* reaction (named after the chemists who discovered the reaction) is shown in Fig. 18-6. Benzene reacts with a primary, secondary, or tertiary alkyl halide, RX (X = F, Cl, Br, or I), in the presence of a Lewis acid catalyst (AlX_3 or FeX_3). The reaction between AlX_3 and the alkyl halide forms a carbocation with secondary and tertiary alkyl halides. This is shown in Reaction (a) in Fig. 18-6. Since primary alkyl carbocations are very unstable, a highly polarized AlX_3–alkyl chloride complex is the likely active reagent. Aromatic

(a)   RCl + AlCl$_3$ $\longrightarrow$ R$^+$ + AlCl$_4^-$ (or $\overset{\delta^+}{R}$ $-$ $\overset{\delta^-}{Cl}$AlCl$_3$)

(b) R$^+$ + [benzene] $\longrightarrow$ [alkylcyclohexadienyl carbocation] $\longrightarrow$ [alkylbenzene with R] + H$^+$

Alkylcyclohexadienyl carbocaton

Two additional resonance forms (not shown)

Fig. 18-6.   Friedel–Crafts alkylation reactions.

halides and vinyl halides do not undergo Friedel–Crafts reactions under these reaction conditions.

The reaction between benzene and the alkyl carbocation (or reactive alkyl complex) initially yields the nonaromatic cyclohexadienyl carbocation intermediate shown in Reaction (b) in Fig. 18-6. This species loses a proton to form an aromatic alkylbenzene.

ADDITIONAL ALKYLATION REACTIONS

The carbocation required for aromatic alkylation reactions can also be generated by the reaction of an alkene with HF. HF protonates an alkene to give a carbocation. The fluoride anion is not a sufficiently strong nucleophile under the reaction conditions to react with the resulting carbocation. The mechanism for the alkylation of benzene is analogous to that shown in Reaction (b) in Fig. 18-6.

$$RCH{=}CH_2 + HF \rightarrow RC^+HCH_3 + F^-$$

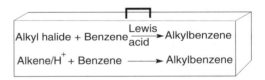

| Alkyl halide + Benzene $\xrightarrow[\text{acid}]{\text{Lewis}}$ Alkylbenzene |
| Alkene/H$^+$ + Benzene $\longrightarrow$ Alkylbenzene |

Toolbox 18-3.

Limitations of the alkylation reaction

Friedel–Crafts alkylation reactions have several limitations. Alkylation of the benzene ring *activates* the ring toward additional alkylation reactions.

Monoalkylbenzenes undergo alkylation reactions more rapidly than the starting nonalkylated benzene. This results in a mixture of mono- and polysubstituted alkylbenzenes. It is difficult to limit the reaction to monoalkylation unless a large excess of benzene is used.

Whenever an alkyl carbocation is formed, one has to consider possible rearrangement reactions (hydride and methyl shifts) to give a more stable carbocation. Rearrangements in secondary carbocations are observed in Friedel–Crafts alkylation reactions. Rearrangements are even observed with primary alkyl halides (where a highly polarized complex and not a free carbocation is the likely alkylating species). Methyl halides and ethyl halides cannot undergo rearrangement reactions but the *n*-propyl group in *n*-propyl halide alkylation reactions undergoes rearrangement to a more stable secondary carbocation.

$$CH_3CH_2CH_2{}^+ \quad \rightarrow \quad (CH_3)_2CH^+$$

$$1° \text{ carbocation} \qquad 2° \text{ carbocation}$$

QUESTION 18-5
Would you expect 2-chloro-3-methylbutane to be a good alkylating agent in a Friedel–Crafts alkylation reaction?

ANSWER 18-5
No, the resulting secondary carbocation would undergo a hydride shift to give a tertiary carbocation.

Substituted benzenes do not undergo Friedel–Crafts alkylation reactions, if the benzene substituent is a deactivating group. Benzenes containing activating substituents and halogens do undergo Friedel–Crafts alkylation reactions. (Deactivating and activating groups are discussed below in the section entitled Multiple Substitution Reactions.)

FRIEDEL–CRAFTS ACYLATION REACTIONS

The reaction of acyl halides (RC(O)X) with Lewis acids (for example AlCl$_3$) forms an *acylium* (ace ill e um) *cation*, Structure 18-7a in Fig. 18-7. The R group can be alkyl or aryl. The halide is typically chloride or bromide. The most stable resonance structure for the acylium cation has a formal plus charge on the oxygen atom, not on the carbon atom. Each second row element in this resonance form is "octet happy". Acylium cations do not undergo rearrangement reactions as do carbocations. The acylium cation undergoes an electrophilic substitution reaction with benzene to give a ketone (18-7b). The reaction mechanism is analogous to that for the alkylation reactions.

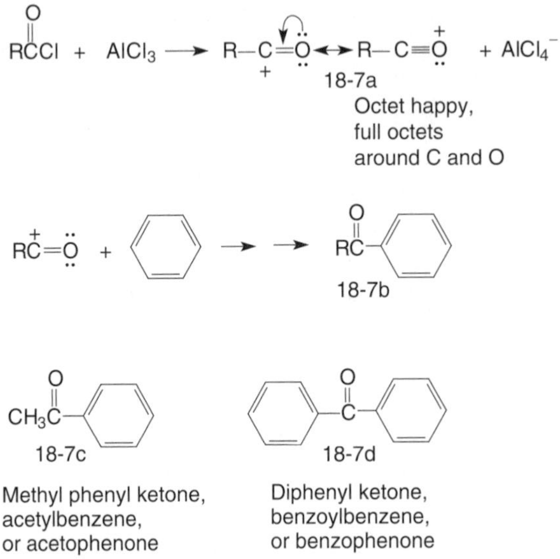

Fig. 18-7.   Friedel–Crafts acylation reactions.

Ketones made from benzene will necessarily be phenyl ketones, also called acyl benzenes or alkyl/arylphenones. Two common phenyl ketones, 18-7c and 18-7d, are shown in Fig. 18-7 along with their multiple names. The acyl group *deactivates the benzene ring* toward further acylation reactions (unlike alkylation reactions) and monoacylated benzene is the major product.

QUESTION 18-6
What product would you expect from the reaction of 2 mol of acetyl chloride with 1 mol of benzene?

ANSWER 18-6
Methyl phenyl ketone. Addition of an acetyl group to a benzene ring deactivates the ring toward further acylation reactions.

Nucleophilic Aromatic Substitution

Nucleophilic aromatic substitution of halide atoms in halobenzenes can occur, if the benzene ring also contains strong electron-withdrawing groups, such as a nitro group. Ring π electrons are withdrawn to such an extent that the ring is attacked by a nucleophile. Treatment of 2,4-dinitro-1-chlorobenzene with ammonia or sodium hydroxide at elevated temperatures results in the substitution of the chlorine atom by an amine group or a hydroxide group, as shown in the two reactions in Fig. 18-8.

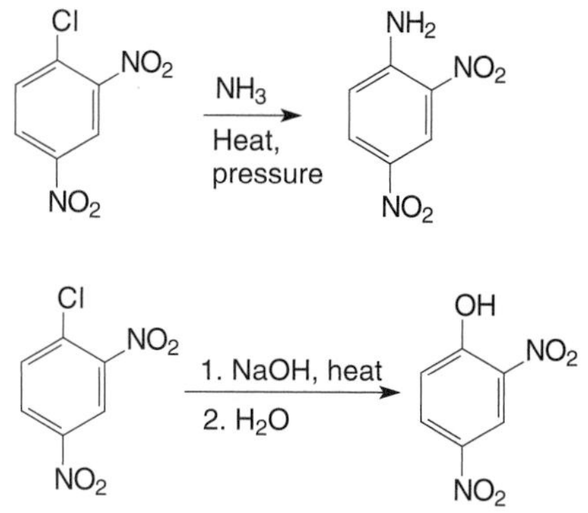

Fig. 18-8. Nucleophilic aromatic substitution reactions.

Chlorobenzene (without additional electron-withdrawing groups) undergoes substitution reactions with NaOH under very vigorous conditions (350 °C) to give phenol (Structure18-9a in Fig. 18-9, Reaction (a)). Reaction (b) in Fig. 18-9 shows the reaction of 4-chlorotoluene (18-9b) with sodium amide (NaNH$_2$), a very strong base, in liquid ammonia at −33°C. Two products, 3-methylaniline (18-9c) and 4-methylaniline (18-9d), are formed in about equal amounts. The intermediate proposed for these reactions is *benzyne* (18-9e).

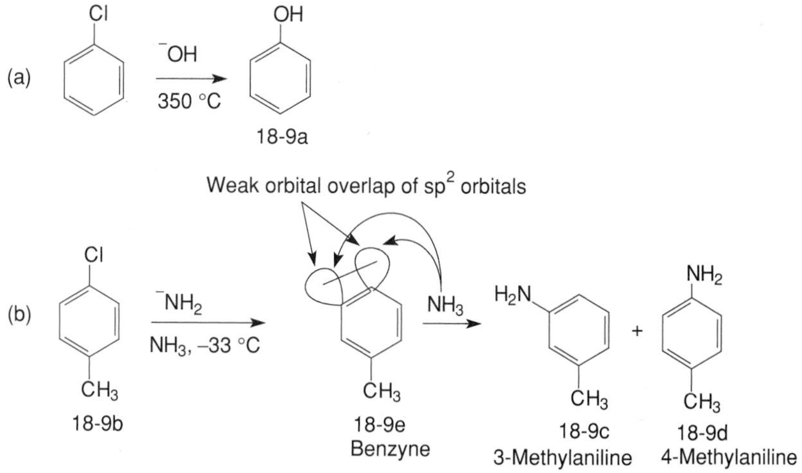

Fig. 18-9. Reactions involving benzyne as an intermediate.

BENZYNE FORMATION

Benzyne is a benzene molecule with an additional "π" bond. Benzyne is often depicted as a benzene molecule containing a triple bond. A compound containing a carbon–carbon triple bond, an alkyne (for a review of *alkynes*, see Chapter 9), has a linear geometry. That is, the carbon atoms in the triple bond and the two atoms bonded directly to each side of the alkyne group have a linear arrangement. The alkyne carbon atoms are sp hybridized. The triple bond is formed by overlapping sp orbitals (the sigma bond) and two pairs of overlapping p orbitals (the π bonds) on adjacent carbon atoms.

This linear arrangement is not possible for any atoms in the six-membered cyclic benzene ring. It is proposed that the carbon atoms in the "triple bond" remain sp^2 hybridized. Two sp^2 orbitals, one on each adjacent carbon atom, overlap to form a sigma bond. The second sp^2 orbital on each carbon atom forms a sigma bond to the other adjacent ring carbon atom. Two p orbitals, one from each adjacent carbon atom, overlap to form a π bond (which is actually part of the π ring system). The remaining third sp^2 orbitals (on adjacent carbon atoms in the triple bond) form a weak second "sigma-like" bond as there is limited overlap of adjacent sp^2 orbitals as shown in Structure 18-9e.

The reaction of *p*-chlorotoluene (18-9b) with base yields the methylbenzyne intermediate 18-9e. Once 18-9e is formed, it can react with ammonia to form an amine. There is an equal probability of ammonia reacting with either carbon atom in the "triple bond" to give the two observed products, 3- and 4-methylaniline (18-9c and 18-9d).

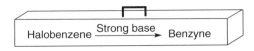

Toolbox 18-4.

QUESTION 18-7
Would you expect the triple bond in benzyne to make the molecule more stable than benzene?

ANSWER 18-7
Two strong sigma bonds are broken and one weak σ-like bond is formed; thus, benzyne is less stable than benzene.

More Derivatives of Benzene

It is difficult to attach groups such as amines ($-NH_2$), carboxyl ($-CO_2H$), and long-chain alkyl groups directly onto benzene by substitution reactions. A more practical approach to making these benzene derivatives is to chemically modify groups, like nitro, alkyl, and acyl groups, that are readily attached to a benzene ring. Examples of bonding amino, alkyl, and carboxyl groups indirectly onto benzene rings are shown in Fig. 18-10.

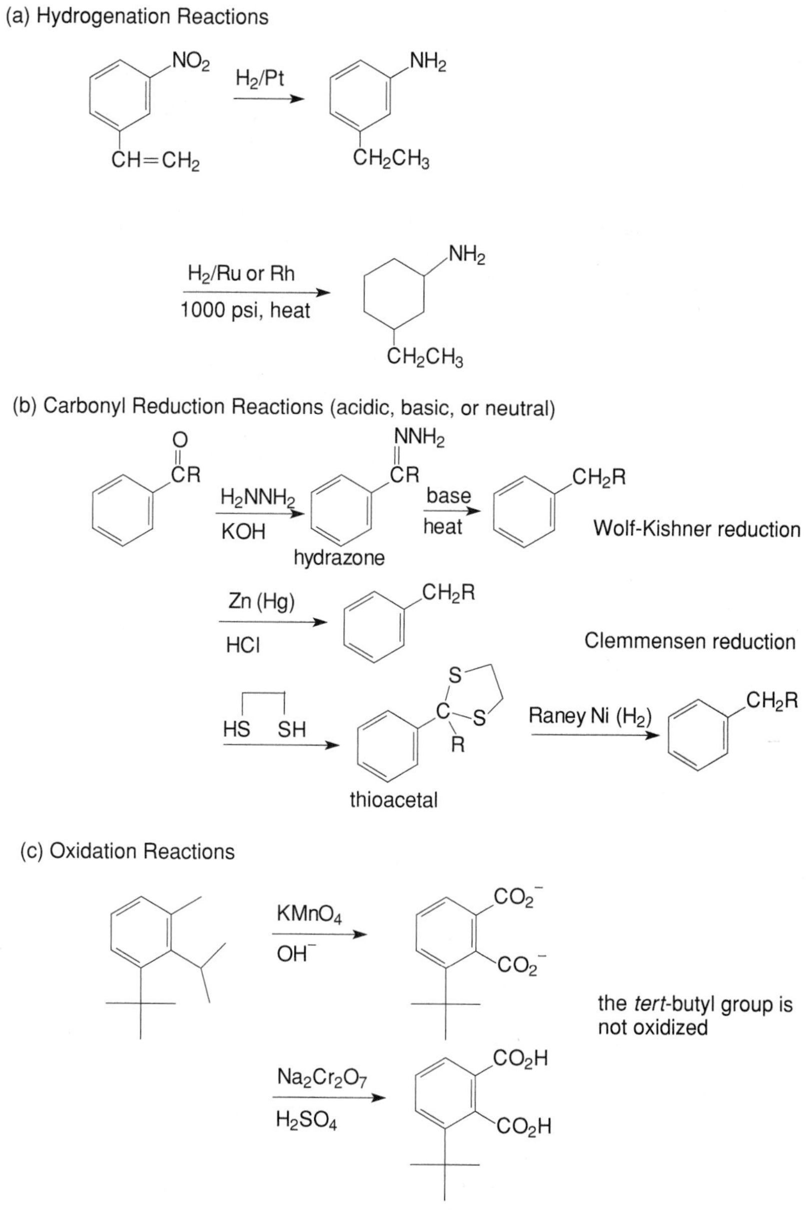

(a) Hydrogenation Reactions

$$\underset{\substack{\text{NO}_2 \\ \text{CH}=\text{CH}_2}}{\qquad} \xrightarrow{\text{H}_2/\text{Pt}} \underset{\substack{\text{NH}_2 \\ \text{CH}_2\text{CH}_3}}{\qquad}$$

$$\xrightarrow[\text{1000 psi, heat}]{\text{H}_2/\text{Ru or Rh}} \underset{\substack{\text{NH}_2 \\ \text{CH}_2\text{CH}_3}}{\qquad}$$

(b) Carbonyl Reduction Reactions (acidic, basic, or neutral)

$$\underset{\overset{\text{O}}{\underset{}{\parallel}}}{\text{CR}} \xrightarrow[\text{KOH}]{\text{H}_2\text{NNH}_2} \underset{\substack{\text{NNH}_2 \\ \text{hydrazone}}}{\overset{\parallel}{\text{CR}}} \xrightarrow[\text{heat}]{\text{base}} \text{CH}_2\text{R} \qquad \text{Wolf-Kishner reduction}$$

$$\xrightarrow[\text{HCl}]{\text{Zn (Hg)}} \text{CH}_2\text{R} \qquad \text{Clemmensen reduction}$$

$$\xrightarrow[\text{HS}\quad\text{SH}]{} \underset{\text{thioacetal}}{\overset{\text{S}}{\underset{\text{R}}{\text{C}}}\overset{\text{S}}{}} \xrightarrow{\text{Raney Ni (H}_2)} \text{CH}_2\text{R}$$

(c) Oxidation Reactions

$$\xrightarrow[\text{OH}^-]{\text{KMnO}_4} \underset{\substack{\text{CO}_2^- \\ \text{CO}_2^-}}{\qquad} \qquad \text{the \textit{tert}-butyl group is not oxidized}$$

$$\xrightarrow[\text{H}_2\text{SO}_4]{\text{Na}_2\text{Cr}_2\text{O}_7} \underset{\substack{\text{CO}_2\text{H} \\ \text{CO}_2\text{H}}}{\qquad}$$

Fig. 18-10.   Attaching groups to benzene by indirect means.

AMINES FROM NITRO GROUPS

Nitro substituents bonded to a benzene ring can be readily reduced by hydrogen, in the presence of a catalyst, to form amine substituents (Reaction (a) in Fig. 18-10). The double bonds in benzene are not hydrogenated under the mild conditions required for reduction of a nitro group. However, functional groups like alkenes or alkynes that may be present in other substituents attached to a benzene ring can be reduced (hydrogenated) under these reaction conditions. Whenever a reaction is planned for one functional group in a molecule, one has to carefully consider other functional groups in the molecule that can also undergo an undesirable reaction at the same time.

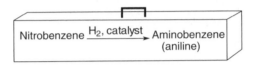

Nitrobenzene $\xrightarrow{H_2, \text{ catalyst}}$ Aminobenzene (aniline)

Toolbox 18-5.

QUESTION 18-8
Can *p*-nitrostyrene (*p*-nitrovinylbenzene) be hydrogenated to give *p*-amino-styrene?

ANSWER 18-8
No, reactions that convert the nitro group to an amine would also convert an alkenyl (vinyl) group into an alkyl group.

INDIRECT ADDITION OF ALKYL GROUPS

One disadvantage of Friedel–Crafts *alkylation* reactions using primary and secondary alkyl halides is the possibility of rearrangement reactions (hydride and alkyl shifts). *A key point: whenever a carbocation is formed, one always has to consider possible rearrangement reactions.* However, rearrangement reactions do not occur with Friedel–Crafts acylation reactions. Thus, the carbonyl group in the ketone product from these acylation reactions can be reduced to a methylene ($-CH_2-$) group. Therefore, an alkyl group can be attached to a benzene ring indirectly (acylation followed by reduction), circumventing the potential rearrangement and multiple alkylation reactions that occur in Friedel–Crafts alkylation reactions.

A carbonyl group bonded directly to a benzene ring can be reduced under basic, acidic, or neutral conditions. Three reduction reactions are shown in Fig. 18-10, Reaction (b). The carbonyl group is reduced to the corresponding methylene group by first forming a hydrazone ($=NNH_2$) by treatment with hydrazine (H_2NNH_2). Heating (200 °C) the hydrazone with base (KOH) replaces the nitrogen atom bonded to the carbonyl carbon atom with two hydrogen atoms. This reaction is called Wolff–Kishner reduction. In an alternate method, Clemmensen reduction, the ketone is reduced to the corresponding methylene group with concentrated HCl and zinc amalgam Zn(Hg). A third method is conducted under neutral conditions. The ketone reacts with a dithiol to give a cyclic thioacetal. The thioacetal is reduced with Raney nickel (which contains adsorbed hydrogen) to the corresponding alkyl group.

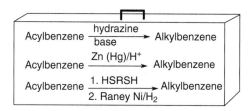

Toolbox 18-6.

BENZENECARBOXYLIC ACIDS

Primary and secondary alkyl substituents on benzene rings can be oxidized to carboxylic acids. Tertiary substituents are not oxidized under these conditions. The reaction can be carried out under basic conditions ($KMnO_4$, OH^-, heat) or acidic conditions ($Na_2Cr_2O_7$, H_2SO_4, heat). These reactions are shown in Fig. 18-10, Reaction (c).

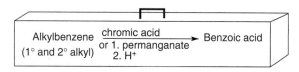

Toolbox 18-7.

QUESTION 18-9
What product results from the chromic acid ($Na_2Cr_2O_7$, H_2SO_4) oxidation of mesitylene (1,3,5-trimethylbenzene)?

ANSWER 18-9
Benzene-1,3,5-tricarboxylic acid.

REDUCTION OF BENZENE

Under vigorous hydrogenation conditions benzene can be reduced to cyclohexane. This reaction requires high pressures (1000 psi) and temperatures (100 °C) in the presence of a catalyst (powdered Ru, Rh, Pt, Pd, or Ni). The reaction is shown in Reaction (a) in Fig. 18-10. Alkenes, alkynes, and nitro groups will also be reduced (hydrogenated) under these conditions. Milder hydrogenation conditions are used to reduce alkenes, alkynes, and nitro groups without reducing the benzene ring.

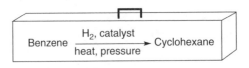

Toolbox 18-8.

ADDITIONAL REACTIONS

Additional derivatives of benzene can be made by chemically modifying substituents. Alkyl, alkenyl, alkynyl, hydroxyl, amino, carboxylic acid, etc., substituents undergo numerous addition and substitution reactions. The reader is referred to the various chapters in this text that discuss these functional groups and the reactions they undergo.

Multiple Substitution Reactions

When a substituent is attached to a benzene ring it may activate the ring toward further substitution reactions or deactivate it toward further substitution reactions. Substituents are classified as activating or deactivating. A substituent also influences the position of attachment (ortho, meta, or para) of additional substituents to the benzene ring, and substituents are therefore classified as ortho- and para-directing or meta-directing. Substituents preferentially direct the incoming substituent to one or more of these positions. Table 18-1 classifies substituents as activating, deactivating, o/p-directing, and m-directing. Substituents that direct ortho also direct para and vice versa.

Substitution reactions often give mixtures of o-, m-, and p-substituted products. Ortho/para-directing groups usually give a mixture of both isomers and a lesser amount of the meta-substituted product. A meta-directing group gives predominately the meta product and a lesser amount of the ortho and para products.

Table 18-1.   Directing effects of substituents.

Ortho/para-directing groups

Strongly activating:
(π donating)
$-\ddot{N}H_2, -\ddot{N}HR, -\ddot{N}R_2, -\ddot{O}H, -\ddot{O}R$

Moderately activating:
(sigma donating)
$-\overset{\overset{\displaystyle :O:}{\|}}{\ddot{N}H\!C}R, -\overset{\overset{\displaystyle :O:}{\|}}{\ddot{N}H\!C}Ar, -\overset{\overset{\displaystyle :O:}{\|}}{\ddot{O}\!C}R, -\overset{\overset{\displaystyle :\ddot{O}:}{\|}}{\ddot{O}\!C}Ar$

Reference:	$-H$

Weakly activating:
(sigma donors)
(π donors)
$-R, -Ar$ (aromatic),
$-R$
$-Ar$

Weakly deactivating:
$-\ddot{F}:, -\ddot{C}l:, -\ddot{B}r:, -\ddot{I}:$

Meta-directing groups

Moderately deactivating:
$-\overset{\overset{\displaystyle :O:}{\|}}{C}H, -\overset{\overset{\displaystyle :O:}{\|}}{C}R, -\overset{\overset{\displaystyle :O:}{\|}}{C}\ddot{O}H, -\overset{\overset{\displaystyle :O:}{\|}}{C}\ddot{O}R, -\overset{\overset{\displaystyle :O:}{\|}}{C}\ddot{N}H_2, -\overset{\overset{\displaystyle :O:}{\underset{\underset{\displaystyle :O:}{\|}}{\|}}}{S}\ddot{O}H, -C\equiv\ddot{N}$

Strongly deactivating:
$-NO_2, -NH_3^+, -CCl_3$

DIRECTING EFFECTS

Atoms and groups attached to a benzene ring can donate electrons to or attract electrons from the benzene ring by two processes: an inductive effect and a resonance effect.

Inductive effects

An inductive effect involves the donation or attraction of electrons through a sigma bond. Atoms or groups that are more electronegative than the benzene ring will attract electrons from the ring and atoms or groups that are less electronegative than benzene will donate electrons to the benzene ring. In electrophilic substitution reactions, the electrophilic species seeks electrons in the benzene ring. Substituents already on the ring that donate electrons to the benzene ring activate the ring toward electrophilic substitution reactions, because they increase the electron density of the benzene ring. Substituents that attract electrons from the benzene ring deactivate electrophilic substitution reactions, because the groups decrease electron density in the benzene ring.

QUESTION 18-10
Is a trichloromethyl substituent an electron-attracting or electron-donating species?

ANSWER 18-10
It is a sigma electron-attracting (and deactivating) substituent.

Resonance effects

Atoms, with nonbonding electron pairs, bonded directly to a benzene ring, including nitrogen and oxygen atoms, can donate these electrons to the benzene ring through π bonds, a resonance effect. These atoms can also withdraw electrons by an inductive effect. Both effects are considered in classifying the groups shown in Table 18-1.

ORTHO- AND PARA-DIRECTING GROUPS

A substituent on a benzene ring that donates electrons to the ring activates the ring toward electrophilic substitution and directs the incoming electrophile toward the ortho and para positions (relative to the existing substituent).

Sigma electron donors

Figure 18-11 shows the directing effects of the methyl group in toluene. The three reactions, a, b, and c, show bonding of the electrophile in the ortho, para, and meta positions. The methyl (or any unsubstituted alkyl) group is a sigma (inductive) electron-donating group. If the incoming electrophilic group bonds to the ring in the ortho position (Reaction (a)), three resonance structures for the allylic carbocation reactive intermediate can be drawn.

In one of the resonance structures (18-11a), the positive charge is on the ring carbon atom attached to the electron-donating methyl group. The electron-donating methyl group stabilizes (lowers the energy of) this resonance structure by donating electrons to the electron-deficient carbon atom. When the electrophile is bonded in the para position (Reaction (b)), three resonance structures can again be drawn. One of these resonance structures (18-11b) has the positive charge on the carbon atom bonded directly to the methyl group. Donation of electrons by the methyl group to this carbon atom stabilizes this resonance structure. If the electrophile is attached in the meta position (Reaction (c)), none of the three resonance structures has the positive charge on the carbon atom bonded to the electron-donating methyl group and no special stability is expected.

Consider the stability of the three resonance structures for the carbocation intermediates formed in each of the three reactions shown in Fig. 18.11. One would expect the ortho and para substitution intermediates to be more stable than the meta substitution intermediate since one of the resonance structures for the ortho and para reactions is particularly stable. The resonance hybrid

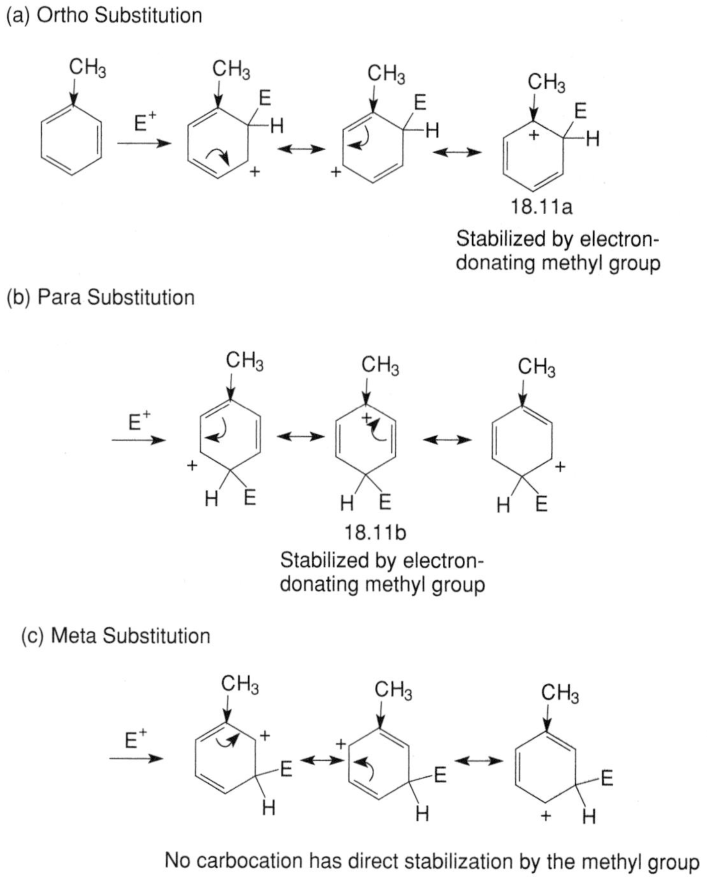

(a) Ortho Substitution

18.11a

Stabilized by electron-donating methyl group

(b) Para Substitution

18.11b

Stabilized by electron-donating methyl group

(c) Meta Substitution

No carbocation has direct stabilization by the methyl group

Fig. 18-11. Directing effects of inductive electron donation.

(average of the resonance structures) should be lower in energy for the ortho and para intermediates than the resonance hybrid structure for meta intermediate. According to the Hammond postulate, the energy of the transition state resembles the energy of the corresponding reactive intermediate in an endothermic reaction. Lowering the energy of the intermediate (by an electron-donating group) also lowers the energy of the corresponding transition state, increasing the rate of that reaction.

Resonance structures similar to those shown in Fig. 18-11 can be drawn for any electron-donating group. Hence electron-donating groups activate the benzene ring toward electrophilic substitution and preferentially direct the incoming electrophile toward the ortho and para positions.

QUESTION 18-11
Is an *n*-propyl group electron-donating or electron-withdrawing?

ANSWER 18-11
Alkyl groups, like the *n*-propyl group, donate electrons by an inductive effect.

Pi electron donors

Substituents that contain a nitrogen or oxygen atom bonded directly to the benzene ring are ortho/para-directing groups (see Table 18-1). Oxygen and nitrogen atoms withdraw electrons from the benzene ring by an inductive effect but also donate electrons to the ring by a pi electron-donating or resonance-donating effect. *A key point: the pi-electron-donating effect is more important than the sigma electron-withdrawing effect for oxygen and nitrogen atoms.* Figure 18-12 shows the resonance structures for the carbocation intermediate, when an electrophile, −OR, is bonded to the ortho, meta, and para positions on a benzene ring.

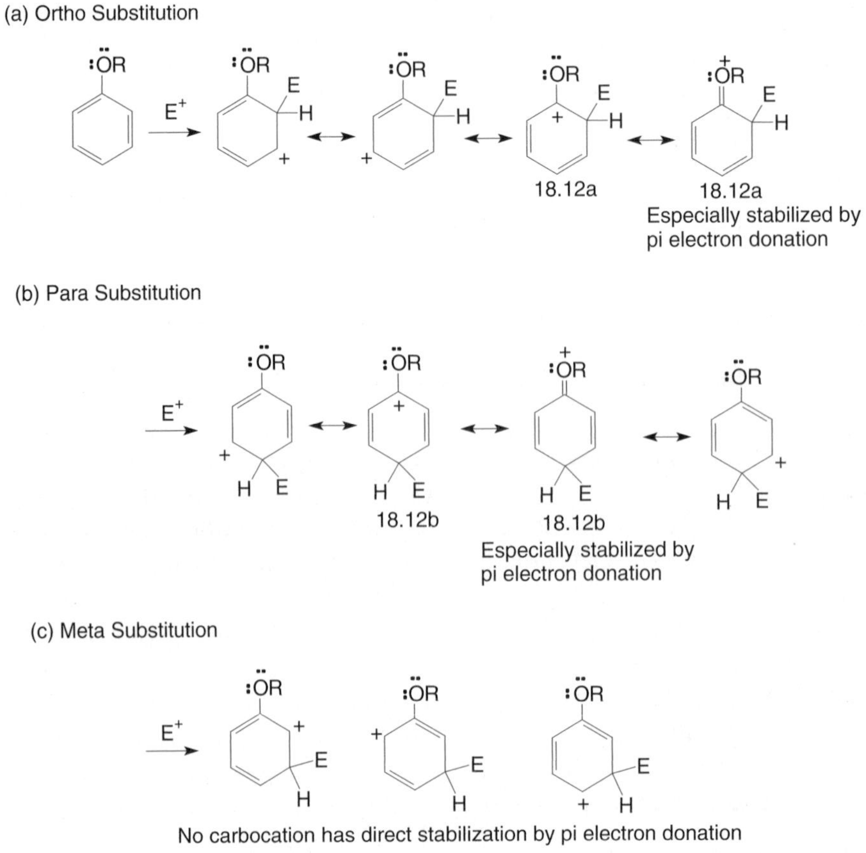

(a) Ortho Substitution

18.12a

18.12a
Especially stabilized by
pi electron donation

(b) Para Substitution

18.12b

18.12b
Especially stabilized by
pi electron donation

(c) Meta Substitution

No carbocation has direct stabilization by pi electron donation

Fig. 18-12. Directing effects of resonance electron donation.

One of the resonance structures (18-12a) for ortho substitution (Reaction (a)) is stabilized by pi electron donation to the ring carbon atom bearing the positive charge. A similar resonance-stabilized structure (18-12b in Reaction (b)) can be drawn when the electrophile is bonded in the para position. A resonance-stabilized structure cannot be drawn when the electrophile is bonded in the meta position (Reaction (c)). The resonance hybrid structures for the reaction intermediates for ortho and para substitution reactions are more stable (lower in energy) than the resonance hybrid structure for the meta substitution reaction. The energy requirements for the ortho and para substitution reactions are therefore lower than the energy requirement for the meta substitution reaction, and the major products are the ortho- and para-substituted isomers.

QUESTION 18-12

Is the N,N-dimethylamino substituent [$(CH_3)_2N-$] ortho/para- or meta-directing?

ANSWER 18-12

Ortho/para directing, since N is a pi electron donor.

META-DIRECTING GROUPS

Substituents that are electron withdrawing are predominately meta-directing and deactivating. The reaction of nitrobenzene with an electrophile is shown in Fig. 18-13. Three resonance structures are shown for the carbocation intermediate, when an electrophile is bonded in the ortho, para, and meta positions. One resonance structure (18-13a in Reaction (a)) for ortho substitution and one resonance structure (18-13b in Reaction (b)) for para substitution is particularly *unstable*. The ring carbon atom with the positive charge is bonded directly to the electron-withdrawing nitro group. The electron-withdrawing nitro group decreases the electron density (increases the positive charge) on the ring carbon atom it is bonded to. This makes that resonance structure less stable relative to the other resonance structures. A similarly unfavorable resonance structure *cannot be drawn* when the electrophile is bonded in the meta position (Reaction (c)).

Electron-withdrawing substituents deactivate the benzene ring toward electrophilic substitution since they decrease electron density in the entire ring. The electron-withdrawing group also makes reaction at the ortho and para positions less favorable (there is one particularly high-energy resonance structure) than reaction at the meta position. Hence these electron-withdrawing groups are meta-directing and deactivating toward electrophilic aromatic substitution reactions.

QUESTION 18-13

Is the $-NHC(O)CH_3$ group an ortho/para or meta director?

ANSWER 18-13
The N atom donates its nonbonding electrons through π bonds and is an ortho/para director. The electron-withdrawing carbonyl group has a weaker deactivating effect since it is not directly bonded to the benzene ring.

DEACTIVATING AND ORTHO/PARA-DIRECTING GROUPS

The halogen atoms belong to a special group that deactivates the benzene ring toward electrophilic aromatic substitution reactions, but directs incoming

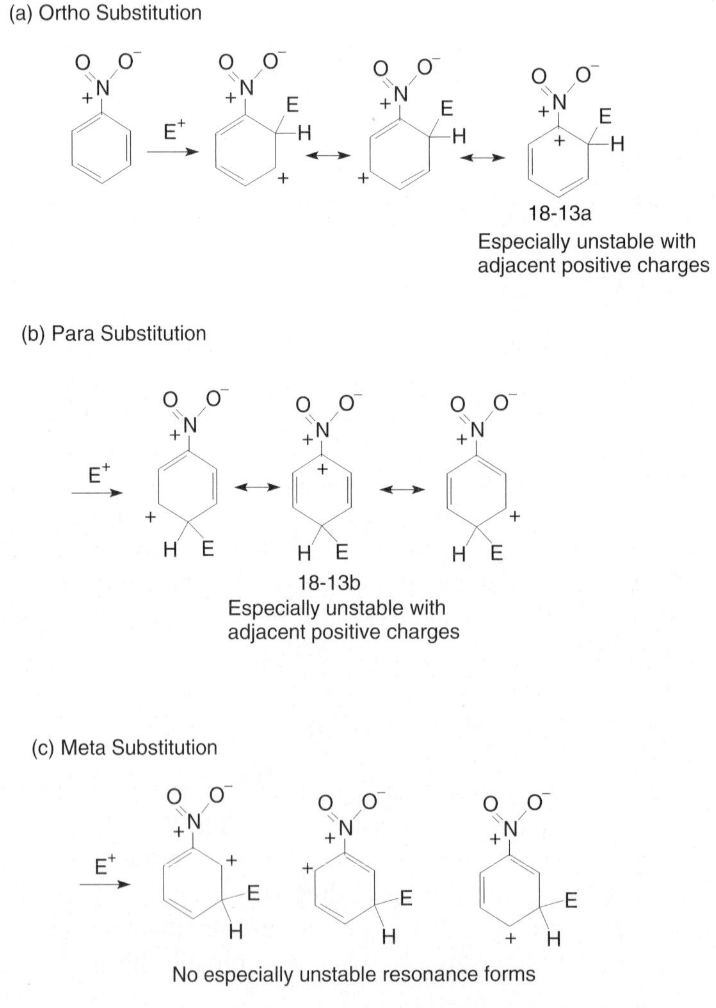

(a) Ortho Substitution

18-13a
Especially unstable with
adjacent positive charges

(b) Para Substitution

18-13b
Especially unstable with
adjacent positive charges

(c) Meta Substitution

No especially unstable resonance forms

Fig. 18-13. Directing effects of electron-withdrawing groups.

electrophiles toward the ortho and para positions. Halogen atoms withdraw electrons from the benzene ring by an inductive effect and donate electrons by a pi-donating effect. Withdrawing electrons from the benzene ring deactivates the ring toward further electrophilic substitution reactions. The resonance-donating effect stabilizes the carbocation intermediate when the electrophile is bonded to the ortho and para positions, but does not stabilize the intermediate when the electrophile is bonded in the meta positions.

QUESTION 18-14
Is the iodine atom an ortho/para or meta director?

ANSWER 18-14
The iodine atom is an ortho/para director. It deactivates the ring by an electron-withdrawing inductive effect and directs ortho/para by a pi electron-donating effect.

Electrophilic Substitution in Disubstituted Benzenes

Each substituent on a benzene ring influences the position of attachment of an incoming electrophile. Two substituents may complement each other and direct the incoming electrophile to the same position(s) on the benzene ring or they may direct the incoming electrophile to different ring positions.

The methyl groups in 1,3-dimethylbenzene (*m*-xylene) are ortho/para directing. Each methyl group directs to the same three ortho and para positions (C–2, C–4, and C–6) as shown in Structure 18-14a in Fig. 18-14. A mixture of three products is obtained. Steric hindrance inhibits the addition of an electrophile between the two methyl substituents that are on alternate (1 and 3) positions.

The ortho/para-directing methyl group in 4-nitrotoluene (*p*-nitrotoluene, 4-nitro-1-methylbenzene, Structure 18-14b) directs an incoming electrophile to the carbon atoms (C–2 and C–6) ortho to it. The para position is blocked (unreactive) by the nitro group. The nitro group directs to the meta positions (C–2 and C–6), and therefore both groups complement each other and the electrophile is directed to the positions ortho to the methyl group. This is shown in Fig. 18-14, Reaction (b).

The methyl and nitro substituents 3-nitrotoluene (*m*-nitrotoluene, Structure 18-14c) do not direct to the same positions. The methyl group directs to positions C–2, C–4, and C–6 that are ortho and para to itself. The nitro group directs to position C–5 that is meta to itself. A mixture of products results.

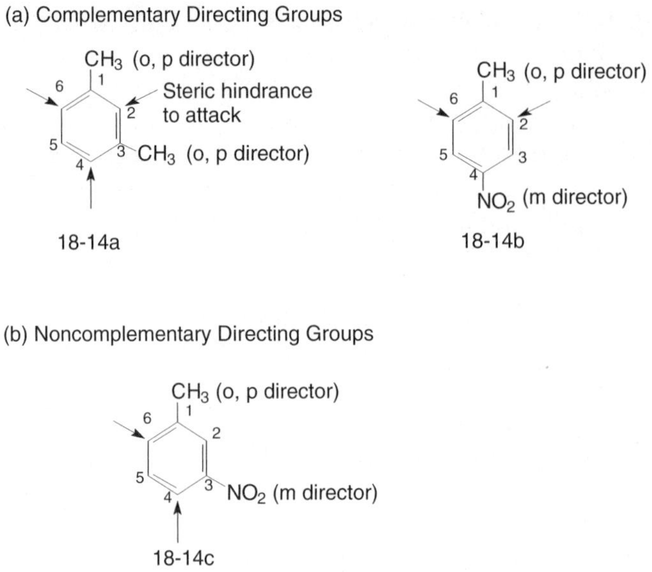

(a) Complementary Directing Groups

18-14a

18-14b

(b) Noncomplementary Directing Groups

18-14c

Fig. 18-14.   Directing effects of multiple groups.

A key point: general rules to follow when the directing effect of the groups do not complement each other are:

(a)   Activating groups are stronger directors than deactivating groups.
(b)   Stronger activating groups predominate over weaker activating groups.
(c)   The general order of activation: $-NR_2$, $-OH$, $-OR > -R$, $-X > C(O)R$, NO_2.

QUESTION 18-15
What product would you expect from the nitration of *m*-nitrobenzoic acid?

ANSWER 18-15
Both groups are meta directors to give 3,5-dinitrobenzoic acid.

Quiz

1.   The intermediate in an electrophilic substitution reaction is
     (a) an anion
     (b) a radical
     (c) an unpaired electron
     (d) a carbocation

2. Benzene reacts with Br^+ because Br^+ is
   (a) a nucleophile
   (b) a radical
   (c) an electrophile
   (d) a bromophile

3. The addition of an electrophile to benzene gives
   (a) a non-allylic carbocation
   (b) an aromatic carbocation
   (c) a diallyl carbocation
   (d) an aromatic anion

4. An electrophilic nitration substitution reaction involves
   (a) NO^+
   (b) NO_2^+
   (c) NO_3^+
   (d) NO_2

5. A sulfonation electrophilic substitution reaction involves
   (a) SO^+
   (b) SO_2^+
   (c) SO_3^+
   (d) SO_3

6. The best Friedel–Crafts alkylating agents are _____ carbocations.
   (a) primary
   (b) secondary
   (c) tertiary
   (d) quaternary

7. A limitation of Friedel–Crafts alkylation reactions is
   (a) rearrangement reactions
   (b) adding deactivating groups to the benzene ring
   (c) spontaneous reactions without a Lewis acid
   (d) multiple acylation reactions

8. An advantage of Friedel–Crafts acylation reactions is that
   (a) rearrangement reactions readily occur
   (b) trialkylbenzenes are easily made
   (c) spontaneous reactions occur without a Lewis acid
   (d) multiple acylation reactions do not readily occur

9. Nucleophilic substitution reactions require a benzene ring to have
   (a) strong electron-donating groups

(b) no substituents

(c) strong electron-withdrawing groups

(d) strong o/p-directing groups

10. Benzyne is a molecule with
    (a) all single bonds
    (b) all double bonds
    (c) three single bonds and three double bonds
    (d) one "triple" bond

11. The amino (H_2N-) group in aniline is ———— substituent.
    (a) an activating
    (b) a deactivating
    (c) neither an activating nor a deactivating
    (d) a meta-directing

12. A Clemmensen reduction reaction
    (a) converts a ketone into an aldehyde
    (b) oxidizes an alcohol to a carboxylic acid
    (c) reduces a ketone to an alcohol
    (d) reduces a carbonyl group to a methylene ($-CH_2-$) group

13. A benzoic acid can be prepared by oxidizing ———— substitutent on a benzene ring.
    (a) a hydroxyl
    (b) an amino
    (c) a primary alkyl
    (d) a tertiary alkyl

14. Benzene can be reduced by reaction with ———— in the presence of a catalyst.
    (a) hydrogen
    (b) oxygen
    (c) nitrogen
    (d) helium

15. An inductive effect refers to the movement of electrons through
    (a) π bonds
    (b) p orbitals
    (c) nonpolar bonds
    (d) sigma bonds

16. Meta-directing groups ———— of electrophilic substitution reactions.
    (a) increase the rates

(b) decrease the rates

(c) do not affect the rates

17. ———— is deactivating and ortho/para directing.

(a) —NO_2

(b) —OH

(c) —H

(d) —Cl

18. The carboxylic acid group in benzoic acid directs the incoming elec-trophile to the ———— position(s).

(a) ortho

(b) meta

(c) para

(d) ortho and para

19. The nitro group in nitrobenzene directs the incoming electrophile to the ———— position(s).

(a) ortho

(b) meta

(c) para

(d) ortho and para

20. The chloro group in chlorobenzene directs the incoming electrophile to the ———— position(s).

(a) ortho

(b) meta

(c) para

(d) ortho and para

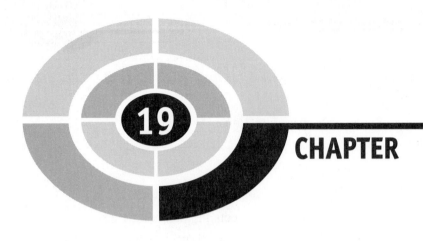

CHAPTER

Aldehydes and Ketones

Introduction

A large number of natural and manmade compounds are *aldehydes* (al-dah-hides) and *ketones* (key-tones). Both families contain the *carbonyl* (car-bow-kneel), C=O, group. The carbonyl group is of particular interest since it is a polar group and undergoes many types of reactions including nucleophilic addition, substitution, oxidation, and reduction reactions.

The simplest aldehyde is formaldehyde, HC(O)H or HCHO. All other aldehydes have the generic formula RCHO, where R can be an alkyl or aryl group. The aldehyde group is often written as —CHO where the double bond between the carbon and oxygen atoms may not be obvious. Formaldehye is a gas at room temperature. A 37 wt % solution of formaldehyde in water is called *formalin*. It is most familiar because of its use as a preservative for biological specimens and as an embalming fluid. Trioxane is a cyclic ether formed from three formaldehyde

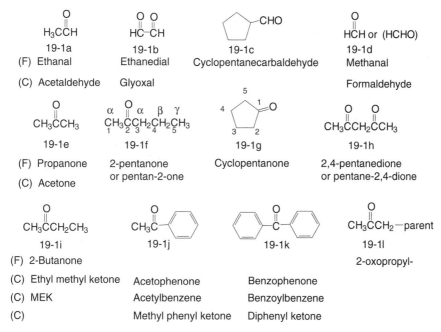

Fig. 19-1. Structures and formal (F) and common (C) names of aldehydes and ketones.

molecules. It is a solid at room temperature and a good source of pure, dry formaldehyde since it decomposes to formaldehyde when heated.

Ketones have the generic structure RC(O)R, where R can be an alkyl or aryl group. Acetone (R=CH$_3$) is a good solvent that is commonly used in organic chemistry laboratories for cleaning glassware. Ethyl methyl ketone is also a good solvent that is sold under the trade name MEK. Structures of these and additional aldehydes and ketones are shown in Fig. 19-1.

Nomenclature

ALDEHYDES

Formal names

The nomenclature of aldehydes and ketones follows the same general rules given for alkanes. The longest carbon–carbon chain containing the aldehyde function is chosen as the parent. The carbonyl carbon atom in aldehydes is necessarily C-1. The carbonyl group has to be at the end of the parent chain for the molecule to be an aldehyde. The parent chain name is based on that of the corresponding alkane, alkene, or alkyne where the terminal -e is replaced by -al. The aldehyde based on the parent alkane ethane is ethanal (Fig. 19-1, Structure 19-1a). If there is an aldehyde at each end of the parent chain, the molecule is a -dial (Structure 19-1b). The two-carbon dialdehyde is called ethanedial. Note that the terminal -e in the alkane name is not dropped when naming di- or polyaldehydes. When the aldehyde group is bonded to a ring, it is named as a carbaldehyde, for example cyclopentanecarbaldehyde (Structure 19-1c). When formaldehyde is a substituent on a molecule it is called a formyl group (−CHO).

QUESTION 19-1
How many different aldehydes can be made based on the isobutane backbone?

ANSWER 19-1
One, a carbonyl group on each of the three identical primary carbon atoms gives the same compound, $(CH_3)_2CHCHO$.

Common names

Lower molecular weight aldehydes have *common names* based on the corresponding carboxylic acid. Formaldehyde (HCHO, Structure 19-1d) is named after formic acid. Acetaldehyde (CH_3CHO, Structure 19-1a) is named after acetic acid. The -ic or -oic suffix in the acid name is replaced by *aldehyde* and the word acid is dropped. Acet*ic acid* becomes acet*aldehyde*.

KETONES
Formal names

The parent name of a ketone is based on the corresponding alkane, alkene, or alkyne where the terminal -e is replaced by -one. The name of the simplest ketone, propanone (Structure 19-1e), is derived from the three-carbon atom alkane, propane. It is not necessary to state the position of the carbonyl group in propanone because it can only be on the center carbon atom. If the carbonyl group was on either of the terminal carbon atoms of propane, the molecule would be an aldehyde, propanal. The position of the carbonyl group must usually be specified for linear ketones having more than four carbon atoms (see Structure 19-1f). The carbonyl carbon atom of a cyclic ketone is always C−1, as shown in Structure 19-1g. Molecules with more than one ketone group (polyketones) are called -diones, -triones, etc. The terminal -e of the alkane is not dropped in naming polyketones. Note the "e" in 2,4-pentanedione (Structure 19-1h).

QUESTION 19-2

Can a ketone be made from the isobutane backbone?

ANSWER 19-2

No, the carbonyl group would have to be on the central carbon atom, which is already bonded to three carbon atoms. One carbon–carbon bond would have to be broken to form a carbonyl group and one would no longer have the isobutane backbone.

Common names

Common names are often used for lower molecular weight ketones. When the two R groups in the ketone are not identical $(RC(O)R')$, they are named alphabetically and the carbonyl group is called ketone. For example, $CH_3C(O)CH_2CH_3$ is ethyl methyl ketone (Structure 19-1i). If the R groups are identical, ketones are named as dialkyl or diaryl ketones. Additional common names include acetone (19-1e), acetophenone, (19-1j), and benzophenone (19-1k).

If a substituent contains a carbonyl group and the carbonyl group is bonded directly to the parent chain or ring (parent–C(O)R), the carbonyl-containing substituent is called an acyl (ā syl) group. The substituent name is based on the name of the corresponding acid. The *-ic* or *-oic* suffix in the name of the corresponding acid is replaced by *-yl* and the word acid is dropped. The acyl group derived from acet*ic acid* $(CH_3C(O)OH)$ is called an acet*yl* group $(CH_3C(O)–)$. Acetylbenzene (19-1j) and benzoylbenzene (19-1k) are examples of this nomenclature. If the carbonyl group in a substituent is not directly bonded to the parent, the oxygen atom is given the name oxo-, as shown in Structure 19-1l.

The formal (IUPAC) nomenclature assigns position numbers (1, 2, 3, etc.) to the parent carbon atoms in aldehydes and ketones in such a way as to give the lowest position number to the carbonyl group. However, in common names, the carbon atom(s) bonded directly to the carbonyl carbon atom (on both sides for ketones) is/are called the alpha (α) carbon atom(s). The next carbon atom is called the beta (β) carbon atom. The third is called the gamma (γ) carbon atom, etc., following the order of letters in the Greek alphabet. See Structure 19-1f for an example of this nomenclature.

QUESTION 19-3

What is the formal name for $CH_3C(O)CH_2CH_2C(O)CH_2CH_3$?

ANSWER 19-3

2,5-Hepanedione or heptane-2,5-dione.

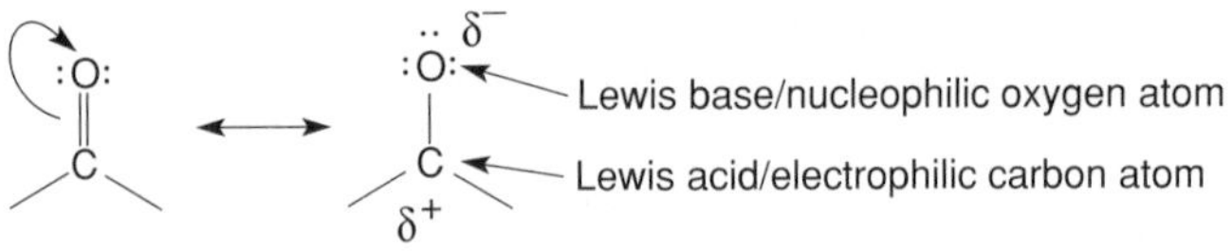

Fig. 19-2.   Resonance forms of carbonyl groups.

Physical Properties

Aldehydes and ketones contain a polar carbonyl bond. The oxygen atom carries a partial negative charge and the carbon atom a partial positive charge, as shown in Fig. 19-2. These partial charges result in polar interactions between molecules. This polar interaction causes ketones and aldehydes to have higher boiling points than nonpolar alkanes of comparable molecular weight. Ketones and aldehydes have lower boiling points than do alcohols of comparable molecular weight. Alcohols can form intermolecular hydrogen bonding, resulting in stronger intermolecular forces and relatively higher boiling points than aldehydes and ketones.

The oxygen atom in the carbonyl group of aldehydes and ketones can bond to other molecules that are capable of hydrogen bonding. Water and alcohols are examples of molecules capable of hydrogen bonding to aldehydes and ketones. Lower molecular weight aldehydes and ketones are soluble in water due to hydrogen bonding and polar interactions between the molecules of water and those of the aldehyde or ketone.

QUESTION 19-4
Which would you predict to have a greater intermolecular interaction with water, $CH_3(CH_2)_5CHO$ or $CH_3(CH_2)_5CH_2OH$?

ANSWER 19-4
$CH_3(CH_2)_5CH_2OH$, since the alcohol has hydrogen atoms that can form hydrogen bonds to water and water has hydrogen atoms that can form hydrogen bonds to the alcohol. The aldehyde has no hydrogen atoms capable of hydrogen bonding to water although water can form hydrogen bonds to the aldehyde oxygen atom.

Chemical Properties

Aldehydes tend to be more chemically reactive than ketones for two principal reasons. The carbonyl carbon atom in ketones is more sterically hindered from nucleophilic attack than is the carbonyl group of an aldehyde. Compare Structure

Ketones

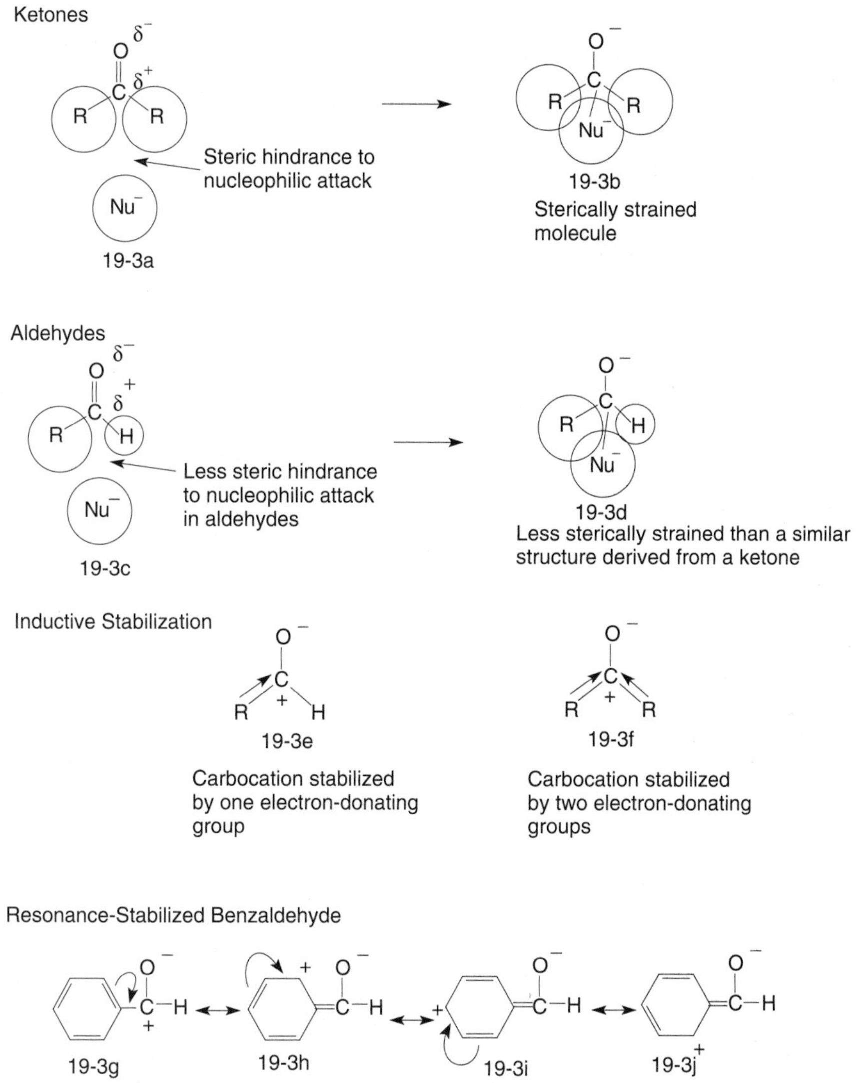

Steric hindrance to
nucleophilic attack

19-3a

19-3b
Sterically strained
molecule

Aldehydes

Less steric hindrance
to nucleophilic attack
in aldehydes

19-3c

19-3d
Less sterically strained than a similar
structure derived from a ketone

Inductive Stabilization

19-3e

Carbocation stabilized
by one electron-donating
group

19-3f

Carbocation stabilized
by two electron-donating
groups

Resonance-Stabilized Benzaldehyde

19-3g   19-3h   19-3i   19-3j

Fig. 19-3. Reactivity of aldehydes and ketones.

19-3a with Structure 19-3c in Fig. 19-3. Circles represent the van der Waals radii and electron clouds. The tetrahedral intermediate resulting from the addition of the nucleophile is more sterically strained (has higher potential energy) for a ketone (Structure 19-3b) than for an aldehyde (Structure 19-3d).

Secondly, alkyl aldehydes have a greater partial positive charge on the carbonyl carbon atom due to reduced inductive electron donation from a second alkyl group. There is only one alkyl group to reduce the partial positive charge on (and stabilize) the carbonyl carbon atom by inductive electron donation, as shown in Structure 19-3e. Dialkyl ketones have two alkyl groups to decrease the partial positive charge on (and help stabilize) the carbonyl carbon atom by electron donation, as shown in Structure 19-3f. The less stabilized carbon atom in an aldehyde has a higher potential energy and is thus more reactive.

QUESTION 19-5
Which compound is more reactive toward electrophilic attack, $CH_3C(O)CH_3$ or CH_3CHO?

ANSWER 19-5
CH_3CHO, since there is less steric hindrance and the carbonyl carbon atom is more reactive.

Aryl aldehydes are more stable (less reactive) than alkyl aldehydes toward nucleophilic addition because of resonance stabilization. The positive charge on the carbonyl carbon atom can be shared with the benzene ring as shown by the resonance-stabilized structures 19-3g–j. Since the positive charge is "spread throughout" the molecule, the positive charge on the carbonyl carbon atom is reduced. The lower partial charge on the carbon atom decreases the attraction toward an attacking nucleophile.

Preparation of Aldehydes and Ketones

OXIDATION REACTIONS

Primary alcohols can be oxidized to aldehydes. Since primary alcohols can also be oxidized to carboxylic acids, specific oxidizing reagents are used in order to stop oxidation at the aldehyde stage. The most common oxidizing agent for aldehyde formation is pyridinium chlorochromate (PCC). The reaction needs to be carried out under anhydrous conditions or the alcohol will be oxidized to an acid. This is shown in Fig. 19-4, Reaction (a).

Secondary alcohols are oxidized to ketones. PCC oxidizes secondary alcohols to ketones. The more common oxidizing agents, aqueous permanganate anion (MnO_4^-) or aqueous chromic acid (shown as $HCrO_4^-$, $Cr_2O_7^{-2}$, or CrO_3/H_2SO_4), also oxidize secondary alcohols to ketones. Carbon–carbon bonds are not cleaved under these oxidation conditions and the oxidation reaction stops at the ketone stage. This is shown in Fig. 19-4, Reaction (b).

QUESTION 19-6
What product results from the PCC oxidation of isopropyl alcohol (2-propanol)?

Primary Alcohol

(a) RCH_2OH $\xrightarrow[CH_2Cl_2]{PCC}$ $R\overset{O}{\overset{\|}{C}}H$ PCC = $\langle\text{pyridine}\rangle NH^+ CrO_3Cl^-$

Secondary Alcohol

(b) $R\overset{OH}{\underset{|}{C}}HR'$ $\xrightarrow[\substack{\text{or } CrO_4^{2-},\ H^+ \\ \text{or } KMnO_4,\ H^+}]{PCC,\ CH_2Cl_2}$ $R\overset{O}{\overset{\|}{C}}R'$

Fig. 19-4. Synthesis of aldehydes and ketones by oxidation of alcohols.

ANSWER 19-6
Propanone (acetone).

Primary alcohol	$\xrightarrow{PCC}$	Aldehyde
Secondary alcohol	$\xrightarrow[\substack{\text{or chromic acid} \\ \text{or permanganate ion}}]{PCC}$	Ketone

Toolbox 19-1.

OZONOLYSIS OF ALKENES

Ozone is used to cleave carbon–carbon double bonds. This reaction is shown in Fig. 19-5. This process is called *ozonolysis*. The initial product of the ozonolysis reaction is a cyclic *ozonide*. The ozonide is cleaved with reducing agents such as dimethyl sulfide [$(CH_3)_2S$] or zinc in aqueous acid. If a carbon atom in the carbon–carbon double bond is bonded to one hydrogen atom, it will be converted

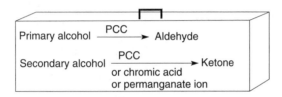

Oxidized to a ketone   Oxidized to an aldehyde or carboxylic acid

Trisubstituted alkene

Ozonide

Ketone   Aldehyde

Ketone   Acid

Fig. 19-5. Bond cleavage with ozone.

into an aldehyde. If a carbon atom in the double bond is bonded to two alkyl and/or aryl groups, it will be converted into a ketone. The ozonide can also be cleaved with an oxidizing agent such as with hydrogen peroxide. In this case, a carbon atom bonded to one hydrogen atom in the carbon–carbon double bond is oxidized to a carboxylic acid. A carbon atom bonded to two alkyl or aryl groups is oxidized to to a ketone.

Ozonolysis can be used to characterize complex molecules by degrading them to lower molecular weight (smaller) molecules that are easier to characterize. The resulting carbonyl groups in the smaller molecules indicate where the carbon–carbon double bonds were in the parent molecule. Ozonolysis reactions take place readily at room temperature, but care must be taken as lower molecular weight ozonides can be explosive. Also, ozone is toxic. (Your city may declare ozone action days during the summer, when ozone levels in the atmosphere exceed acceptable levels.)

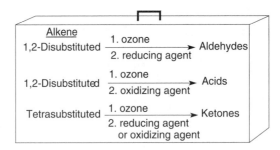

Toolbox 19-2.

QUESTION 19-7
What alkene gives only acetaldehyde upon ozonolysis and subsequent reduction with dimethyl sulphide?

ANSWER 19-7
2-Butene is oxidized to give 2 equiv. of acetaldehyde.

Hydration of Alkynes

MARKOVNIKOV ADDITION

Alkynes undergo acid-catalyzed Markovnikov addition of water to give ketones. The reaction is shown in Fig. 19-6. Mercury (Hg^{2+}) salts are used to increase the reaction rate. Terminal alkynes give methyl ketones (Reaction (a)). Symmetrical internal alkynes give one ketone product (Reaction (b)), while unsymmetrical

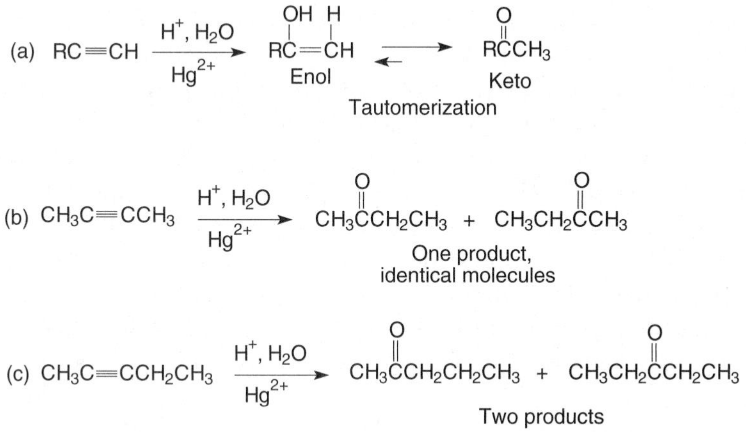

(a) $RC\equiv CH$ $\xrightarrow[Hg^{2+}]{H^+, H_2O}$ $RC\!\!=\!\!CH$ (OH, H) Enol $\rightleftarrows$ $RCCH_3$ (O) Keto

Tautomerization

(b) $CH_3C\equiv CCH_3$ $\xrightarrow[Hg^{2+}]{H^+, H_2O}$ $CH_3\overset{O}{\overset{||}{C}}CH_2CH_3$ + $CH_3CH_2\overset{O}{\overset{||}{C}}CH_3$

One product,
identical molecules

(c) $CH_3C\equiv CCH_2CH_3$ $\xrightarrow[Hg^{2+}]{H^+, H_2O}$ $CH_3\overset{O}{\overset{||}{C}}CH_2CH_2CH_3$ + $CH_3CH_2\overset{O}{\overset{||}{C}}CH_2CH_3$

Two products

Fig. 19-6.   Hydration of alkynes (Markovnikov addition).

internal alkynes give a mixture of two products (Reaction (c)). In these reactions, the initially formed enol tautomerizes to the more thermodynamically stable keto form.

QUESTION 19-8
What product results from the acid-catalyzed addition of water to acetylene, $HC\equiv CH$?

ANSWER 19-8
Only one product, acetaldehyde, CH_3CHO.

ANTI-MARKOVNIKOV ADDITION

Borane (BH_3) and substituted borane compounds (RBH_2 and R_2BH) are used to give anti-Markovnikov addition of 1 equiv. of water (as an H and an OH) to alkynes. This reaction is shown in Fig. 19-7. A hindered boron compound, such as *di(secondary-isoamyl)borane* (also called *disiamylborane* or *(sia)$_2$BH*;

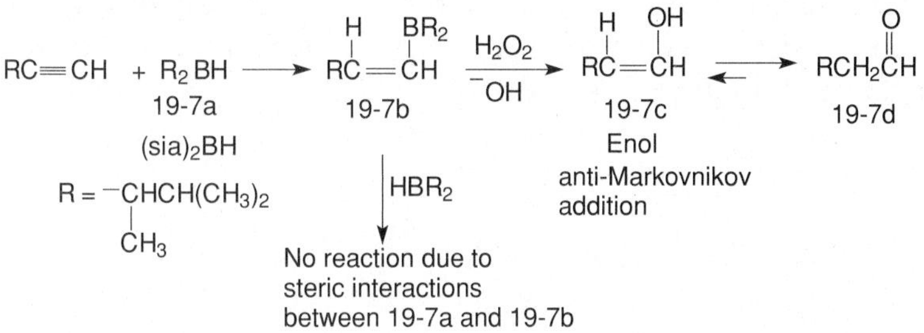

$RC\equiv CH$ + R_2BH $\longrightarrow$ $RC\!\!=\!\!CH$ (H, BR$_2$) $\xrightarrow[^-OH]{H_2O_2}$ $RC\!\!=\!\!CH$ (H, OH) $\rightleftarrows$ $RCH_2\overset{O}{\overset{||}{C}}H$

19-7a   (sia)$_2$BH $R = ^-CHCH(CH_3)_2$, CH_3

19-7b

19-7c   Enol   anti-Markovnikov addition

19-7d

$\downarrow$ HBR$_2$

No reaction due to steric interactions between 19-7a and 19-7b

Fig. 19-7.   Aldehydes from terminal alkynes (anti-Markovinkov addition).

Structure 19-7a), reacts with an alkyne to give the monoadditon product 19-7b. Structure 19-7b contains a carbon–carbon double bond that could react with another molecule of (sia)$_2$BH. However, steric hindrance between bulky 19-7a and 19-7b prevents the addition of a second molecule of (sia)$_2$BH.

When the initially formed boron compound (19-7b) reacts with aqueous basic hydrogen peroxide, the $-BR_2$ is replaced by a hydroxyl group. The resulting hydroxyalkene (19-7c), called an enol, tautomerizes to the corresponding aldehyde (19-7d). Terminal alkynes undergo this reaction to give aldehydes and internal alkynes give ketones. Hydration of internal alkynes by both acid-catalyzed hydration and (sia)$_2$BH addition-oxidation give identical ketone products, and thus the less expensive acid-catalyzed hydration is a more economical synthetic route.

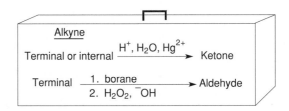

Toolbox 19-3.

Reduction of Acid Chlorides

Lithium aluminum hydride (LiAlH$_4$ or LAH) is a very powerful reducing agent. LAH reduces ketones and aldehydes to alcohols. Carboxylic acid chlorides (derivatives of carboxylic acids) are reduced to aldehydes by using a less reactive aluminum hydride, *lithium tri(tert-butoxy)aluminum hydride* [Li(OC(CH$_3$)$_3$)AlH].

This reduction reaction is shown in Fig. 19-8. Carboxylic acids (Structure 19-8a) are easily converted to carboxylic acid chlorides (Structure 19-8b) by treatment with thionyl chloride (SOCl$_2$). Lithium tri(*tert*-butoxy)aluminum hydride transfers a hydride anion to the carbon atom of the carbonyl group, giving intermediate 19-8c. This intermediate undergoes an elimination reaction to form an aldehyde (19-8d) and a chloride anion.

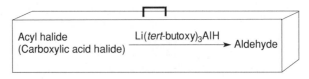

Toolbox 19-4.

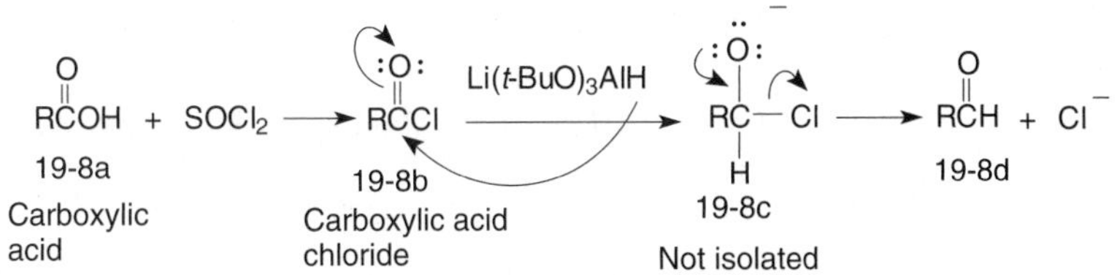

Fig. 19-8. Reduction of carboxylic acids to aldehydes.

QUESTION 19-9
What product results from the reaction of acetyl chloride, $CH_3C(O)Cl$, with Li($tert$-butoxy)$_3$AlH.

ANSWER 19-9
Acetaldehyde, CH_3CHO.

SYNTHESIS OF ALDEHYDES AND KETONES BY ACYLATION REACTIONS

Preparation of ketones

Aromatic compounds undergo *Friedel–Crafts acylation* reactions with carboxylic acid halides to give ketones. Acetyl chloride reacts with benzene, in the presence of a Lewis catalyst, to give methyl phenyl ketone (acetophenone). This reaction is shown in Fig. 19-9. A large variety of ketones are made by reacting various acyl halides with aromatic compounds under Friedel–Crafts conditions. Friedel–Crafts reactions are discussed in detail in the section entitled *Friedel–Crafts Acylation Reactions* in Chapter 18.

QUESTION 19-10
What product results from the reaction of benzoyl chloride, $C_6H_5C(O)Cl$, with benzene?

ANSWER 19-10
Diphenyl ketone (also called benzophenone), $C_6H_5C(O)C_6H_5$.

Organometallic reagents

Acyl chlorides (also called carboxylic acid chlorides or just acid chlorides) react with Grignard reagents (a type of organometallic compound, RMgX) to give,

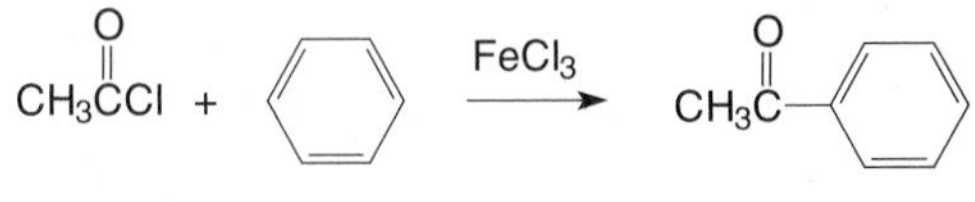

Fig. 19-9. Synthesis of ketones by Friedel–Crafts acylation.

(a) $R'MgX$ + $\overset{\overset{\displaystyle ::\text{O}::}{\parallel}}{R}CCl$ ⟶ $\underset{\underset{\displaystyle R'}{|}}{RC}—Cl$ ⟶ $\overset{\overset{\displaystyle O}{\parallel}}{R}CR'$ $\xrightarrow[\text{2. }H_3O^+]{\text{1. }R'MgX}$ $\underset{\underset{\displaystyle R'}{|}}{R}\overset{\overset{\displaystyle OH}{|}}{C}R'$

Slower reaction   Faster reaction

(b) R'_2CuLi + $\overset{\overset{\displaystyle O}{\parallel}}{R}CCl$ ⟶ $\overset{\overset{\displaystyle O}{\parallel}}{R}CR'$

Gilman reagent   A ketone is the final product

(c) $R'—Li$ + $\overset{\overset{\displaystyle O}{\parallel}}{R}COH$ ⟶ $\overset{\overset{\displaystyle O}{\parallel}}{R}CO^-$ + $R'H$

19-10a $\xrightarrow{R^{\delta-}Li^{\delta+}}$ $\underset{\underset{\displaystyle R'}{|}}{\overset{\overset{\displaystyle O^-}{|}}{R}CO^-}$ $\xrightarrow{H^+}$ $\underset{\underset{\displaystyle R'}{|}}{\overset{\overset{\displaystyle OH}{|}}{R}COH}$ ⟷ $\overset{\overset{\displaystyle O}{\parallel}}{R}CR'$ + H_2O

19-10b   19-10c   19-10d

Fig. 19-10. Organometallic reactions with acyl chlorides.

initially, ketones. Ketones are more reactive toward Grignard reagents than are the starting acyl chlorides, and therefore the ketone is converted into a tertiary alkoxide anion before all of the acyl chloride reacts. Subsequent protonation of the alkoxide gives a tertiary alcohol. This is shown in Fig. 19-10, Reaction (a).

A less reactive organometallic reagent is used to make ketones from acyl chlorides. *Gilman reagents*, lithium dialkylcuprates, LiR_2Cu, are less reactive toward ketones than are Grignard reagents. Gilman reagents react with acyl chlorides to give the corresponding alkyl ketones (Reaction (b) in Fig. 19-10). Gilman reagents react more slowly with the alkyl ketone than with the acyl chloride, and thus the ketone is the major product of the reaction when equivalent amounts of acyl chloride and Gilman reagent are used.

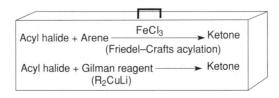

Acyl halide + Arene $\xrightarrow{FeCl_3}$ Ketone
(Friedel–Crafts acylation)

Acyl halide + Gilman reagent ⟶ Ketone
(R_2CuLi)

Toolbox 19-5.

Alkyllithium compounds react with carboxylic acids to give ketones, as shown in Reaction (c) in Fig. 19-10. The lithium carboxylate salt (19-10a) is initially formed in an acid–base reaction. The lithium carboxylate salt then reacts with more alkyllithium to give a dianion, Structure 19-10b. Treatment of

19-10b with acid gives gem diol 19-10c. Gem diols are in equilibrium with their dehydrated ketone form. The equilibrium strongly favors the ketone 19-10d.

Toolbox 19-6.

NITRILES

The nitrile group ($-C\equiv N$) is also called a cyano or cyanide group in organic compounds. Alkyl and aryl nitriles ($RC\equiv N$ or $ArC\equiv N$) react with Girgnard reagents to give ketones. This reaction is shown in Fig. 19-11. The $C\equiv N$ group is polarized with the nitrogen atom having a partial negative charge and the carbon atom having a partial positive charge. The carbon atom in the R group bonded to Mg in the Grignard reagent has a partial negative charge and attacks the carbon atom in the nitrile group. The initial product of the reaction (19-11a) is the magnesium salt of an imine ($R_2C=NH$). Hydrolysis of the imine salt gives the corresponding imine (19-11b). Acid-catalyzed hydrolysis of the imine gives ketone 19-11c.

QUESTION 19-11
What ketone is formed by the reaction of acetonitrile, $CH_3C\equiv N$, with methyl-magnesium chloride and subsequent acidification?

ANSWER 19-11
Acetone, $CH_3C(O)CH_3$.

$$\overset{\delta^+}{R}\overset{\delta^-}{C}\equiv\overset{\delta^-}{N} \ + \ \overset{\delta^+}{R'}\!-\!MgX \longrightarrow \underset{\underset{\text{Imine salt}}{\text{19-11a}}}{RC=\overset{-}{N}\overset{+}{MgX}}\overset{H^+}{\underset{H_2O}{\longrightarrow}}\underset{\underset{\text{Imine}}{\text{19-11b}}}{RC=NH}\overset{H^+}{\underset{H_2O}{\longrightarrow}}\underset{\text{19-11c}}{RC=O}$$

Fig. 19-11. Synthesis of ketones from nitriles.

Alkyl nitrile or Aryl nitrile $\xrightarrow[\text{2. } H_3O^+]{\text{1. } RMgX}$ Imine $\xrightarrow{H_3O^+}$ Ketone

Toolbox 19-7.

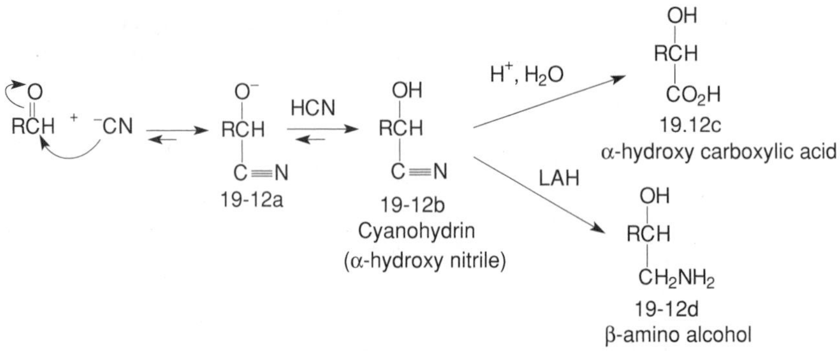

Fig. 19-12. Reaction with cyanide ion and derivatives.

Reactions of Aldehydes and Ketones with Nucleophiles

Nucleophiles react with the carbonyl carbon atom of aldehydes and ketones. The nucleophile can be an anion such as ^-OH, H^-, R^-, $N{\equiv}C^-$, and $HC{\equiv}C^-$. Alkyl anions (R^-) derived from Grignard reagents or alkyl metals (RLi) were discussed above. Nucleophiles can also be neutral species that contain nonbonding electron pairs such as H_2O, ROH, NH_3, or HNR_2.

REACTION OF CYANIDE WITH THE CARBONYL GROUP

The cyanide ion ($-C{\equiv}N$) reacts with aldehydes and ketones to form cyanohydrins (α-hydroxy nitriles). This reaction is shown in Fig. 19-12. The cyano group is also called a nitrile in organic compounds. The reaction is reversible with the equilibrium shifted toward the cyanohydrin for most aldehydes. The equilibrium is not favorable for most ketones for steric reasons. A small amount of base is added to HCN to form the cyanide ion. Reaction of the cyanide ion with the carbonyl group results in an alkoxide (19-12a, an alcohol anion). This anion removes a proton from HCN to form the cyanohydrin (19-12b) and more CN^-, continuing the reaction until all the aldehyde is depleted.

The nitrile group of a cyanohydrin can be hydrolyzed in the presence of aqueous acid to give an α-hydroxy carboxylic acid (19-12c). Reduction of the nitrile group with LAH gives the corresponding primary β-amino alcohol (19-12d). This is an example where carbonyl and nitrile functional groups are converted into other functional groups.

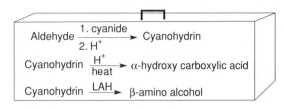

Toolbox 19-8.

QUESTION 19-12
What product results from the reaction of sodium cyanide, NaCN, with acetone, followed by reaction with LAH?

ANSWER 19-12
1-Amino-2-methyl-2-propanol, $(CH_3)_2C(OH)CH_2NH_2$.

HYDRATION REACTIONS

Water reacts with aldehydes and ketones to form hydrates, $RCHO \cdot H_2O$ or $RC(OH)_2H$ for aldehydes and $RC(O)R \cdot H_2O$ or $RC(OH)_2R$ for ketones. These hydrates are geminal diols, meaning both hydroxyl groups are on the same carbon atom. The equilibrium expression for the hydration of an aldehyde, shown below, is shifted strongly toward the aldehyde except for the low molecular weight aldehydes. Values for the equilibrium ratio of hydrate to aldehyde for formaldehyde, acetaldehyde, and propanal are 1000/1, 1.4/1, and 0.9/1, respectively. The hydration reactions are catalyzed by acid or base. The mechanism for hydration is analogous to that of hemiacetal and acetal formation, described in the following section:

$$\underset{\text{RCH}}{\overset{\text{O}}{\|}} + H_2O \rightleftharpoons \underset{\text{RCH}}{\overset{\text{OH}}{|}} \overset{|}{\underset{\text{OH}}{}}$$

QUESTION 19-13
What is the structure of formaldehyde hydrate?

ANSWER 19-13
$H_2C(OH)_2$ (or $HC(OH)_2H$ or $HCHO \cdot H_2O$)

FORMATION OF HEMIACETALS AND ACETALS

Aldehydes and ketones react with alcohols to form *hemiacetals* and *acetals*. The mechanism is best understood by following the reaction sequence shown

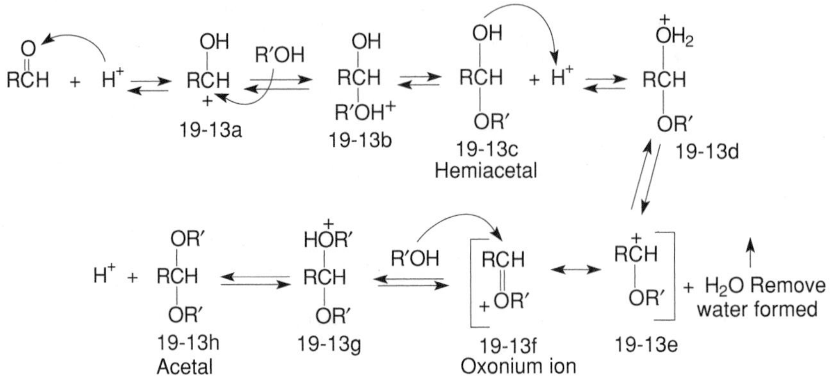

Fig. 19-13.   Acetal formation.

in Fig. 19-13. The first step in the acid-catalyzed reaction is protonation of the aldehyde to give carbocation 19-13a. An alcohol molecule reacts with 19-13a to form a protonated hemiacetal, 19-13b. Another alcohol molecule removes the proton from 19-13b to give hemiacetal 19-13c.

The acetal is made by replacing the remaining hydroxyl group with an alkoxy group. Since the hydroxyl group is a poor leaving group it is protonated in this acid-catalyzed reaction to give 19-13d. Structure 19-13d loses a molecule of water (water is a better leaving group than the hydroxyl group) to give carbocation 19-13e. This cation is stabilized by formation of the octet-happy oxonium ion, resonance structure 19-13f. Since all the steps in acetal formation are equilibrium reactions, the water produced during the formation of 19-13e/19-13f is removed from the reaction, forcing the equilibrium to favor the acetal product. Another alcohol molecule reacts with 19-13e, giving a protonated acetal, 19-13g. An alcohol molecule removes a proton from 19-13g, giving the acetal 19-13h.

Hemiacetal formation may be acid- or base-catalyzed, but acetal formation must be acid-catalyzed. The term acetal refers to the reaction products of aldehydes and alcohols. The reaction products of alcohols and ketones are also called acetals, although the term ketals is often used instead.

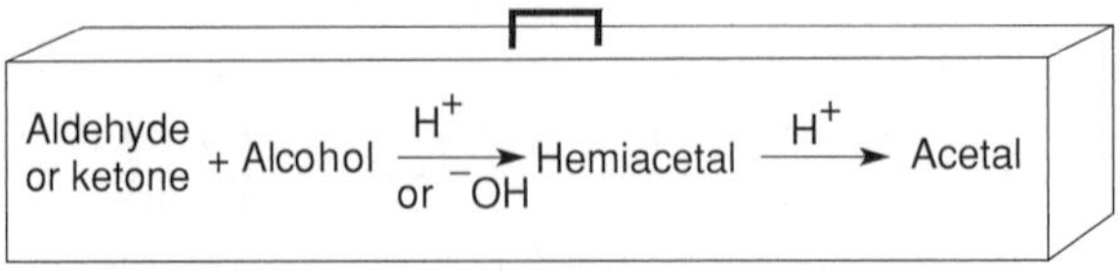

Toolbox 19-9.

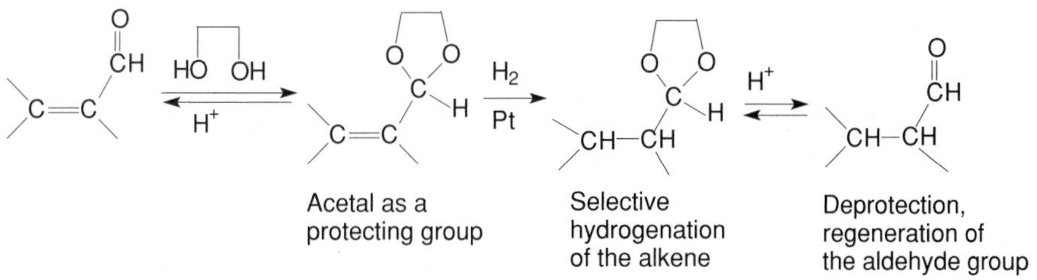

Fig. 19-14. Cyclic acetal protecting group.

QUESTION 19-14

What are the structures of the hemiacetal and acetal formed by the reaction of acetaldehyde with methanol?

ANSWER 19-14

$CH_3CH(OH)OCH_3$ and $CH_3CH(OCH_3)_2$.

Cyclic acetals

An acetal can be made by the reaction of two molecules of a monoalcohol with a carbonyl group or by the reaction of one molecule of a diol with one carbonyl group. A particularly useful acetal is formed by the reaction of ethylene glycol (a diol) with an aldehyde or ketone. This reaction is shown in Fig. 19-14.

An acetal may be formed to act as a *protecting group* for the original carbonyl group. For example, it may be desirable to reduce a carbon–carbon double bond in a molecule that also contains an aldehyde or ketone group. The reduction reaction (metal-catalyzed hydrogenation) may also reduce the carbonyl group to an alcohol. To prevent reduction of the carbonyl group, it is converted to a cyclic acetal, as shown in Fig. 19-14. Acetals are not susceptible to metal-catalyzed hydrogenation. After the double bond is hydrogenated, the protecting group is removed in an acid-catalyzed reaction and the carbonyl group is regenerated (deprotected).

Equilibrium positions

All the steps in the acid-catalyzed acetal formation are equilibration reactions as shown in Fig. 19-13. Sterically hindered aldehydes and ketones do not form acetals in significant quantities. The equilibrium can be shifted toward the acetal by removing water as it is being formed. Carbohydrates (e.g., starch and cellulose) are examples of molecules containing stable acetal groups as a key structural component.

QUESTION 19-15

A chemist reacted an aqueous solution of acetone with methanol under acid-catalyzed conditions, but could not detect the desired acetal. Why was he not successful?

ANSWER 19-15

This equilibrium reaction is shifted toward the starting materials. Removing water as it is formed (if run under anhydrous conditions) would help shift the equilibrium toward the acetal, but since the reaction is run in water, little, if any, acetal will be formed.

REACTION WITH AMINES

Imine formation

Aldehydes and ketones react with ammonia and primary amines to form imines (−C=NH or −C=NR). Imines are also known by an older name, *Schiff bases*. Imine formation is acid-catalyzed. A pH of 4.5 is an optimum condition. The reaction mechanism is shown in Fig. 19-15. A pH in this range protonates the carbonyl oxygen atom forming a carbocation (19-15a) that is more susceptible to nucleophilic attack by the amine than is the unprotonated carbonyl group. A lower pH (higher acidity) would protonate the amine to a greater extent, reducing its nucleophilicity. Higher pH levels result in less protonation of the carbonyl oxygen atom. Reaction of an amine with the carbocation gives a protonated amino alcohol (19-15b).

Loss of a proton from 19-15b gives a carbinolamine (amino alcohol, 19-15c). Protonation of this alcohol and loss of a molecule of water gives the resonance-stabilized carbocation 19-15d. Loss of a proton from 19-15d gives

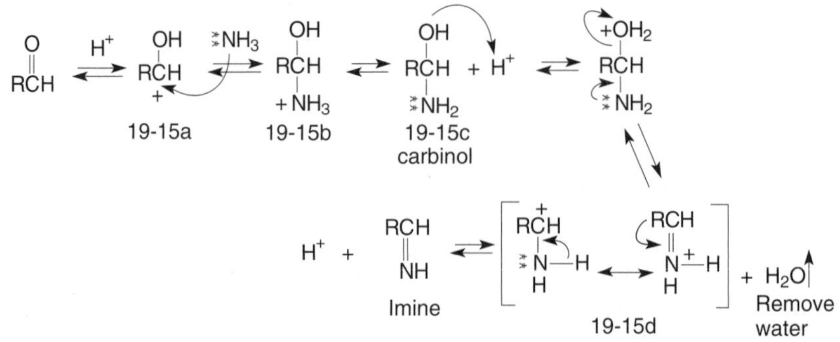

Fig. 19-15. Imine formation.

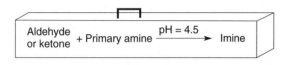

Fig. 19-16. Imine derivatives of aldehydes and ketones.

an imine. Imine formation is a reversible reaction. Removal of water as it is being formed shifts the equilibrium toward imine formation. Most imines are readily converted to the corresponding amine and aldehyde or ketone by the reverse of the reaction shown.

| Aldehyde or ketone + Primary amine $\xrightarrow{pH = 4.5}$ Imine |

Toolbox 19-10.

A variety of amine compounds are used to make derivatives of aldehydes and ketones. Examples of derivatives are given in Fig. 19-16. Reaction with hydroxylamine gives oximes (Reaction (a)), with hydrazine gives hydrazones (Reaction (b)), and with semicarbazide gives semicarbazones (Reaction (c)). If one is trying to confirm the structure of an unknown aldehyde or ketone, and the measured properties cannot differentiate between two (or more) possible structures, imine derivatives of the unknown aldehyde or ketone can be made. The derivatives of different compounds usually have different properties (e.g., melting points), allowing one to identify the unknown. Nowadays, NMR and IR spectroscopies are more efficient ways of confirming structures rather than the time-consuming synthesis and characterization of derivatives.

Enamine formation

Primary amines react with carbonyl compounds to form imines. *Secondary amines*, however, react with aldehydes and ketones to form *enamines (vinyl amines)*. The product is an alkene and amine, and hence the name enamine. This reaction is shown in Fig. 19-17. The nitrogen atom of a secondary amine does not have a second hydrogen atom bonded to it that can be removed to

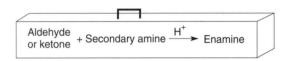

Fig. 19-17. Enamine formation.

give an imine. Instead, a hydrogen atom bonded to an α carbon atom in the aldehyde or ketone is removed, resulting in an enamine.

Aldehyde or ketone + Secondary amine $\xrightarrow{\ H^+\ }$ Enamine

Toolbox 19-11.

Wittig Reactions

The *Wittig reaction* is a method of making carbon–carbon bonds by converting a carbon–oxygen double bond into a carbon–carbon double bond. A Wittig reagent is a phosphorus-containing compound called an *ylide* (ill id), in which the major resonance form contains positive and negative charges on adjacent atoms. The reaction mechanism is shown in Fig. 19-18.

Wittig reagents are prepared by reacting triphenylphosphine (Ph_3P; Ph represents a phenyl group) with an unhindered (usually primary) alkyl halide. This is shown in Reaction (a). An S_N2 reaction results in a phosphonium salt, Structure 19-18a. Reaction of this salt with a strong base (butyllithium) gives a

$$Ph_3\ddot{P} + RCH_2{-}X \longrightarrow Ph_3\overset{+}{P}{-}CH_2R + X^- \xrightarrow{BuLi} Ph_3P{=}CHR \longleftrightarrow Ph_3\overset{+}{P}{-}\overset{-}{C}HR$$

19-18a
Phosphonium salt

19-18b
Ylide (major form)

$$Ph_3\overset{+}{P}{-}\overset{-}{C}HR \longrightarrow \quad Ph_3\overset{+}{P}{-}CHR \longrightarrow Ph_3P{-}CHR \longrightarrow Ph_3P + CHR$$

19-18b

19-18c
Betaine

19-18d
Oxaphosphetane

Alkene

Fig. 19-18. Wittig reaction.

phosphorus ylide, whose major resonance form is the charge-separated structure (19-18b).

The ylide (19-18b) reacts with the carbonyl group of an aldehyde or ketone (Reaction (b)), giving a *betaine* (bay-tuh-ene), 19-18c, a species with a positive-charged phosphorus atom and a negative-charged oxygen atom bonded to adjacent carbon atoms. The betaine closes to a four-membered ring, an *oxaphosphetane*, 19-18d. The ring reopens in a different way, giving an alkene and triphenylphosphine oxide.

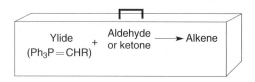

Toolbox 19-12.

QUESTION 19-16
What product results when the ylide of triphenylphosphine and methyl bromide is reacted with acetone?

ANSWER 19-16
2-Methylpropene, $(CH_3)_2C=CH_2$.

Oxidation and Reduction Reactions

OXIDATION OF ALDEHYDES

Aldehydes can be oxidized to carboxylic acids by a variety of oxidizing reagents, such as chromic acid, permanganate anion, or silver salts, as shown in Fig. 19-19. Silver ion (Ag^+) is reduced to silver metal $(Ag°)$, which forms a silver "mirror" on the side of the glass reaction vessel or a black precipitate of finely divided silver metal. This reaction is often used as a quick laboratory (Tollen's) test for the presence of an aldehyde group. The reaction of aldehydes with cupric salts (Fehling's or Benedict's reagent) oxidizes the aldehyde to a carboxylic acid and reduces the cupric ion (Cu^{2+}) to the cuprous ion (Cu^+). Under basic conditions, red cuprous oxide (Cu_2O) precipitates. This reaction was once used by diabetic patients to check for glucose (an aldehyde) levels in their urine. These reactions are shown in Fig. 19-19.

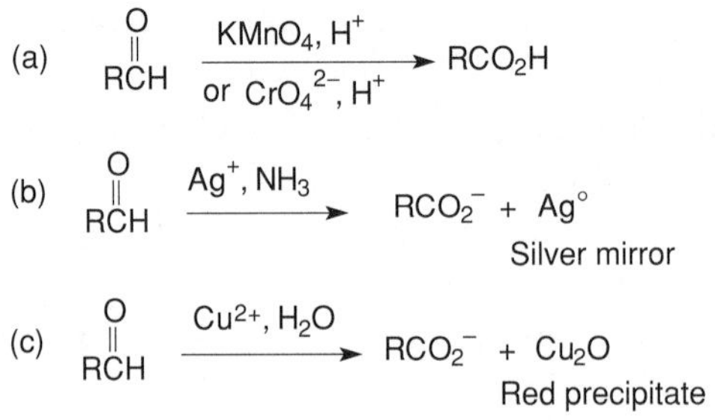
(a)
O
RCH
KMnO4, H+
or CrO42−, H+
RCO2H
(b)
O
RCH
Ag+, NH3
RCO2− + Ag°
Silver mirror
(c)
O
RCH
Cu2+, H2O
RCO2− + Cu2O
Red precipitate

Fig. 19-19. Oxidation of aldehydes.

OXIDATION OF KETONES

Ketones are not oxidized by the same reagents under the reaction conditions that oxidize aldehydes, since a carbon–carbon bond would have to be broken in the ketone for oxidation to occur. However, carbon–carbon bonds in ketones can be cleaved by vigorous heating of the ketone with aqueous basic $KMnO_4$. Either bond adjacent to the carbonyl group can be broken. The reaction is most useful for symmetrical ketones, otherwise a mixture of acids is obtained, as shown in Fig. 19-20.

REDUCTION OF ALDEHYDES AND KETONES

Aldehydes and ketones are reduced by LAH or sodium borohydride ($NaBH_4$) to primary and secondary alcohols. These reactions are discussed in more detail in the section entitled *Reactions with LAH and* $NaBH_4$ in Chapter 13. Carbonyl groups can also be reduced by catalytic hydrogenation. However, catalytic hydrogenation also reduces carbon-carbon double bonds. LAH and $NaBH_4$ reduce

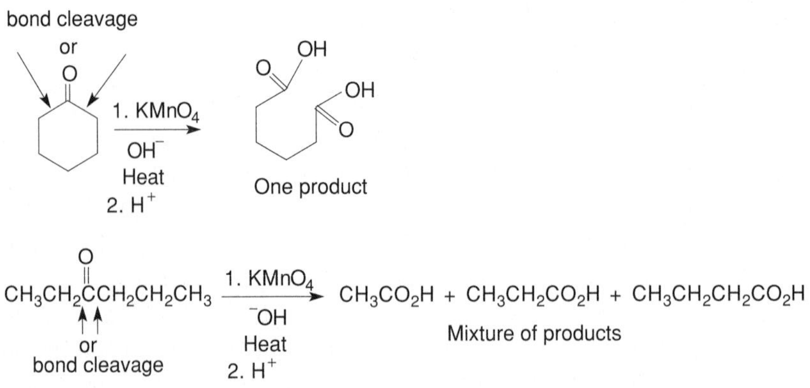
bond cleavage
or
O
1. KMnO4
OH−
Heat
2. H+
OH
O
OH
O
One product
O
CH3CH2CCH2CH2CH3
or
bond cleavage
1. KMnO4
−OH
Heat
2. H+
CH3CO2H + CH3CH2CO2H + CH3CH2CH2CO2H
Mixture of products

Fig. 19-20. Oxidation of ketones.

carbonyl groups more rapidly than they reduce alkenes and therefore the carbonyl is selectively reduced in the presence of a carbon-carbon double bond.

Clemmensen and Wolf–Kishner reactions are used to reduce carbonyl groups to the corresponding alkane, $RC(O)R' \rightarrow RCH_2R'$. These reactions are discussed in more detail in the section entitled *Alkyl Groups* in Chapter 18.

Quiz

1. The formal name for aldehydes ends in
   (a) -one.
   (b) -ide
   (c) -al.
   (d) -ane

2. The formal name for ketones ends in
   (a) -ane
   (b) -ide
   (c) -al
   (d) -one

3. Aldehydes have ———— boiling points than/as alcohols of comparable molecular weight.
   (a) higher
   (b) lower
   (c) about the same

4. Acetone is the common name for
   (a) 1-butanone
   (b) 1-propanone
   (c) 2-propanone
   (d) 2-butanone

5. Which material is most reactive toward nucleophilic addition?
   (a) acetaldehyde
   (b) dimethyl ketone
   (c) diisopropyl ketone
   (d) di-*tert*-butyl ketone

6. The oxidation of 1-propanol with chromic acid yields
   (a) an alcohol
   (b) an aldehyde
   (c) a ketone
   (d) an acid

7. The oxidation of 2-propanol with chromic acid yields
    (a) an alcohol
    (b) an aldehyde
    (c) a ketone
    (d) an acid

8. The oxidation of 1-propanol with PCC gives
    (a) an alcohol
    (b) an aldehyde
    (c) a ketone
    (d) an acid

9. The oxidation of _____ with ozone followed by reduction with $(CH_3)_2S$ yields one product, 1,5-pentanedial.
    (a) 1,5-pentanediol
    (b) 2,4-pentanedione
    (c) cyclopentene
    (d) cyclopentane

10. The acid-catalyzed hydration of cyclodecyne yields
    (a) cyclodecanone
    (b) cyclodecadiene
    (c) cyclodecanediol
    (d) cyclodecanedial

11. Reduction of acetic acid with Li(tri-*tert*-butoxy)AlH yields
    (a) an acid
    (b) an aldehyde
    (c) an alcohol
    (d) a ketone

12. The reaction of acetyl chloride with excess Grignard reagent followed by acidification yields
    (a) a ketone
    (b) an ester
    (c) a secondary alcohol
    (d) a tertiary alcohol

13. Lithium dimethylcuprate (Gilman reagent) reacts with acetyl chloride to give
    (a) a ketone
    (b) an ester
    (c) a secondary alcohol
    (d) a tertiary alcohol

14.   The cyanide ion $^-C\equiv N$ reacts with a ketone to give
(a) an acid
(b) a cyanohydrin
(c) an aldehyde
(d) an amine

15.   The acid-catalyzed addition of water to formaldehyde gives
(a) a ketal
(b) a hemiacetal
(c) an acetal
(d) a hydrate

16.   The acid-catalyzed addition of 2 mol of alcohol to 1 mol of aldehyde yields
(a) a ketal
(b) a hemiacetal
(c) an acetal
(d) a hydrate

17.   Ammonia reacts with an aldehyde to give an
(a) alcohol
(b) amide
(c) imine
(d) amine

18.   The reaction of a Wittig reagent with ketones give
(a) alkenes
(b) ketones
(c) aldehydes
(d) alkanes

19.   Silver ion, Ag^+, reacts with aldehydes to form
(a) a red precipitate
(b) a silver "mirror"
(c) a blue solution
(d) there is no reaction

20.   Dimethylamine reacts with a ketone to give an
(a) amide
(b) imide
(c) amine
(d) enamine

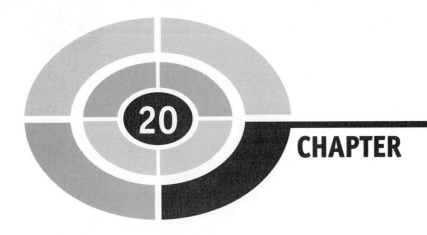

Carboxylic Acids

Introduction

A large variety of aliphatic and aromatic carboxylic acids are found in nature and many more are synthesized in laboratories. Carboxylic acids contain a *carboxyl* group that is made of a *carb*onyl and a hydr*oxyl* group. The carboxyl group is usually represented in one of the three following three ways:

$$-\overset{\overset{\textstyle O}{\|}}{C}OH \qquad -CO_2H \qquad -COOH$$

The structure $-COOH$ may suggest (incorrectly) that there is an oxygen–oxygen peroxide bond. Remember, an uncharged carbon atom has four bonds. The oxygen atom following the carbon atom is doubly bonded to that carbon atom, as shown in the first structure.

Carboxylic acids containing about 12 or more carbon atoms are called fatty acids. Many of these acids were originally obtained from animal fats, and hence the name fatty acids. Acids derived from animal fats typically contain 18 carbon atoms. Carboxylic acids are often just called acids. This term can be confused with inorganic acids (HCl etc.) that are also just called acids. Organic acids should really be called carboxylic acids to avoid this confusion.

Lower molecular weight carboxylic acids do not have pleasant aromas. The most familiar acid, acetic acid (found in vinegar), is used to flavor foods (e.g., in salad dressing) but most of us would not consider it to have a pleasant aroma. Aromas of other acids are compared to putrid butter (butanoic acid) and dirty socks (hexanoic acid).

Nomenclature

FORMAL NAMES

Monocarboxylic acids

The formal (IUPAC) names for carboxylic acids are based on the longest, continuous carbon–carbon (the parent) chain containing the carboxyl group as the terminal group. Several examples are given in Fig. 20-1. The terminal -e in the name of the corresponding alkane, alkene, or alkyne is replaced by -oic followed by the word acid. Thus ethane becomes ethanoic acid (Structure 20-1a). Butane becomes butanoic acid (Structure 20-1b). Substituents (branches) are named alphabetically and the point of attachment to the parent chain is indicated by a number, as shown in Structure 20-1c. The carbon atom in the carboxyl group is always carbon atom number one (C–1) in acyclic carboxylic acids.

QUESTION 20-1
Give the formal name of a linear carboxylic acid that contains six carbon atoms with a carboxyl group in the terminal position.

ANSWER 20-1
Hexanoic acid, from hexan(e), oic, and acid.

When the carboxyl group is bonded to a ring, the ring is named followed by the suffix carboxylic acid. An example is cyclopentanecarboxylic acid, Structure 20-1d. The terminal -e of the cyclic alkane is not dropped. The ring atom bonded to the carboxyl group is carbon atom (C-1) in cyclic carboxylic acids.

Polycarboxylic acids

Formal names for di- and polycarboxylic acids are based on the name of the parent chain, and the suffix dicarboxylic (or polycarboxylic) acid is added without dropping the terminal -e in the parent name. The carbon atoms in the carboxyl group are counted as members of the parent chain. Examples are shown in Fig. 20-1 (Structures 20-1e–j).

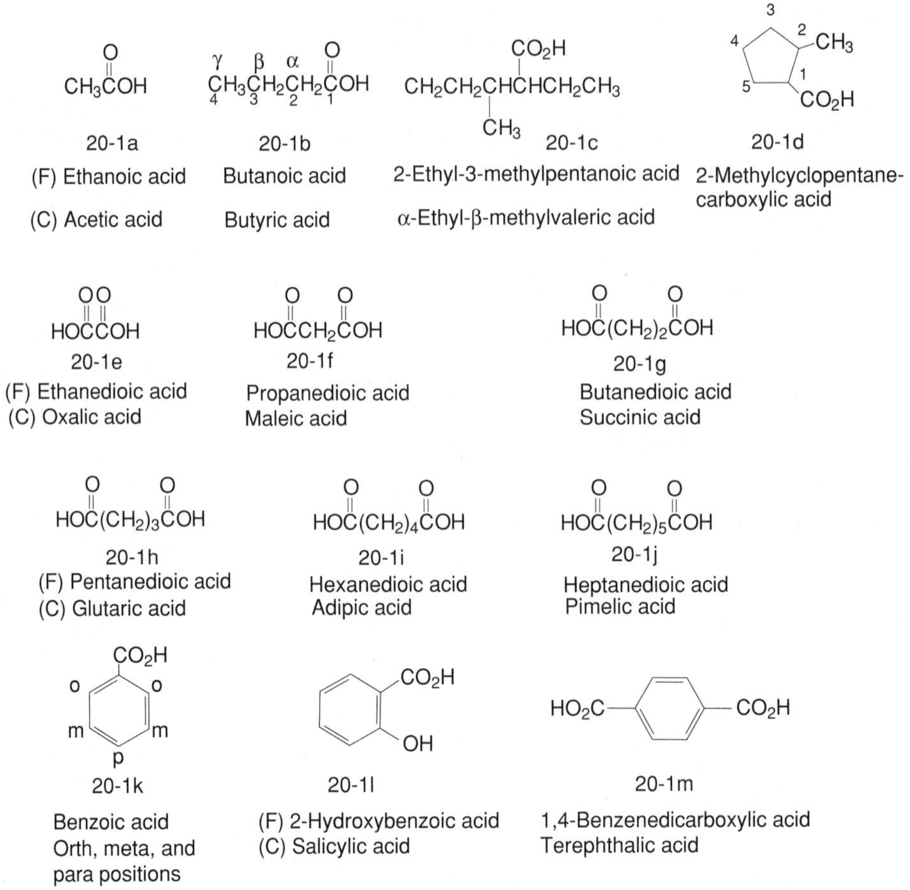

Fig. 20-1.   Formal (F) and common (C) names for carboxylic acids.

COMMON NAMES

Aliphatic carboxylic acids

Lower molecular weight carboxylic acids were initially isolated from natural sources and were given names (in Latin or Greek) based on their source. Common names of these acids are given in Table 20-1. The positions of substituents in common names are preceded by the Greek letters α, β, γ, δ, etc. The carbon atom next to the carboxyl group is the α-carbon atom (see Structure 20-1b).

QUESTION 20-2
Is the hydroxyl group in 2-hydroxyheptanoic acid on the α-, β-, γ-, or δ-carbon atom?

ANSWER 20-2

It is on the α-carbon atom.

The mnemonic "*oh my such good apple pie*" can be used to remember the common names of the first five dicarboxylic acids: *o*xalic, *m*alonic, *s*uccinic, *g*lutaric, *a*dipic, and *p*imelic acids. Or, make up your own mnemonic that is easy for you to remember.

Aromatic carboxylic acids

The simplest aromatic carboxylic acid is benzoic acid (20-1k). Ring substituents can be assigned numbers to indicate their position on the ring (the carboxyl group is always on ring carbon atom C–1). If benzoic acid contains one substituent, its position can be indicated by the prefix ortho- (o-), meta- (m-), or para- (p-). The ortho position corresponds to ring atoms C–2 or C–6, the meta position corresponds to ring atoms C–3 or C–5, and the para position corresponds to ring atom C–4. You may use the mnemonic "*oh my papa*" to remember the ortho, meta, and para prefixes. Structure 20-1k shows the ortho, meta, and para positions. If benzoic acid contains more than one substituent, the numbering system (1, 2, etc.) must be used.

QUESTION 20-3

Is the chloro atom in the ortho, meta, or para position in 3-chlorobenzoic acid?

ANSWER 20-3

It is in the meta position.

ortho-Hydroxybenzoic (*o*-hydroxybenzoic acid or 2-hydroxybenzoic) acid (Structure 20-1l) has the common name salicylic acid. It was originally obtained from the bark of willow trees. A derivative of this material, acetyl salicylic acid, is more commonly called aspirin. Terephthalic acid (Structure 20-1m) has the formal name 1,4-benzenedicarboxylic acid. It is a primary component in polyester fibers.

Nomenclature hierarchy

If there is more than one functional group in a molecule, there is a hierarchy for assigning the parent. The hierarchy order, highest to lowest, is: carboxylic acid, ester, amide, nitrile, aldehyde, ketone, alcohol, thiol, amine, alkene, alkyne, alkane, ether, and halide. The highest priority group present is the parent in the name. If a compound contains a carboxylic acid and a ketone group, the compound would be named as an acid and the ketone oxygen atom would be named as a substituent.

CARBOXYLATE SALTS

Salts of carboxylic acids are made by replacing the acid hydrogen ($-CO_2H$) with a metal cation or ammonium ion. The salts are named by first naming the metal (or ammonium) followed by the name of the *carboxylate anion*. The anion is named by replacing the suffix -ic in the name of the parent acid with -ate. The sodium salt of ace*tic* acid (CH_3CO_2H) is called sodium ace*tate* ($CH_3CO_2^- Na^+$).

Three-Dimensional Structure

The carbon atom in the carboxyl group ($-CO_2H$) is sp^2 hybridized as is the carbonyl oxygen atom. All bond angles around the carboxyl carbon atom are about 120°, and the molecular geometry is trigonal planar. Three Lewis resonance structures, 20-2a, 20-2b, and 20-2c shown in Fig. 20-2, can be drawn for the carboxyl group. The structures with charge separation (Structures 20-2b and 20-2c) are less important resonance structures, but their contribution is shown by the polarity of the carbonyl group in the resonance hybrid.

QUESTION 20-4
Why is Structure 20-2a the best resonance structure?

ANSWER 20-4
The best resonance structures have atoms that are octet or duet happy and have no charge separation, which is the case for Structure 20-2a.

Molecular Orbital (MO) Description

An MO diagram can also be used to describe the electronic properties of the carboxylate anion. Localized valence bonds are usually used to describe the sigma bond framework and MOs are used to describe the π bonds. Three atomic p orbitals, one on the sp^2 hybridized carbon atom and one on each of the oxygen

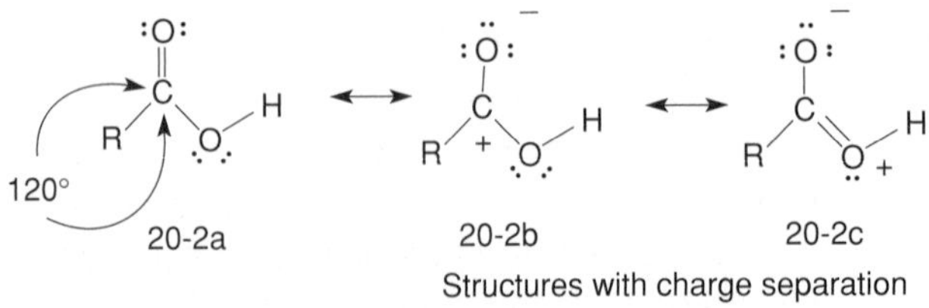

Fig. 20-2.   Resonance structures of the carboxyl group.

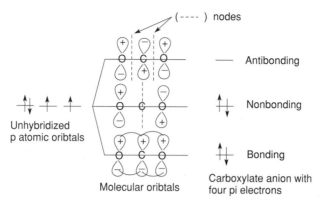

Fig. 20-3.   MO description of a carboxylate anion.

atoms, are used to form the MOs shown in Fig. 20-3. Three atomic p orbitals form three π MOs: a bonding, nonbonding, and antibonding orbital. Two of the π electrons go into the lowest (bonding) MO. This bonding MO has constructive orbital overlap (all overlapping lobes have the same sign) of all three p orbitals and the two electrons in this MO are shared with all three atoms. This suggests a π bond exists to some extent between all adjacent atoms. The other two π electrons go into the nonbonding MO. Since there is a node on the carbon atom, the electrons spend their time on the two oxygen atoms, resulting in a negative charge on these atoms. This is in agreement with Lewis resonance structures that show a negative charge on each oxygen atom (as shown in Fig 20-6).

The carbon atom in the carbonyl group in carboxylic acids is less reactive (electrophilic) than the carbon atom in the carbonyl group in aldehydes and ketones. The electrons on the —OH oxygen atom are shared with the carbonyl carbon atom, making it less electropositive than the carbonyl carbon atom in aldehydes or ketones. (There is no —OH group bonded to the carbonyl group in aldehydes and ketones.)

Physical Properties

SOLUBILITIES

The carboxyl group is polar and hydrophilic. Low molecular weight alkyl carboxylic acids containing four or fewer carbon atoms are miscible (completely soluble) in water. As the hydrophobic (alkyl) group of the carboxylic acid increases in length, water solubility decreases. Decanoic acid is only 0.2 wt % soluble in water.

QUESTION 20-5
Which acid is more soluble in water, $CH_3(CH_2)_8CO_2H$ or $HO_2C(CH_2)_8CO_2H$?

ANSWER 20-5
$HO_2C(CH_2)_8CO_2H$, since it contains two hydrophilic groups per 10 carbon atoms, while $CH_3(CH_2)_8CO_2H$ contains one hydrophilic group per 10 carbon atoms.

Soluble salts of carboxylic acids

Loss of the acidic proton from a carboxylic acid results in a carboxylate anion. The anion is more hydrophilic than its conjugate acid. Group I and ammonium carboxylate salts are soluble in water, although solubility decreases with increasing molecular weight of the alkyl group. The hydrophobic alkyl segments in higher molecular weight acids (containing about 18 carbon atoms) aggregate to form spherical or cylindrical micelles composed of 100 to 200 carboxylate molecules. The structure of a spherical micelle is shown in Fig. 20-4. The interior of these micelles is hydrophobic and absorbs/attracts hydrophobic (oleophilic) materials (oils and dirt). The outer surface is hydrophilic, allowing the micelle to be suspended in water. Hence these carboxylate salts are used as soaps for dispersing non-water-soluble materials.

Insoluble salts

Calcium and magnesium salts of carboxylic acids are insoluble in water. When calcium and magnesium ions are added to aqueous solutions of sodium or potassium carboxylates, a calcium or magnesium carboxylate salt precipitates. This is the familiar "soap scum" or "bathtub ring." Water softeners are devices that replace the calcium and magnesium (hard) ions in water with sodium (soft) ions eliminating the precipitation problem.

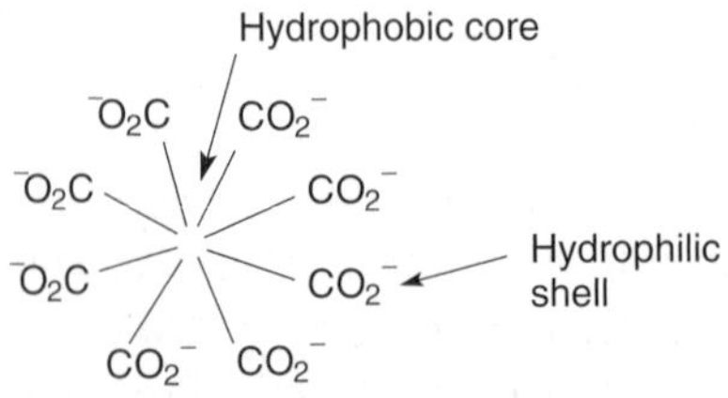

Fig. 20-4. Micelle structure.

Fig. 20-5. Dimeric structure of carboxylic acids.

BOILING POINTS

Boiling points of carboxylic acids are higher than those of alcohols, ketones, and aldehydes of comparable molecular weight. The boiling points of carboxylic acids are even higher than one would predict based on their structure. This apparently results from the dimeric interaction between molecules effectively doubling their molecular weight. As shown in Fig. 20-5, an eight-membered ring results from hydrogen bonding between the carbonyl oxygen atom in one molecule and the acidic hydrogen atom in another molecule.

QUESTION 20-6
Which has a higher boiling point, $CH_3CH_2CO_2H$ or $CH_3CH_2CH_2OH$?

ANSWER 20-6
$CH_3CH_2CO_2H$, since it forms a dimeric structure.

Acidity of Carboxylic Acids

Aqueous solutions of carboxylic acids are much weaker acids than inorganic acids, such as HCl, but are much stronger acids than most other types of hydrocarbons (alcohols, ketones, alkynes, etc.). Carboxylic acids typically have pK_a values of 4 to 5. Inorganic acids have negative pK_a values. Alcohols have pK_a values of about 15 to 16.

Carboxylic acids and alcohols can both act as Brønsted–Lowry acids and ionize to give a proton (H^+) and an anionic conjugate base. Each base has a negative charge on an oxygen atom. The carboxylate anion is more stable (is a weaker base) than an alkoxide anion. One reason for this stability is that the electrons in the carboxylate anion can be delocalized over both oxygen atoms. Delocalization increases stability. As the stability of an anion increases, the acid dissociation equilibrium shifts to produce more of the anion, concurrently increasing the hydrogen ion concentration (acidity). (For a review, the properties of acids are discussed in Chapter 3.) Acid dissociation equations and Lewis resonance structures of the two anions (carboxylate and alkoxide) are shown in

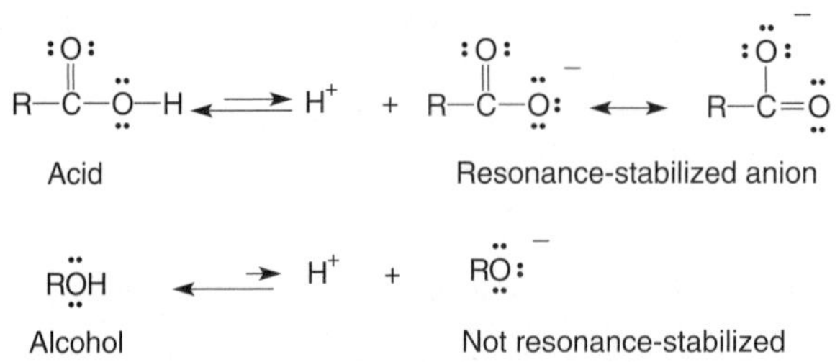

Fig. 20-6. Conjugate Base Stability.

Fig. 20-6. The MO description of the carboxylate anion (Fig. 20-3) also shows that the negative charge is equally dispersed (delocalized) on both oxygen atoms, in agreement with the Lewis strucures.

INDUCTIVE EFFECTS

Decreasing the electron density of an anion, by delocalizing electrons, increases the stability of that anion. Electron withdrawal through sigma bonds is called an inductive effect. Electron-withdrawing atoms or groups bonded to the carbon atom α (adjacent) to the carboxylate group inductively reduce the electron density in, and increase the stability of, the carboxylate anion. The more stable the carboxylate anion, the further the acid–base equilibrium will shift to produce more carboxylate anions and protons, resulting in an increase in acidity. This effect is seen in a series of chlorinated acetic acids. The chlorine atom is more electronegative than a carbon atom and withdraws electrons inductively. The pK_a values of acetic, monochloroacetic, dichloroacetic, and trichloroacetic acids are 4.75, 2.85, 1.48, and 0.68, respectively. Smaller pK_a values indicate stronger acids. Electron-withdrawing atoms bonded to the β-carbon atom decrease the electron density in the carboxylate anion but to a lesser extent.

QUESTION 20-7
Which is a stronger acid, 2-bromopropionic acid or 3-bromopropionic acid?

ANSWER 20-7
2-Bromopropionic acid; bromine is more effective in withdrawing electrons the closer it is to the carboxylate group.

RESONANCE EFFECTS

An increase or decrease of acidity can also be influenced by the resonance delocalization of electrons. A series of para-substituted benzoic acids and their

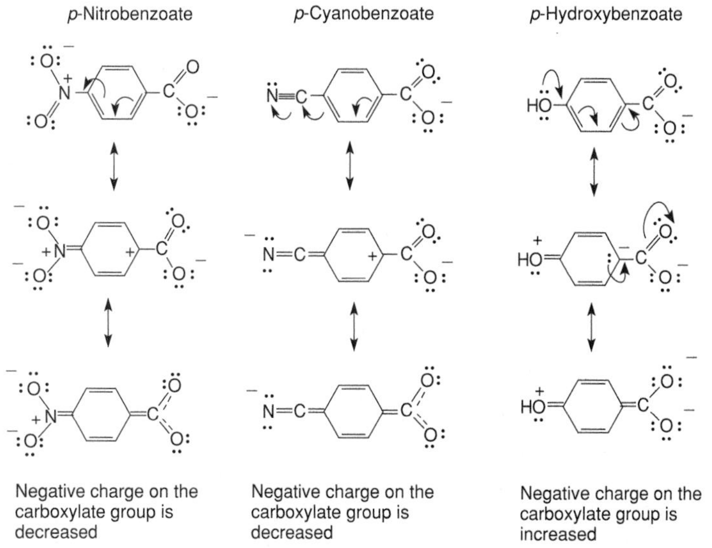

Fig. 20-7.   Resonance stabilized.

corresponding pK_a values is shown in Fig. 20-7. Substituted benzoic acids containing electron-withdrawing groups, such as NO_2 and CN, in the para position are more acidic than benzoic acid (with an H in the para position). Substituted benzoic acids containing electron-donating groups, such as OH, in the para position are less acidic than benzoic acid. (For a review of electron-donating and -attracting groups, see the section entitled *Multiple Substitution Reactions* in Chapter 18.) Increasing the electron density in a carboxylate group decreases its stability and its tendency to form, while decreasing the electron density increases its stability and favors its formation. In acid dissociation equilibrium reactions, factors that increase the stability of the carboxylate anion favor its formation and consequently the proton concentration (acidity).

Preparation of Carboxylic Acids

OXIDATION OF ALCOHOLS

Carboxylic acids can be made by oxidizing various functional groups. Some common oxidation reactions are shown in Fig. 20-8. Primary alcohols can be oxidized to carboxylic acids with chromic acid (represented by CrO_3, $HCrO_4^-/H^+$, or $Cr_2O_7^{2-}/H^+$). Basic solutions of permanganate anion will also oxidize primary alcohols to carboxylate salts. The alcohol is first oxidized to an aldehyde,

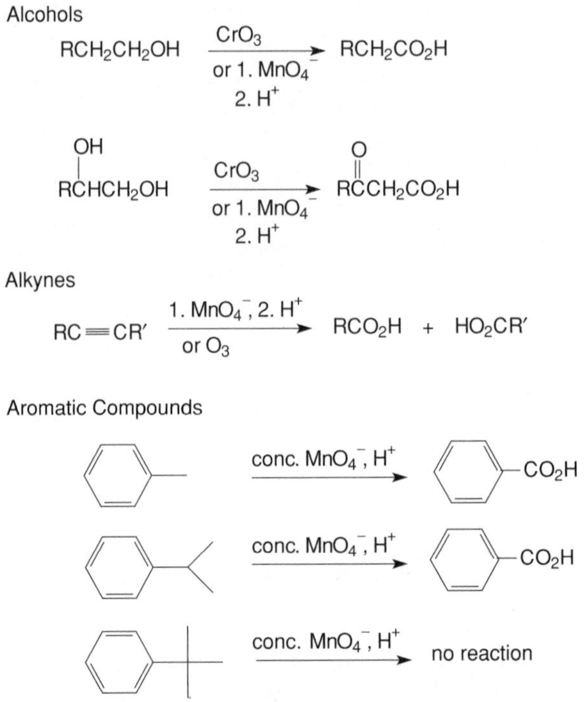

Alcohols

$$RCH_2CH_2OH \xrightarrow[\substack{\text{or 1. } MnO_4^- \\ 2.\ H^+}]{CrO_3} RCH_2CO_2H$$

$$\underset{\underset{OH}{|}}{RCHCH_2OH} \xrightarrow[\substack{\text{or 1. } MnO_4^- \\ 2.\ H^+}]{CrO_3} \underset{\underset{O}{\|}}{RCCH_2CO_2H}$$

Alkynes

$$RC\equiv CR' \xrightarrow[\text{or } O_3]{1.\ MnO_4^-,\ 2.\ H^+} RCO_2H\ +\ HO_2CR'$$

Aromatic Compounds

Fig. 20-8. Oxidation reactions leading to carboxylic acids.

which is rapidly oxidized to the carboxylic acid. Hence aldehydes can also be directly oxidized to carboxylic acids.

QUESTION 20-8
What product results from the chromic acid oxidation of $CH_3CH(OH)CH_2CH_2OH$?

ANSWER 20-8
$CH_3C(O)CH_2CO_2H$, the primary alcohol is oxidized to a carboxylic acid and the secondary alcohol is oxidized to a ketone, not an acid.

OXIDATION OF ALKENES

Hot permanganate solutions will cleave and oxidize alkenes to carboxylic acids. If a carbon atom in a carbon–carbon double bond is monosubstituted (it contains just one hydrogen atom), it is oxidized to a carboxylic acid. If the carbon atom in the double bond is disubstituted (it is not bonded to a hydrogen atom), it is oxidized to a ketone. Carbon–carbon sigma bonds are more difficult to break in oxidation reactions and are stable under the conditions used to leave double bonds.

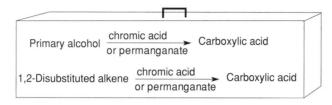

Toolbox 20-1.

QUESTION 20-9
What product results from the permanganate oxidation of cyclopentene?

ANSWER 20-9
Pentanedioic acid.

OXIDATION OF ALKYNES

Alkynes can be oxidized to carboxylic acids by treatment with concentrated aqueous permanganate solutions or with ozone as shown in Fig. 20-8. One method of determining where alkene and alkyne functions are in large, complex molecules is to oxidize the molecule. The resulting smaller fragments with ketone and acid groups are easier to characterize. Like a puzzle, one can then determine how the smaller pieces would fit together to reconstruct the larger starting molecule.

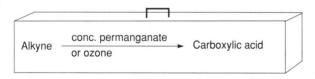

Toolbox 20-2.

OXIDATION OF ALKYL BENZENES

Primary and secondary alkyl substituents on a benzene ring are also oxidized to carboxylic acids by hot, concentrated permanganate or hot chromic acid solutions. The carbon atom *bonded directly* to the benzene ring is oxidized to a carboxyl group. This carbon atom needs to be bonded to at least one hydrogen atom for the oxidation to occur. A tertiary carbon atom bonded to the benzene ring is not oxidized by these reagents, since no hydrogen atom is bonded to it.

Very rigorous oxidation reaction conditions are required, and no other oxidizable groups can be present on the benzene ring (unless one also wants to oxidize these groups). These reactions are shown in Fig. 20-8.

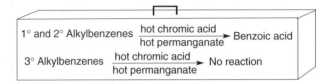

Toolbox 20-3.

QUESTION 20-10
What product results from the permanganate oxidation of 1,4-diethylbenzene?

ANSWER 20-10
1,4-Benzenedicarboxylic acid (terephthalic acid).

ORGANOMETALLIC REAGENTS

Organometallic reagents such as alkyllithium and Grignard reagents (discussed in the section entitled *Organometallic Compounds* in Chapter 13) react with carbon dioxide to form carboxylate salts. Examples are shown in Fig. 20-9. The carbon atom in the alkyl group bonded to the lithium or magnesium atom has a partial negative charge and is nucleophilic. The carbon atom in carbon dioxide has a partial positive charge and is electrophilic. The nucleophilic carbon atom in the organometallic compound attacks and forms a bond with the electrophilic carbon atom in carbon dioxide.

The resulting carboxylate anion is protonated in a subsequent step. In predicting reactions, the most logical mechanism involves the interaction of an atom with a positive (+) or partial positive (δ^+) charge with an atom with a negative (−) or partial negative (δ^-) charge.

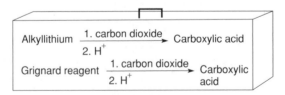

Toolbox 20-4.

QUESTION 20-11
What product results from the reaction of ethylmagnesium chloride with carbon dioxide followed by acidification?

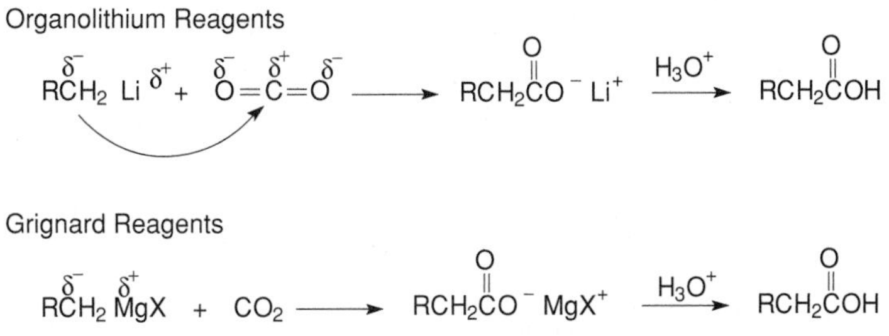

Organolithium Reagents

Grignard Reagents

Fig. 20-9.   Organometallic reactions.

ANSWER 20-11
Propionic acid. The alkyl (ethyl) group is increased in length by one carbon atom.

HYDROLYSIS OF NITRILES

The nitrile group consists of a highly polarized triple bond with a partial positive charge (δ^+) on the carbon atom and a partial negative charge (δ^-) on the nitrogen atom. Acid- or base-catalyzed hydrolysis (addition of water) gives initially an imine which is further hydrolyzed to an amide and then to a carboxylate group. This reaction is shown in Fig. 20-10.

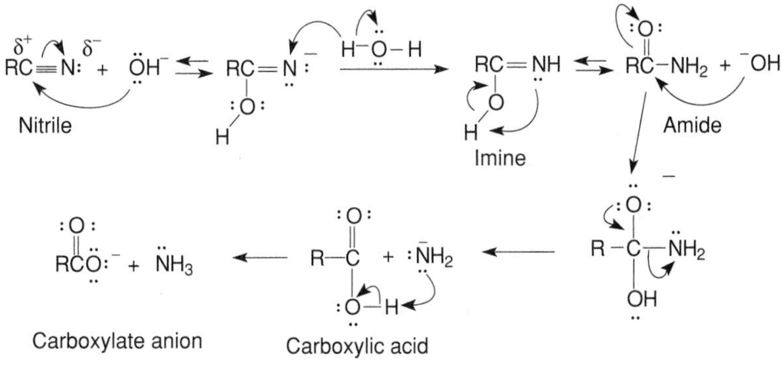

Fig. 20-10.   Hydrolysis of nitriles.

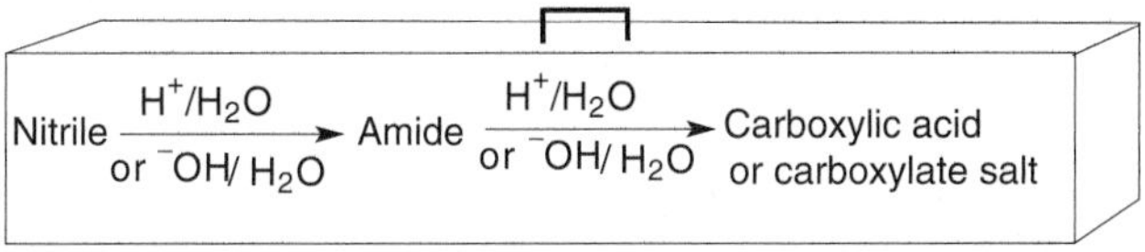

Toolbox 20-5.

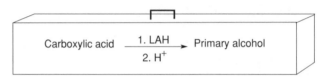

Fig. 20-11. Reduction of carboxylic acids.

REDUCTION OF CARBOXYLIC ACIDS

Carboxylic acids are reduced to alcohols by LAH. The initial reduction product is an aldehyde which is subsequently reduced to an alkoxide anion. In a second step, the alkoxide anion is protonated to give the corresponding primary alcohol. This reaction is shown in Fig. 20-11. Note that both carbonyl groups are reduced.

Carboxylic acid $\xrightarrow[\text{2. H}^+]{\text{1. LAH}}$ Primary alcohol

Toolbox 20-6.

QUESTION 20-12
What product results from LAH treatment of oxalic acid?

ANSWER 20-12
Ethylene glycol (1,2-ethanediol).

Derivatives of Carboxylic Acids

Carboxylic acids are excellent starting materials to make a variety of compounds containing other functional groups. The most common derivatives of carboxylic acids are carboxylic acid halides, carboxylic anhydrides, esters, and amides. These materials are discussed in Chapter 21.

Quiz

1. Cyclohexanetricarboxylic acid contains _____ carboxylic acid groups.
   (a) one
   (b) two
   (c) three
   (d) four

2. α-Bromobutyric acid is also called
   (a) 1-bromobutanoic acid
   (b) 2-bromobutanoic acid
   (c) 2-bromobutanoate acid
   (d) 3-bromobutanoic acid

3. The magnesium salt of acetic acid is called magnesium
   (a) formate
   (b) acetate
   (c) ethanoic acid
   (d) acetic acid

4. _____ is most soluble in water.
   (a) Sodium octanoate
   (b) Octanoic acid
   (c) Sodium decanoate.
   (d) Decanoic acid

5. The strongest acid is
   (a) 2-chloroacetic acid
   (b) 2,2-dichloroacetic acid
   (c) 3-chloroacetic acid
   (d) 2,3-dichloroacetic acid

6. The strongest acid is
   (a) p-nitrobenzoic acid
   (b) m-nitrobenzoic acid
   (c) p-hydroxybenzoic acid
   (d) m-hydroxybenzoic acid

7. Chromic acid oxidation of $HOCH_2CH_2CHO$ gives
   (a) $HOCH_2CH_2CO_2H$
   (b) HO_2CCH_2CHO
   (c) $HO_2CCH_2CO_2H$
   (d) $HOCH_2CH_2CHO$

8. Potassium permanganate oxidation of 3-hexene gives
   (a) potassium propionate
   (b) potassium hexanate
   (c) potassium acetate
   (d) potassium ethanoate

9. Chromic acid oxidation of isobutylbenzene gives
   (a) benzoic acid

(b) dimethylbenzoic acid

(c) isopropylbenzoic acid

(d) no reaction

10. Butylmagnesium chloride reacts with carbon dioxide to give
    (a) butyl alcohol
    (b) butyric acid
    (c) butyl chloride
    (d) pentanoate anion

11. Reaction of 2-methylpropanoic acid with LAH gives
    (a) ethylene glycol
    (b) 1,3-butylenediol
    (c) 2-methyl-1,2-propanediol
    (d) 2-methylpropanol

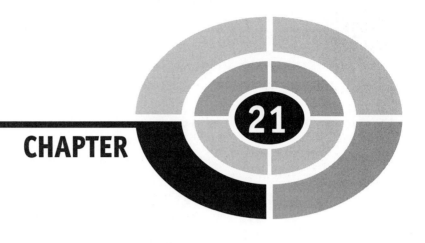

Derivatives of Carboxylic Acids

Introduction

Carboxylic acids have a terminal OH group bonded to a carbonyl (C=O) group. Derivatives of carboxylic acids are compounds where the OH group has been replaced by another group. The derivatives are not necessarily made from carboxylic acids, rather one derivative is often made from another derivative. The derivatives contain the RC(O) group, called an acyl (ā sil) group. The derivatives are prepared by addition–elimination reactions, also called acyl transfer reactions or nucleophilic acyl substitution reactions. Derivatives of carboxylic acids include carboxylic acid halides, anhydrides, esters, amides, and nitriles. A generic structure of a member of each family is shown in Fig. 21-1.

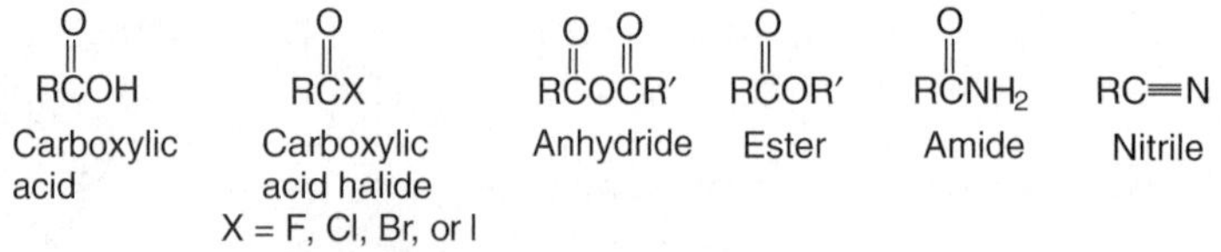

$$\underset{\substack{\text{Carboxylic} \\ \text{acid}}}{\overset{\overset{\displaystyle O}{\parallel}}{RCOH}} \qquad \underset{\substack{\text{Carboxylic} \\ \text{acid halide} \\ \text{X = F, Cl, Br, or I}}}{\overset{\overset{\displaystyle O}{\parallel}}{RCX}} \qquad \underset{\text{Anhydride}}{\overset{\overset{\displaystyle O\ \ O}{\parallel\ \ \parallel}}{RCOCR'}} \quad \underset{\text{Ester}}{\overset{\overset{\displaystyle O}{\parallel}}{RCOR'}} \quad \underset{\text{Amide}}{\overset{\overset{\displaystyle O}{\parallel}}{RCNH_2}} \quad \underset{\text{Nitrile}}{RC\!\equiv\!N}$$

Fig. 21-1. Derivatives of carboxylic acid.

REACTION MECHANISMS

The reactivities of the various derivatives are dependent on steric and electronic factors. The generic reaction mechanism is diagrammed in Fig. 21-2. Resonance structure 21-2a has a positive charge on the carbonyl carbon atom as a result of the more electronegative oxygen atom withdrawing the bonding electrons. Other groups attached to the carbonyl carbon atom can increase or decrease the extent of positive charge on this carbon atom and thus influence its reactivity. A nucleophile attacks the electron-deficient carboxyl carbon atom and the X group is subsequently eliminated. The reaction mechanisms are analogous to that shown in Fig.21-2 and therefore, are not repeated.

INDUCTIVE EFFECTS

X in Structure 21-2a can be a halogen, an oxygen, or a nitrogen atom. These atoms withdraw electrons inductively (through a sigma bond), increasing the positive charge on the carbon atom and making it even more susceptible to attack by a nucleophile.

RESONANCE EFFECTS

Halogen, oxygen, and nitrogen atoms can also (back-) donate nonbonding π electrons *to* the carbon atom by a resonance effect. (A more detailed description of inductive and resonance effects is given in the section entitled *Directing Effects* in Chapter 19.) A nitrogen atom donates its nonbonding electrons to the carbon atom to a greater extent than does an oxygen atom. Oxygen donates its nonbonding electrons to a greater extent than do halogen atoms. This resonance effect decreases the electropositive character of the carbonyl carbon atom. Thus

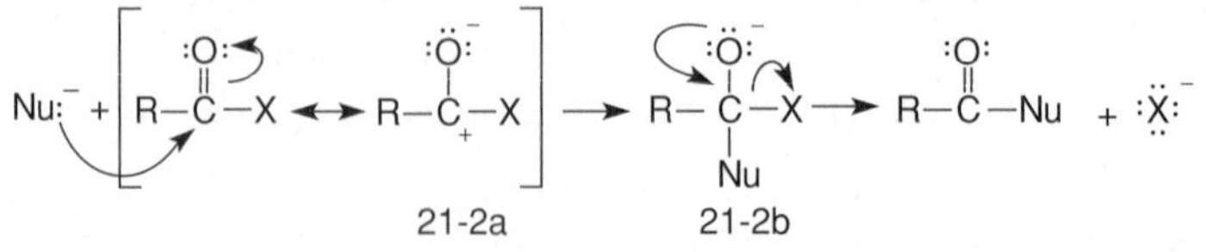

$$Nu\!:^- + \left[R\!-\!\overset{\overset{\displaystyle :O:}{\parallel}}{C}\!-\!X \longleftrightarrow R\!-\!\overset{\overset{\displaystyle :\overset{..}{O}:^-}{|}}{\underset{+}{C}}\!-\!X \right] \longrightarrow R\!-\!\overset{\overset{\displaystyle :\overset{..}{O}:^-}{|}}{\underset{\underset{\displaystyle Nu}{|}}{C}}\!-\!X \longrightarrow R\!-\!\overset{\overset{\displaystyle :O:}{\parallel}}{C}\!-\!Nu \ + \ :\overset{..}{\underset{..}{X}}:^-$$

$$\qquad\qquad\qquad\qquad\qquad 21\text{-}2a \qquad\qquad\qquad 21\text{-}2b$$

Fig. 21-2. Nucleophilic substitution reactions.

amides (which back-donate electrons to the greatest extent) are less reactive to-ward nucleophilic substitution reactions than are esters, which are less reactive than anhydrides, which are less reactive than acid halides.

QUESTION 21-1

Which compound acts as a better electrophile, a carboxylic acid chloride or an amide?

ANSWER 21-1

The carbonyl carbon atom in an acid chloride has a greater partial positive charge and is a better electrophile. Nitrogen reduces the partial positive charge on the carbon atom to a greater extent than does chlorine.

BASICITY OF LEAVING GROUP

A nucleophile attacks the carbonyl carbon atom forming a tetrahedral sp^3 carbon atom now bonded to two electron-withdrawing groups (X and O in Structure 21-2b). These tetrahedral structures tend to be unstable due to steric repulsion between the four groups bonded to the carbon atom. A more stable structure results if one of the electron-withdrawing groups leaves. If the incoming nu-cleophile leaves, the original molecule is obtained and there is no reaction. If the electron-withdrawing group initially bonded to the carbonyl group (X in Structure 21-2b) leaves, a new derivative is formed.

Weak bases are good leaving groups. Weak bases do not want to share their nonbonding electron pairs with other species. The order of basicity for the atoms mentioned above is N > O > halogen (weakest). Halogens are the best leaving groups in this series.

QUESTION 21-2

Which is the best leaving group, a chloride anion or an ethoxide anion?

ANSWER 21-2

A chloride anion because it is the weaker base of the two. (Consider the acidity of the two conjugated acids.)

REACTIVITY OF CARBOXYLIC ACID DERIVATIVES

The order of reactivity for the various derivatives is based on three factors: electron-withdrawing/donating effects, basicity of the leaving group, and steric effects.

The order of reactivity of carboxylic acid derivatives is shown in Fig. 21-3. Carboxylic acid halides are readily converted into anhydrides, esters, and amides. Anhydrides are readily converted into esters and amides. Esters are readily converted into amides. Amides are the most stable type of derivative. The

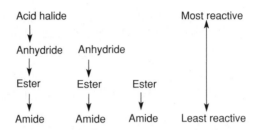

Fig. 21-3. Reactivity order of carboxylic acid derivatives.

derivatives can be made in the reverse order, but rigorous reaction conditions are generally required. Acids are not included in this series as they often are the starting materials for the various derivatives.

Carboxylic Acid Halides

Carboxylic acid halides are not found in nature. They are sufficiently reactive that they are readily converted into the parent acid or one of the other acid derivatives.

NOMENCLATURE

Carboxylic acid halides are often just called acid halides or acyl halides. The halide is usually chloride or bromide. Formal names are derived from the corresponding acid by replacing the terminal *-oic acid* in the formal name with *-oyl halide*. Ethanoic acid ($CH_3C(O)OH$) becomes ethanoyl chloride ($CH_3C(O)Cl$). Common names are derived by replacing the *-ic acid* ending in the common name with *-yl halide*. Acetic acid ($CH_3C(O)OH$) becomes acetyl chloride ($CH_3C(O)Cl$). Cyclic carboxylic acids become carbonyl halides. Cyclopentanecarboxylic acid becomes cyclopentanecarbonyl chloride.

QUESTION 21-3
What is the structure of propionyl chloride?

ANSWER 21-3
$CH_3CH_2C(O)Cl$.

PREPARATION OF CARBOXYLIC ACID HALIDES

Carboxylic acids are usually the starting materials for making acid halides. Since the OH group in the carboxylic acid is not a good leaving group, it has to be converted into a better leaving group. Thionyl chloride ($SOCl_2$) is commonly used to convert acids to acid chlorides. Figure 21-4 shows this reaction

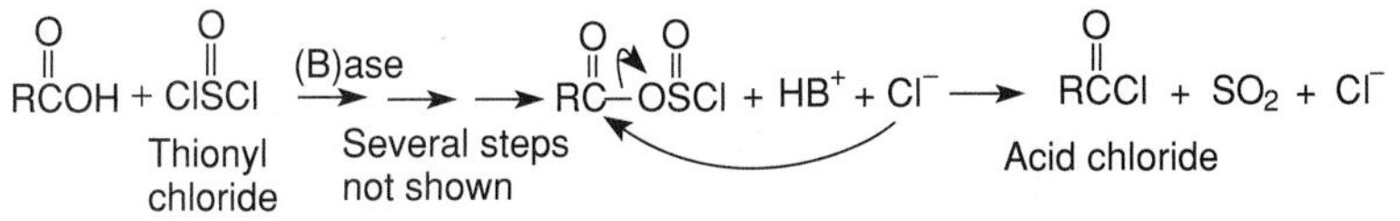

Fig. 21-4. Synthesis of an acid chloride.

mechanism. The poor OH leaving group is converted into a chlorosulfite group (OS(O)Cl), a good leaving group. The chloride anion acts as the nucleophile in an S_N2 reaction to form the acid chloride. Thionyl bromide ($SOBr_2$), phosphorous tribromide (PBr_3), phosphorous trichloride (PCl_3), and oxalyl chloride (ClC(O)C(O)Cl) are other reagents used to make acid halides. These reactions take place readily at room temperature. A base (B:) such as pyridine or a trialkyl amine is used to neutralize the hydrogen halide byproduct formed in the reaction.

REACTIONS OF CARBOXYLIC ACID HALIDES

Acid halides are the most reactive of the various acid derivatives. Acid halides are often the starting materials used to prepare other acid derivatives. The various reactions are shown in Fig. 21-5.

Reaction with alcohols

Acid halides react readily with alcohols to give esters (Structure 21-5a). A base (B:) is usually present to neutralize the hydrogen halide byproduct. As a result of steric effects, primary alcohols react more rapidly than do secondary alcohols, which react more rapidly than do tertiary alcohols. Esters can be made from all three types of alcohols, but it is more difficult to prepare esters from bulky tertiary alcohols.

QUESTION 21-4
What product results from the reaction of acetyl chloride with ethanol?

ANSWER 21-4
Ethyl acetate, $CH_3CO_2CH_2CH_3$.

Reaction with water

Acid halides react readily with water to give the corresponding carboxylic acid (Structure 21-5b). This is generally not a useful reaction since carboxylic acids are usually used to prepare acid halides. However, it is important to note that if

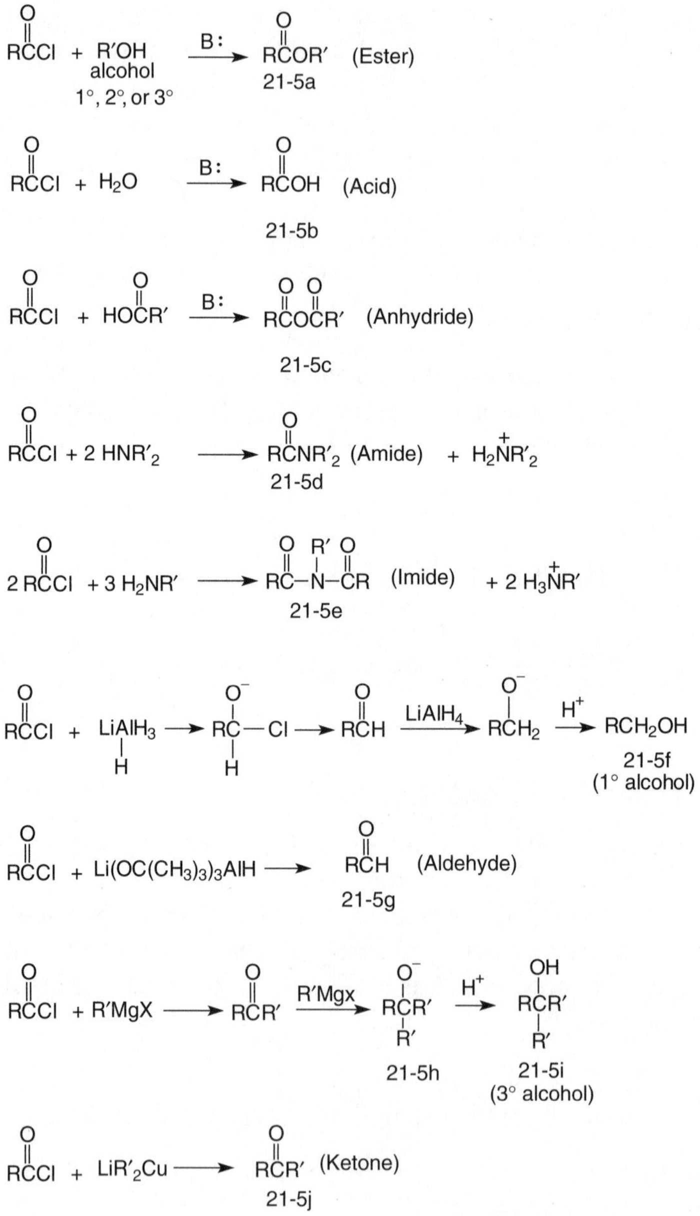

$$\underset{\text{RCCl}}{\overset{\text{O}}{\parallel}} + \underset{\substack{\text{alcohol}\\1°,\,2°,\,\text{or}\,3°}}{\text{R'OH}} \xrightarrow{\text{B:}} \underset{\text{RCOR'}}{\overset{\text{O}}{\parallel}} \;(\text{Ester})$$
21-5a

$$\underset{\text{RCCl}}{\overset{\text{O}}{\parallel}} + \text{H}_2\text{O} \xrightarrow{\text{B:}} \underset{\text{RCOH}}{\overset{\text{O}}{\parallel}} \;(\text{Acid})$$
21-5b

$$\underset{\text{RCCl}}{\overset{\text{O}}{\parallel}} + \underset{\text{HOCR'}}{\overset{\text{O}}{\parallel}} \xrightarrow{\text{B:}} \underset{\text{RCOCR'}}{\overset{\text{O}\;\;\text{O}}{\parallel\;\;\parallel}} \;(\text{Anhydride})$$
21-5c

$$\underset{\text{RCCl}}{\overset{\text{O}}{\parallel}} + 2\,\text{HNR}'_2 \longrightarrow \underset{\text{RCNR}'_2}{\overset{\text{O}}{\parallel}} (\text{Amide}) + \text{H}_2\overset{+}{\text{N}}\text{R}'_2$$
21-5d

$$2\,\underset{\text{RCCl}}{\overset{\text{O}}{\parallel}} + 3\,\text{H}_2\text{NR}' \longrightarrow \underset{\text{RC}-\text{N}-\text{CR}}{\overset{\text{O}\quad\text{R}'\;\text{O}}{\parallel\qquad\;\parallel}} (\text{Imide}) + 2\,\text{H}_3\overset{+}{\text{N}}\text{R}'$$
21-5e

$$\underset{\text{RCCl}}{\overset{\text{O}}{\parallel}} + \underset{\overset{|}{\text{H}}}{\text{LiAlH}_3} \longrightarrow \underset{\overset{|}{\text{H}}}{\overset{\text{O}^-}{\text{RC}-\text{Cl}}} \longrightarrow \underset{\text{RCH}}{\overset{\text{O}}{\parallel}} \xrightarrow{\text{LiAlH}_4} \underset{\text{RCH}_2}{\overset{\text{O}^-}{|}} \xrightarrow{\text{H}^+} \text{RCH}_2\text{OH}$$
21-5f
(1° alcohol)

$$\underset{\text{RCCl}}{\overset{\text{O}}{\parallel}} + \text{Li(OC(CH}_3)_3)_3\text{AlH} \longrightarrow \underset{\text{RCH}}{\overset{\text{O}}{\parallel}} \;(\text{Aldehyde})$$
21-5g

$$\underset{\text{RCCl}}{\overset{\text{O}}{\parallel}} + \text{R'MgX} \longrightarrow \underset{\text{RCR'}}{\overset{\text{O}}{\parallel}} \xrightarrow{\text{R'Mgx}} \underset{\overset{|}{\text{R}'}}{\overset{\text{O}^-}{\text{RCR'}}} \xrightarrow{\text{H}^+} \underset{\overset{|}{\text{R}'}}{\overset{\text{OH}}{\text{RCR'}}}$$
21-5h   21-5i
(3° alcohol)

$$\underset{\text{RCCl}}{\overset{\text{O}}{\parallel}} + \text{LiR}'_2\text{Cu} \longrightarrow \underset{\text{RCR'}}{\overset{\text{O}}{\parallel}} \;(\text{Ketone})$$
21-5j

Fig. 21-5. Reactions of acid halides.

water is present as an impurity it can react with the acid chloride and compete with the intended nucleophilic acyl substitution reaction.

Reaction with carboxylic acids

Acid halides react with carboxylic acids to give anhydrides (Structure 21-5c). Since hydrogen halide is a byproduct, a base (pyridine or a tertiary amine) is usually present to neutralize this acid.

Reaction with amines

Acid halides react with *secondary amines* to give amides (Structure 21-5d). At least one hydrogen atom must be bonded to the nitrogen atom of the amine for amide formation. *Tertiary amines* do not form amides since no hydrogen atom is bonded to the nitrogen atom. *Primary amines* form imides (Structure 21-5e). Primary amines can react with acid halides to form amides, if a large excess of the primary amine is used.

Unlike other reactions discussed above, 2 mol of a secondary amine are required to react with 1 mol of acid halide. The reaction of an acid halide and an amine gives a hydrogen halide as a byproduct. The hydrogen halide protonates the amine, giving an alkyl ammonium cation. The ammonium cation is not nucleophilic and does not react with the acid halide. This in effect "wastes" 1 equiv. of amine reagent for each equivalent of amide formed. Two equivalents of acid chloride require three equivalents of primary amine to form one equivalent of imide. See this reaction in Fig. 21-5.

QUESTION 21-5
What product results from the addition of triethylamine to acetyl chloride?

ANSWER 21-5
There is no reaction since the nitrogen atom in the amine must be bonded to at least one hydrogen atom for a reaction to occur.

Bases like pyridine or a trialkylamine are used in the above reactions to neutralize the HX byproduct. These bases do not interfere with the intended reaction. They do not contain a hydrogen atom bonded to the nitrogen atom and hence do not react with an acid halide (to form an amide).

Friedel–Crafts reactions

Acid chlorides undergo Friedel–Crafts reactions with benzene (and other aromatic rings), in the presence of Lewis acids, to give ketones. These reactions are discussed in the section entitled *Friedel–Crafts Acylation Reactions* in Chapter 18.

REDUCTION REACTIONS

Acid chlorides are reduced with lithium aluminum hydride (LiAlH$_4$ or LAH) to the corresponding primary alcohol (Structure 21-5f). A less active hydride reducing reagent, lithium tri-(*tert*-butoxy)aluminum hydride reduces an acid halide to the corresponding aldehyde (Structure 21-5g) and not to the alcohol. The mechanism for the reduction reaction is shown in Fig.19-8 in Chapter 19.

REACTION WITH ORGANOMETALLIC REAGENTS

Grignard reagents (RMgX) react with acid halides to give, initially, ketones. It is difficult to isolate the ketone as it reacts rapidly with more Grignard reagent to give a tertiary alkoxide anion. These reactions are carried out with 2 equiv. of Grignard reagent to go directly to the tertiary alkoxide anion (Structure 21-5h). Subsequent addition of acid protonates the alkoxide anion, giving the corresponding tertiary alcohol (Structure 21-5i). The mechanism for these reactions are discussed in the section entitled Organometallic Compounds in Chapter 13.

QUESTION 21-6

What product forms when excess methylmagnesium chloride reacts with acetyl chloride?

ANSWER 21-6

A tertiary alcohol, *tert*-butyl alcohol (after treatment with dilute acid).

Gilman reagents, lithium diorganocopper (LiR$_2$Cu) compounds, are less reactive than Grignard reagents and are used to convert acid halides to the corresponding ketone (Structure 21-5j). Reactions are carried out at low temperatures and generally give excellent yields. Gilman reagents react only with acid halides and not with esters, anhydrides, or amides.

Carboxylic Acid Anhydrides

NOMENCLATURE

Anhydrides are named by *replacing the word acid with anhydride* in both the formal and common names. Examples are shown in Fig. 21-6. The anhydride of ethanoic acid is ethanoic anhydride. The anhydride of acetic acid is acetic anhydride. Symmetrical anhydrides are sometimes called bis anhydrides, e.g. bisacetic anhydride (Structure 21-6a). Usually the prefix bis is eliminated and, for example, acetic anhydride refers to the symmetrical anhydride. Mixed

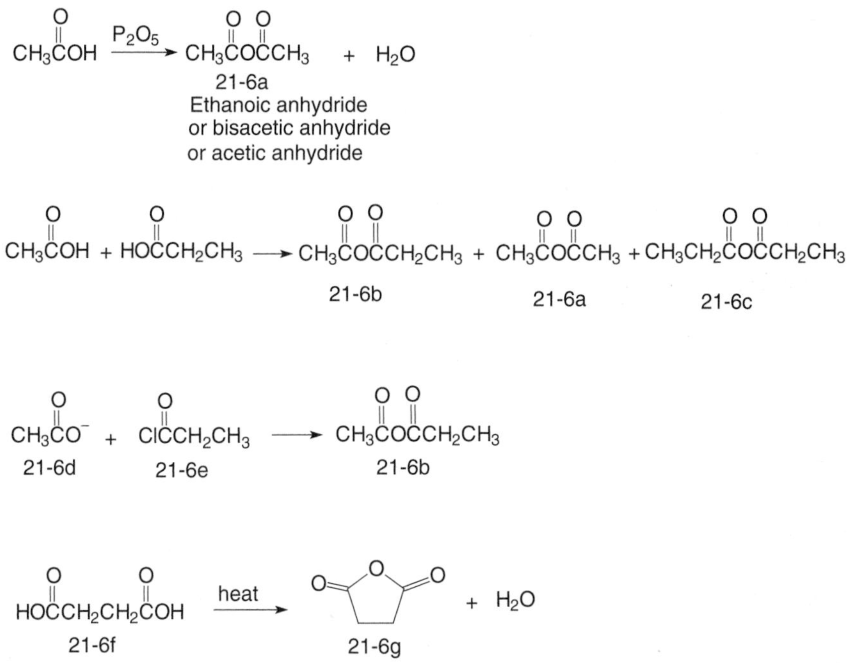

$$\underset{\text{CH}_3\overset{\text{O}}{\overset{\|}{\text{C}}}\text{OH}}{} \xrightarrow{P_2O_5} \underset{\text{CH}_3\overset{\text{O}}{\overset{\|}{\text{C}}}\text{O}\overset{\text{O}}{\overset{\|}{\text{C}}}\text{CH}_3}{} + \text{H}_2\text{O}$$

21-6a
Ethanoic anhydride
or bisacetic anhydride
or acetic anhydride

$$\underset{\text{CH}_3\overset{\text{O}}{\overset{\|}{\text{C}}}\text{OH}}{} + \underset{\text{HO}\overset{\text{O}}{\overset{\|}{\text{C}}}\text{CH}_2\text{CH}_3}{} \longrightarrow \underset{\text{CH}_3\overset{\text{O}}{\overset{\|}{\text{C}}}\text{O}\overset{\text{O}}{\overset{\|}{\text{C}}}\text{CH}_2\text{CH}_3}{} + \underset{\text{CH}_3\overset{\text{O}}{\overset{\|}{\text{C}}}\text{O}\overset{\text{O}}{\overset{\|}{\text{C}}}\text{CH}_3}{} + \underset{\text{CH}_3\text{CH}_2\overset{\text{O}}{\overset{\|}{\text{C}}}\text{O}\overset{\text{O}}{\overset{\|}{\text{C}}}\text{CH}_2\text{CH}_3}{}$$

21-6b   21-6a   21-6c

$$\underset{\text{CH}_3\overset{\text{O}}{\overset{\|}{\text{C}}}\text{O}^-}{} + \underset{\text{Cl}\overset{\text{O}}{\overset{\|}{\text{C}}}\text{CH}_2\text{CH}_3}{} \longrightarrow \underset{\text{CH}_3\overset{\text{O}}{\overset{\|}{\text{C}}}\text{O}\overset{\text{O}}{\overset{\|}{\text{C}}}\text{CH}_2\text{CH}_3}{}$$

21-6d   21-6e   21-6b

$$\underset{\text{HO}\overset{\text{O}}{\overset{\|}{\text{C}}}\text{CH}_2\text{CH}_2\overset{\text{O}}{\overset{\|}{\text{C}}}\text{OH}}{} \xrightarrow{\text{heat}} \quad + \text{H}_2\text{O}$$

21-6f   21-6g

Fig. 21-6. Synthesis of anhydrides.

anydrides, made from two different acids, are named alphabetically. The formal name for Structure 21-6b is ethanoic methanoic anhydride and the common name is acetic formic anhydride.

SYNTHESIS OF ANHYDRIDES

Anhydride means "without water." Two carboxylic acid molecules can react, eliminating a molecule of water, to give an anhydride. A dehydrating reagent, such as P_2O_5, is used in the synthesis of anhydrides. Two different acids can be used to make a mixed (unsymmetrical) anhydride. This latter method is not very efficient as one obtains the two symmetrical anhydrides in addition to the desired mixed anhydride. The preparation of several anhydrides is shown in Fig. 21-6. The dehydration of a mixture of acetic acid and propionic acid gives three anhydrides: acetic anhydride (Structure 21-6a), acetic propionic anhydride (Structure 21-6b), and propionic anhydride (Structure 21-6c).

A better method of making mixed anhydrides is to react an acid halide with the salt of a carboxylic acid. This method can be used to make symmetrical and mixed anhydrides. The reaction of acetate anion (21-6d) with propionyl chloride

(21-6e) gives acetic propionic anhydride (21-6b). Five- and six-membered cyclic anhydrides can be made from the corresponding dicarboxylic acids. Heating succinic acid (21-6f) to about 200 °C gives succinic anhydride (21-6g).

QUESTION 21-7
What anhydride forms when maleic acid (HO$_2$CCH=CHCO$_2$H) is heated at 200 °C?

ANSWER 21-7
Maleic anhydride.

REACTIONS OF ANHYDRIDES

Reactions of anhydrides with various nucleophiles are shown in Fig. 21-7. Anhydrides, like acid chlorides, are quite reactive and few anhydrides are found in nature. Acid halides are not easily made by reacting an anhydride with halide anion. The reaction of a chloride anion with a symmetrical anhydride (21-7a) would give a tetrahedral alkoxide (21-7b) intermediate. Either the chloride or the carboxylate anion could be the leaving group in Structure 21-7b. Since chloride anion is a better leaving group, the anhydride (21-7a) is regenerated.

Reaction with alcohols

Esters are prepared by reacting anhydrides with alcohols. Anhydrides react readily with primary, secondary, or tertiary alcohols to form an ester (21-7c) and an acid (21-7d). These reactions are inefficient since half of the anhydride is "wasted," forming a carboxylic acid (only 1 equiv. of ester is formed from 1 equiv. of anhydride).

QUESTION 21-8
What product results from the reaction of 1 mol of succinic anhydride (21-6g) and 1 mol of ethanol?

ANSWER 21-8
Monoethyl succinate, CH$_3$CH$_2$O$_2$CCH$_2$CH$_2$CO$_2$H. An ester and an acid are formed in this reaction.

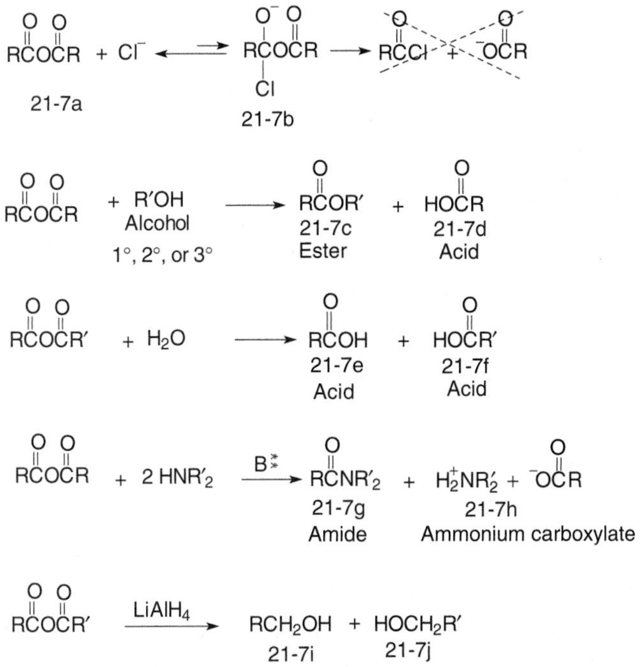

Fig. 21-7. Reactions of anhydrides.

REACTION WITH WATER

Water reacts with mixed anhydrides to form 1 equiv. of two different carboxylic acids (21-7e and 21.7f). If a symmetrical anhydride is hydrolyzed, 2 equiv. of the same carboxylic acid are produced. This is not a very useful reaction since carboxylic acids are often the starting materials for making anhydrides.

Reaction with amines

Anhydrides react with secondary amines to give amides. Again, half of the anhydride is "wasted." One equivalent of anhydride reacts with 2 equiv. of amine. One equivalent of amine forms the amide (21-7g) and the other forms the ammonium carboxylate salt (21-7h).

Reduction reaction

Anhydrides are reduced with LAH to give primary alcohols from each "half" of the anhydride. Symmetrical anhydrides (where R = R′) give one product (Structure 21-7i), and mixed anhydrides give two different primary alcohols (Structures 21-7i and 21.7j).

The reaction mechanisms are consistent with that shown in Fig. 21-2. Practice drawing mechanisms using curved arrows to show electron movement.

Carboxylic Esters

Many carboxylic esters exist in nature and many more have been prepared in laboratories. Many esters have pleasant aromas and are used in perfumes. Isopentyl acetate smells like bananas and methyl butyrate has the aroma of pineapples.

NOMENCLATURE

Esters are named by first giving the name of the alkyl (or aryl) group bonded to the noncarbonyl oxygen atom and then naming the acyl segment. The acyl segment name is based on the corresponding acid, changing *-ic acid to -ate.* The ethyl ester (21-8a) of ethan*oic acid* is called ethyl ethan*oate* (formal name). The common name, ethyl acet*ate*, is derived from acet*ic acid*. Esters of cyclic carboxylic acids are named alkyl cycloalkanecarboxylates, for example ethyl cyclohexanecarboxylate.

QUESTION 21-9
What is the name of an ester made from formic acid and ethanol?

ANSWER 21-9
Ethyl methanoate or more commonly ethyl formate, $HCO_2CH_2CH_3$.

Cyclic esters are called *lactones*. The ester oxygen atom (not the carbonyl oxygen atom) is a member of the ring. In the formal name, the ring is named as a cyclic ketone. The carbonyl carbon atom is always atom C-1. The term *oxa* is used to name the ring oxygen atom. For example, Structure 21-8b is called 2-oxacyclopentanone. Note the oxygen atom, not the α carbon atom, is ring atom number 2. One common name for this material, made from 4-hydroxybutanoic acid, is 4-hydroxybutanoic acid lactone. Another common naming system

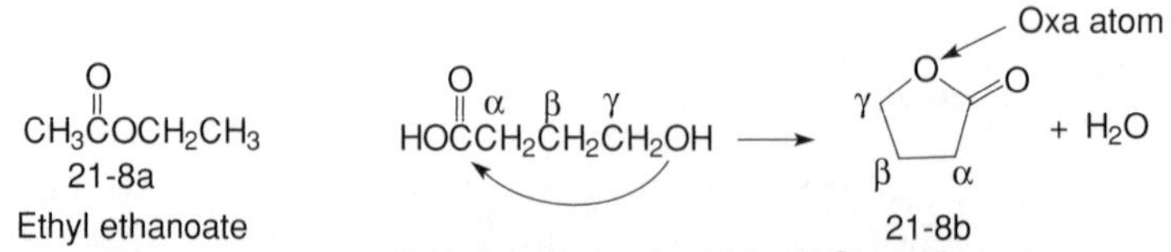

21-8a
Ethyl ethanoate
or ethyl acetate

4-Hydroxybutanoic acid
or γ-Hydroxybutyric acid

21-8b
2-Oxacyclopentanone
or γ-Butyrolactone

Fig. 21-8. Ester nomenclature.

replaces *-ic acid* or *-oic acid by -olactone*. Greek letters are used to indicate ring size. The α position is the carbon atom not the oxygen atom. In 21-8b, the gamma (γ) atom is bonded to the ring oxygen atom and this compound is called γ-butyrolactone.

QUESTION 21-10
What is the structure of the lactone made from 5-hydroxypentanoic acid?

ANSWER 21-10
5-Hydroxypentanoic acid lactone also called δ-pentanolactone or δ-valerolactone. Valeric acid is a common name for pentanoic acid.

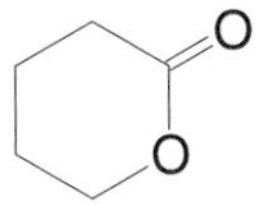

SYNTHESIS OF ESTERS

Esters can be made from acid halides (see Fig. 21-5) and anhydrides (see Fig. 21-7). Figure 21-9 shows the synthesis of esters from carboxylic acids. Carboxylic acids (21-9a) react with primary, secondary, or tertiary alcohols to give esters (21-9d). The reaction is catalyzed by strong acids. The protonated carboxylic acid (21-9b) reacts with an alcohol, giving a tetrahedral cation (21-9c). The OH oxygen is protonated (OH_2^+), converting OH into a better leaving group. This intermediate eliminates a molecule of water, giving an ester (21-9d) after protonation. Acid-catalyzed ester formation is called *Fisher esterification.*

The best known ester is probably ascorbic acid, Vitamin C. Why is it called an acid if it is a cyclic ester? Ascorbic acid is a polyhydroxy carboxylic acid that is in equilibrium with its internal cyclic ester and the equilibrium favors the ester (a lactone).

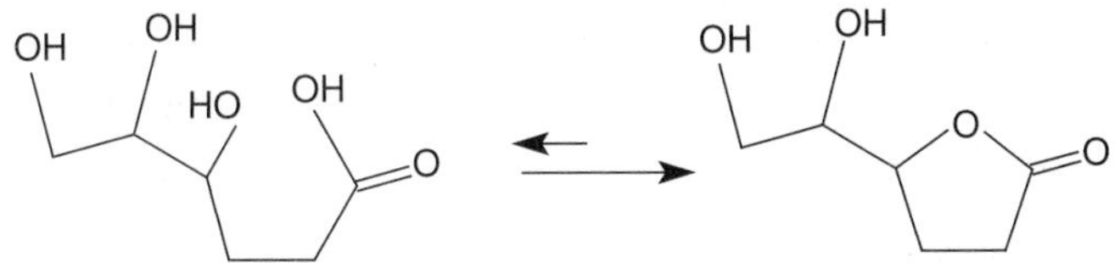

ascorbic acid/Vitamin C

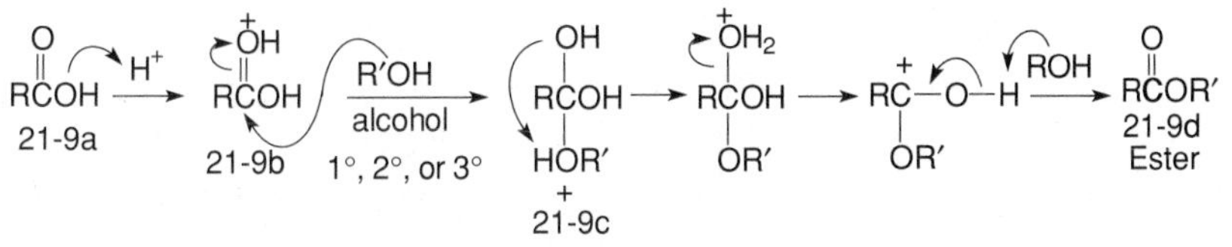

Fig. 21-9. Synthesis of esters.

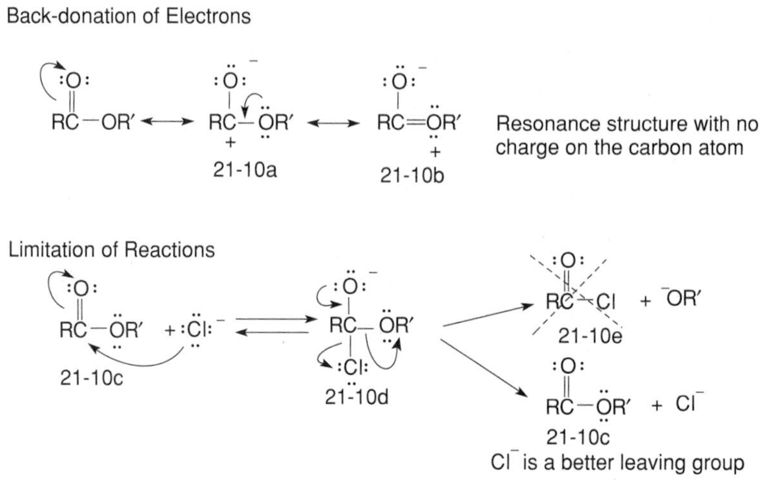

Back-donation of Electrons

Resonance structure with no charge on the carbon atom

21-10a   21-10b

Limitation of Reactions

21-10c   21-10d   21-10e

21-10c

Cl⁻ is a better leaving group

Fig. 21-10.   Reactivity of esters.

REACTIVITY OF ESTERS

The reactivity of a carbonyl compound toward nucleophilic attack depends on the magnitude of the partial positive charge on the carbonyl carbon atom. The "back-donation" of nonbonding electron pairs (a resonance effect) to the carbonyl carbon atom in an ester decreases the positive charge on the carbonyl carbon atom as shown in Structure 21-10a in Fig. 21-10. Resonance structure 21-10b has no formal positive charge on the carbonyl carbon atom. The hybrid resonance structure depends on the contribution of each resonance form. The chlorine atom in acid chlorides and the oxygen atom in anhydrides do not back-donate electrons to the same extent as observed for esters. Thus, the carbonyl carbon atom is more electropositive and more susceptible to nucleophilic attack in acid chlorides and anhydrides. Esters are therefore less reactive (more stable) than acid halides and anhydrides.

Limitation of reactions

Esters generally do not react with halide or carboxylate anions to form acid halides or anhydrides. If a halide were to react with ester 21-10c, a tetrahedral intermediate (21-10d) would result. This intermediate could eliminate an alkoxide anion (⁻OR′) to give an acid chloride (21-10e) or it could eliminate a chloride anion to give the starting ester (21-10c). Since the chloride anion is a better leaving group than an alkoxide anion, the starting ester is regenerated. A similar explanation is given for the reaction with carboxylate anions.

REACTIONS OF ESTERS

Reaction with water

The reaction of an ester with water is called *hydrolysis*. This reaction is very slow unless catalyzed by acid or base. The reaction with water is shown in Fig. 21-11. The base-catalyzed hydrolysis reaction is called *saponification*. This name is derived from the basic hydrolysis of animal fats (triesters) resulting in the

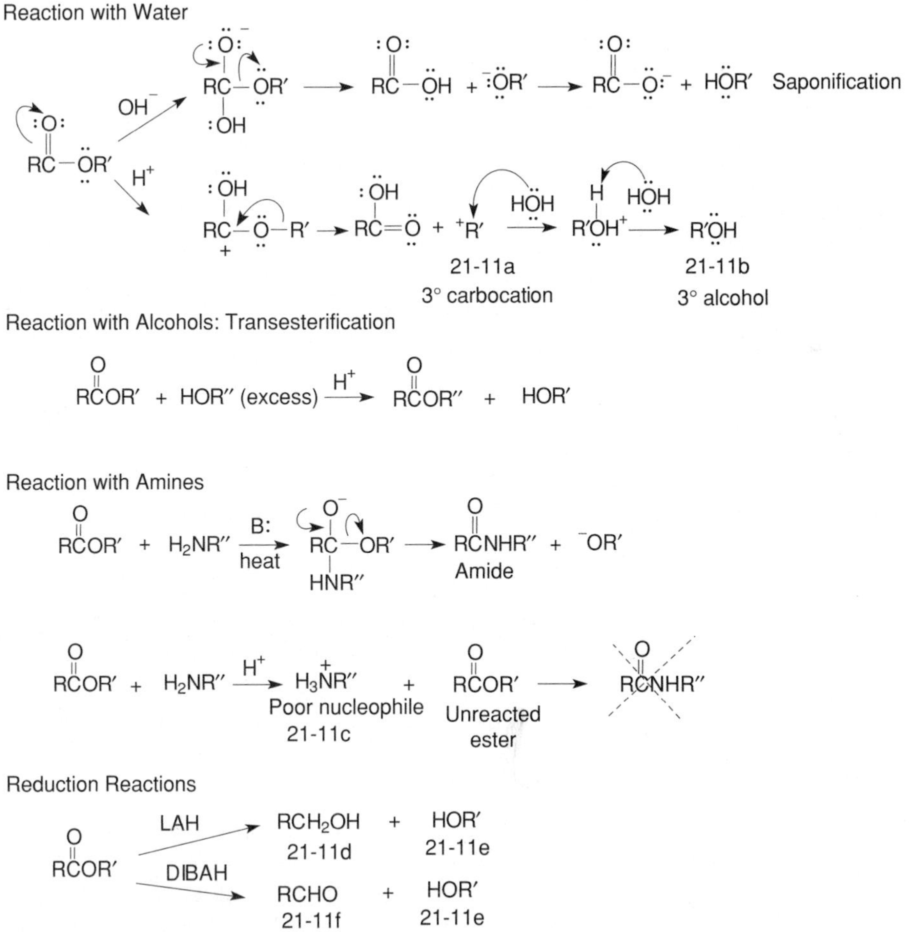

Reaction with Water

Reaction with Alcohols: Transesterification

Reaction with Amines

Reduction Reactions

Fig. 21-11.   Reactions of esters.

formation of the salts of long-chain (about 18 carbon atoms) fatty carboxylic acids. These carboxylic acids are the major component of soap, and hence the term saponification.

Tertiary esters are hydrolyzed more rapidly than secondary or primary esters. One reason is that a bulky ester group is a better leaving group than a less bulky one. In the acid-catalzyed hydrolysis of esters, the tertiary leaving group is a carbocation (21-11a). The carbocation reacts with water giving a protonated alcohol that deprotonates to a tertiary alcohol (21-11b).

QUESTION 21-11
What products result from the acid-catalyzed hydrolysis of *tert*-butyl acetate?

ANSWER 21-11
tert-Butyl alcohol and acetic acid.

Reaction with alcohols

Alcohols react with esters to undergo an exchange reaction of the ester alkoxy or aryloxy group. This reaction is shown in Fig. 21-11. The reaction is acid catalyzed and is called *transesterification*. The equilibrium is controlled by adding a large excess of the alcohol to be exchanged or by removing (by distillation) the alcohol that is formed upon ester exchange. This is a very common way of forming new esters from readily available esters.

Reaction with amines

Esters undergo reactions with primary or secondary amines to form amides. These reactions, shown in Fig. 21-11, are carried out under basic conditions. The mechanism for the reaction of an ester with an amine, under basic conditions, involves the alkoxide anion group as the leaving group. The alkoxide anion is usually not a good leaving group (it is a strong base and good nucleophile). The reaction is endothermic and must be carried out at high temperatures. The reaction is not carried out under acid conditions, because the amine would be protonated (Structure 21-11c) and would no longer be a good nucleophile.

Reduction reactions

Esters are reduced by LAH giving two alcohols, one (21-11d) from the acyl segment and one (21-11e) from the alkoxide segment. When esters are reacted with 1 equiv. of a less reactive reducing reagent, such as diisobutylaluminum hydride (DIBAH), the acyl group is reduced to an aldehyde (21-11f) and the ester group is still converted into an alcohol (21-11e).

QUESTION 21-12
What product results from the reduction of ethyl acetate with $LiAlH_4$?

ANSWER 21-12
Ethanol.

Grignard reactions

Esters react with 2 equiv. of a Grignard reagent to produce tertiary alcohols. The initially formed ketone is more reactive than the starting ester and hence the reaction is carried out with 2 equiv. of Grignard reagent to prevent a product mixture of tertiary alkoxide anion and unreacted ester. The final reaction mixture is acidified to protonate the alkoxy anion to an alcohol. These reactions are discussed in more detail in the section entitled *Grignard Reagents* in Chapter 13.

Amides

Amides are the most stable of all the carboxylic acid derivatives. An amide with the structure $RC(O)NH_2$ is called a primary amide. The nitrogen atom is bonded to only one carbon atom. The nitrogen atom of a secondary amide, $RC(O)NHR'$, is bonded to two carbon atoms. The nitrogen atom in a tertiary amide, $RC(O)NR'_2$, is bonded to three carbon atoms.

Physical properties

Primary amides contain two hydrogen atoms bonded to the nitrogen atom. Secondary amides contain one hydrogen atom bonded to the nitrogen atom. These hydrogen atoms can undergo intermolecular hydrogen bonding with other molecules. As a result of hydrogen bonding, amides have higher melting and boiling points than do esters of comparable molecular weight. Esters do not undergo intermolecular hydrogen bonding with other ester molecules (assuming no other hydrogen bonding group is present in the molecule).

Amides are usually shown as a carbonyl group with a single bond to a nitrogen atom (Structure 21-12a in Fig. 21-12). A resonance structure (21-12b) can be drawn using the nonbonding electron pair on the nitrogen atom to form a double bond with the carbonyl carbon atom. Since the nitrogen atom is not as electronegative as the oxygen atom in esters, resonance structure 21-12b contributes more to the resonance hybrid structure of amides than resonance structure 21-12c contributes to the resonance hybrid structure of esters. There

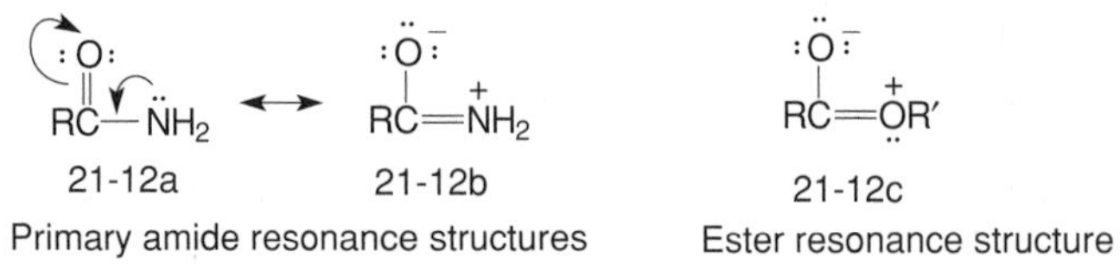

21-12a   21-12b   21-12c

Primary amide resonance structures   Ester resonance structure

Fig. 21-12. Resonance stabilization.

is significant double bond character in the $C-NH_2$ bond of amides, and rotation around this bond is more restricted (of higher energy) than a comparable $C-OR$ bond in esters. The energy input required for rotation around the carbon–nitrogen bond is about 70 to 80 kJ/mol.

Unlike amines, amides are only weakly basic. In the presence of a strong acid, the carbonyl oxygen atom of an amide, not the nitrogen atom, is protonated.

NOMENCLATURE

Amide names are derived from those of the corresponding carboxylic acids. The *-oic acid* suffix in the formal name and the *-ic acid* suffix in the common carboxylic acid name is replaced by *-amide*. The primary amide derived from ethanoic acid is called ethanamide. The same amide is called acetamide based on the common name acetic acid.

The structures of acetamide (Structure 21-13a) and derivatives of acetamide are shown in Fig. 21-13. The alkyl or aryl group bonded to nitrogen atom is indicated by a capital N preceding the alkyl or aryl group name. *N*-Methylformamide (21-13b) and *N*-ethyl-*N*-methylformamide (21-13c) are examples of secondary and tertiary amides. Groups bonded to the nitrogen atom are listed alphabetically.

Amides based on cyclic carboxylic acids are named by replacing the words carboxylic acid with carboxamide. The primary amide derived from cyclopentanecarboxylic acid is called cyclopentanecarboxamide.

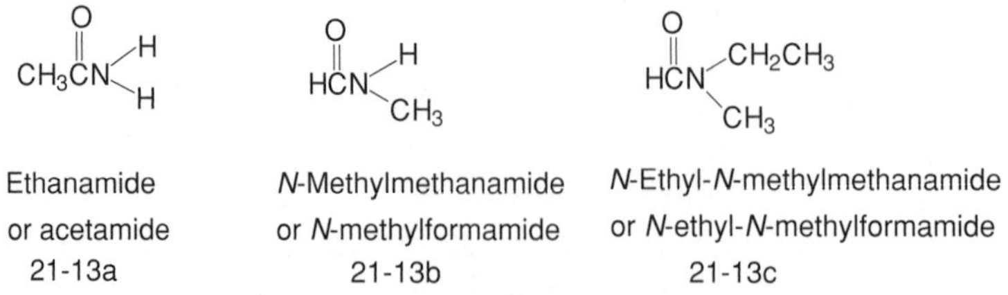

Ethanamide   *N*-Methylmethanamide   *N*-Ethyl-*N*-methylmethanamide
or acetamide   or *N*-methylformamide   or *N*-ethyl-*N*-methylformamide
21-13a   21-13b   21-13c

Fig. 21-13. Amide nomenclature.

QUESTION 21-13
What is the structure of N,N-dimethylacetamide?

ANSWER 21-13

$$\underset{\displaystyle CH_3CN(CH_3)_2}{\overset{\displaystyle \overset{O}{\|}}{}}$$

PREPARATION OF AMIDES

Amides can be made from all of the other acid derivatives. The reaction of a primary or secondary amine with an acid chloride, anhydride, acid, or ester will form an amide. These reactions are shown above in the corresponding acid derivative sections. Tertiary amines, lacking a hydrogen atom bonded to the nitrogen atom, do not form amides.

Amides can also be made by forming a salt from equal molar amounts of carboxylic acid and amine, giving an ammonium carboxylate. Heating this salt (usually higher than 100 °C) results in dehydration and amide formation.

QUESTION 21-14
What is the structure of the amide that results from the reaction of acetyl chloride and dimethylamine?

ANSWER 21-14
N,N-Dimethylacetamide (shown in Question 21-13).

REACTIONS OF AMIDES

Secondary amides are unreactive toward acid halides and anhydrides for steric reasons. Primary amides react with excess acid halides to form imides. Amides react with alcohols (to form esters) and water (to form acids) only at elevated temperatures and in the presence of an acid catalyst. The mechanism of this reaction is shown in Fig. 21-14, Reaction (a).

Reduction to amines

Primary, secondary, and tertiary amides are reduced by LAH to primary, secondary, and tertiary amines. An example is shown in Reaction (b) in Fig. 21-14. Amides do not react with Girgnard reagents or Gilman reagents.

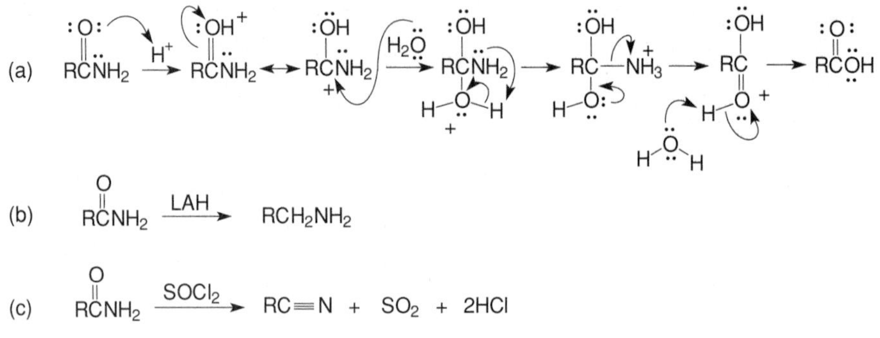

(a) [reaction mechanism diagram showing protonation and hydrolysis of amide]

(b) $\overset{O}{\overset{\|}{R\overset{}{C}NH_2}} \xrightarrow{LAH} RCH_2NH_2$

(c) $\overset{O}{\overset{\|}{R\overset{}{C}NH_2}} \xrightarrow{SOCl_2} RC{\equiv}N + SO_2 + 2HCl$

Fig. 21-14. Reactions of amides.

Dehydration of amides

Dehydrating reagents, like thionyl chloride ($SOCl_2$), remove 1 molecule of water from amides to give nitriles. An example is shown in Reaction (c) in Fig. 21-14.

Cyclic Amides

Cyclic amides are called *lactams*. The most stable lactams contain five or six atoms in the lactam ring. The nitrogen atom is one of the ring atoms. The very important family of penicillin compounds contain a four-membered lactam ring. Lactams are made by an intramolecular condensation reaction between amine and carboxylic acid functional groups. This condensation reaction is shown in Fig. 21-15. Lactams undergo ring-opening polymerization to make polyamides, also known as nylons.

NOMENCLATURE

Lactams are formally named by adding the word lactam to the name of the starting amino acid. The lactam made from 4-aminobutanoic acid (Structure 21-15a) is called 4-aminobutanoic acid lactam (Structure 21-15b). Another

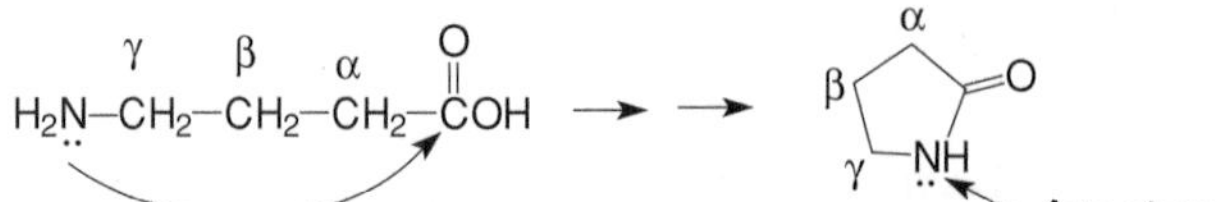

$$H_2N-\overset{\gamma}{C}H_2-\overset{\beta}{C}H_2-\overset{\alpha}{C}H_2-\overset{O}{\overset{\|}{C}}OH \longrightarrow$$

21-15a
4-Aminobutanoic acid
or γ-Aminobutyric acid

[structure of lactam with α, β, γ positions, NH, and Aza atom label]

21-15b
4-Aminobutanoic acid lactam
2-Azacyclopentanone
or γ-Butyrolactam

Fig. 21-15. Lactam formation.

formal name for this material is 2-azacyclopentanone. The ring is named as a cyclic ketone and the nitrogen atom is designated by aza (a historic name referring to nitrogen compounds). The carbonyl carbon atom is atom number one and the nitrogen atom is number two in the ring.

The common name for a lactam is derived from the common name of the corresponding amino acid by dropping the word *amino* and replacing the -*ic or* -*oic acid* by -*olactam*. The lactam derived from γ-*amino*butyr*ic acid* is called γ-butyr*olactam*. Greek letters designate the ring carbon atom bonded to the nitrogen atom as shown in Structure 21-15b. The α carbon is adjacent to the carbonyl group.

Imides

Imides contain nitrogen atoms bonded to two acyl groups and can be cyclic or acyclic. Figure 21-16 shows the structure of two cyclic imides, phthalimide (21-16a) and succinimide (21-16b). Imides are more acidic than amides, as the nitrogen atom is bonded to two electron-withdrawing groups. The negative charge on the nitrogen atom in the conjugate base is delocalized as shown by resonance structures 21-16c–e. Electron delocalization increases the stability of the imide anion. Amides have pK_a values of 15 to 17, while imides have pK_a values of 8 to 10. This is a significant increase in acidity.

NOMENCLATURE

The common name of imides is based on the parent carboxylic acid. The suffix -*ic acid* is dropped and *imide* is added. Phathal*ic acid* becomes phthal*imide* (Structure 21-16a). Succin*ic acid* becomes succin*imide* (Structure 21-16b).

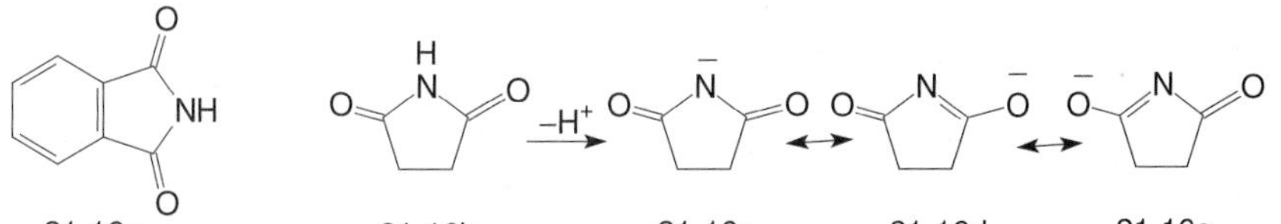

| 21-16a | 21-16b | 21-16c | 21-16d | 21-16e |
| Phthalimide | Succinimide | | | |

Fig. 21-16. Imides.

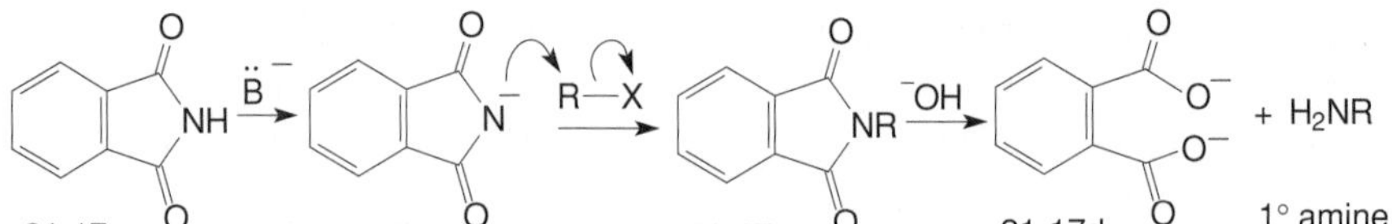

Fig. 21-17. Gabriel amine synthesis.

REACTIONS OF IMIDES

Phthalimide is a starting material for synthesizing *primary amines*. This is called the *Gabriel amine synthesis reaction*. Phthalic acid is heated with ammonia to make phthalimide (Structure 21-17a). A base, like sodium hydroxide, removes the nitrogen proton. The resulting anion (21-17b) undergoes an S_N2 reaction with a primary alkyl halide to give an N-substituted imide (21-17c). This substituted imide is heated with base (or acid) to give the salt of phthalic acid (21-17d) and a primary alkyl amine.

Nitriles

Nitriles are not acyl compounds, but are considered carboxylic acid derivatives because they can be hydrolyzed to carboxylic acids. The nitrile group C≡N is highly polar. Low molecular weight nitriles (RC≡N) are good solvents for polar compounds.

NOMENCLATURE

Formal names are derived from the terms alkane and nitrile. The formal name for $CH_3C≡N$ is ethanenitrile. *A key point: the carbon atom in the C ≡ N group is* included in the alkane name. This compound is not called methanenitrile. The carbon atom in the nitrile group is always atom number one in the alkane segment.

The common name for nitriles is derived from the common name of the carboxylic acid that results from hydrolysis of the nitrile. The suffix *-ic* or *-oic acid* is replaced by *-onitrile*. A two-carbon nitrile, $CH_3C≡N$, is called acetonitrile. Acetic acid is formed upon hydrolysis of acetonitrile. Cyano is also a common name for the C≡N group, when it is present as a substituent. The ⁻C≡N anion is called a cyanide ion.

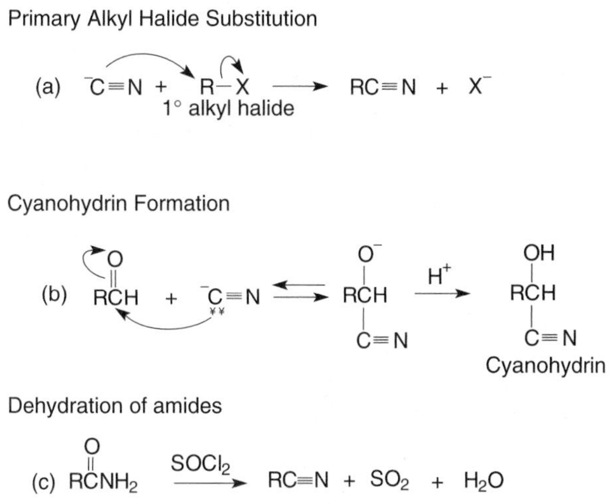

Primary Alkyl Halide Substitution

(a) $^-C\equiv N$ + R—X $\longrightarrow$ RC$\equiv$N + X$^-$
1° alkyl halide

Cyanohydrin Formation

(b)
$$\underset{RCH}{\overset{O}{\|}} + ^-C\equiv N \rightleftharpoons \underset{RCH}{\overset{O^-}{\underset{C\equiv N}{|}}} \xrightarrow{H^+} \underset{RCH}{\overset{OH}{\underset{C\equiv N}{|}}}$$
Cyanohydrin

Dehydration of amides

(c) $\underset{RCNH_2}{\overset{O}{\|}} \xrightarrow{SOCl_2}$ RC$\equiv$N + SO$_2$ + H$_2$O

Fig. 21-18. Preparation of nitriles.

PREPARATION OF NITRILES

Nitriles are made by an S_N2 reaction between a cyanide anion and a primary alkyl halide. This is shown in Fig. 21-18, Reaction (a). Elimination, rather than substitution, tends to be the major reaction between cyanide anions and secondary or tertiary alkyl halides.

Cyanohydrins result from the reaction of cyanide ion with aldehydes or ketones. This is shown in Fig. 21-18, Reaction (b). This is an equilibrium reaction. Formaldehde reacts rapidly and quantitatively with cyanide ion. Bulky ketones give poor yields of cyanohydrins. The general order of reactivity is formaldehyde > other aldehydes > ketones. Steric effects appear to be the main reason for the unfavorable equilibrium position for bulky ketones.

Nitriles are also made by dehydration of (elimination of water from) amides. Heating amides with a dehydrating reagent such as thionyl chloride (SOCl$_2$) gives nitriles. This is shown in Fig. 21-18, Reaction (c).

REACTIONS OF NITRILES

Nitriles are hydrolyzed by heating with strong acids or bases. Stepwise hydrolysis initially forms an amide, which is subsequently hydrolyzed to a carboxylic acid. The hydrolysis can be controlled in some reactions to stop when the amide is formed. The hydrolysis of a nitrile to an amide is shown in Fig. 21-19, Reaction (a).

Grignard reagents react with the polarized nitrile group to form an imine anion. Additional Grignard reagent does not react with the imine anion, since

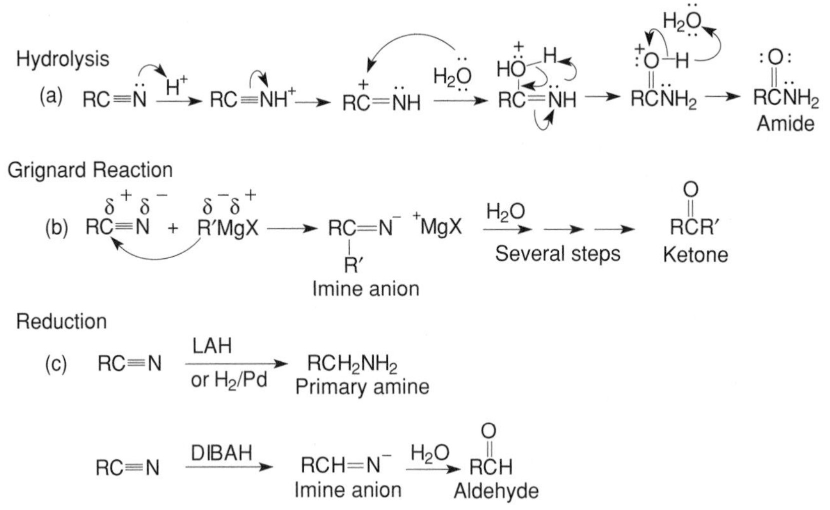

Fig. 21-19. Reactions of nitriles.

the resulting amine dianion would be too unstable (high in energy) to form. The imine magnesium salt is hydrolyzed by the addition of water to give the corresponding ketone. Ketone formation is shown in Fig. 21-19, Reaction (b).

Reduction of nitriles

Nitriles are reduced with LAH to the primary amine. A less vigorous reducing reagent, DIBAH, reduces nitriles to imine anions. Protonation and hydrolysis of the imine gives an aldehyde. These reactions are shown in Fig. 21-19, Reaction (c). Catalytic hydrogenation can also be used to reduce nitriles to primary amines.

Quiz

1. An inductive effect involves
   (a) the withdrawal of electrons through sigma bonds
   (b) the donation of electrons through sigma bonds
   (c) the withdrawal and donations of electrons through sigma bonds
   (d) all of the above

2. Good leaving groups in acyl transfer reactions are
   (a) strong bases
   (b) weak bases

(c) strong acids
(d) weak acids

3. Acid halides are _____ esters.
   (a) more reactive than
   (b) less reactive than
   (c) of equal reactivity to

4. Acid chlorides react with
   (a) alcohols
   (b) amines
   (c) carboxylic acids
   (d) all of the above

5. Acid chlorides react with secondary amines to give
   (a) acids
   (b) amides
   (c) imides
   (d) no reaction

6. Acid chlorides are reduced to _____ with LAH.
   (a) carboxylic acids
   (b) alcohols
   (c) ketones
   (d) aldehydes

7. Acetic acid and propionic acid react to give _____ anhydride(s).
   (a) acetic
   (b) propionic
   (c) acetic propionic
   (d) all of the above

8. Propionyl bromide reacts with sodium acetate to give _____ anhydride(s).
   (a) acetic
   (b) acetic propionic
   (c) propionic
   (d) all of the above

9. Esters are made from the reaction between
   (a) carboxylic acid molecules
   (b) alcohol molecules
   (c) alcohol and carboxylic acid molecules
   (d) anhydride and water molecules

10. The self-condensation reaction of one molecule of 5-hydroxyhexanoic acid gives
    (a) an anhydride
    (b) a lactone
    (c) a ketone
    (d) a lactam

11. Ethyl acetate is hydrolyzed by water to give
    (a) a lactone
    (b) an ester
    (c) an anhydride
    (d) a carboxylic acid and an alcohol

12. Anhydrides are hydrolyzed _____ esters.
    (a) faster than
    (b) slower than
    (c) at the same rate as

13. The reaction of excess ethyl acetate with diethylamine gives an
    (a) amide
    (b) imide
    (c) anhydride.
    (d) amino acid

14. The reaction of excess acetyl chloride with ethylamine gives an
    (a) amide
    (b) imide
    (c) anhydride
    (d) amino acid

15. The reduction of acetamide, $CH_3C(O)NH_2$, with $LiAlH_4$ gives an
    (a) carboxylic acid
    (b) amide
    (c) alcohol
    (d) amine

16. The self-condensation of one molecule of 5-aminopentanoic acid gives
    (a) a lactam
    (b) an ester
    (c) an amide
    (d) a lactone

17. Imides are _____ acidic than amides.
    (a) more
    (b) less

18. Nitriles are hydrolyzed to
    (a) alcohols
    (b) lactones
    (c) imides
    (d) amides and acids

19. Nitriles are reduced to
    (a) alcohols
    (b) carboxylic acids
    (c) amines
    (d) amides

Alpha-Substitution Reactions in Carbonyl Compounds

Introduction

A carbon atom bonded to either side of a carbonyl group is called an *alpha (α) carbon*. Hydrogen atoms bonded to α carbons are called α hydrogens. Alpha-substitution reactions involve the replacement of an α hydrogen with an electrophile. Carbonyl compounds that undergo these reactions are aldehydes, ketones, esters, acids, and amides. This reaction is an important way of making new carbon–carbon bonds. Many new compounds can be prepared by these substitution reactions.

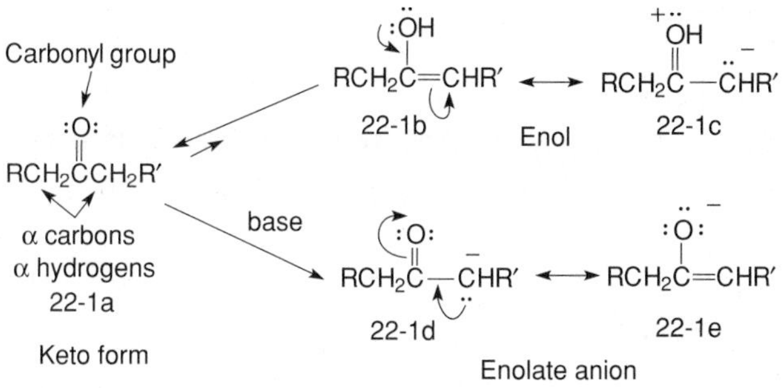

Fig. 22-1.   Enol and enolate structures.

Enol and Enolate Anions

Alpha-substitution reactions involve the reaction of an electrophile with an *enol or an enolate anion*. Enols are compounds that contain alk*ene* and alcoh*ol* functional groups. Enols are in equilibrium with their corresponding ketones or aldehydes. This equilibrium, shown in Fig. 22-1, is usually shifted strongly toward the ketone or aldehyde in monocarbonyl compounds. (For a review of keto–enol equilibria, see the section entitled *Hydration Reactions* in Chapter 9.)

TAUTOMERS

Keto and enol structures are called *tautomers*. The conversion of one tautomer into the other is called *tautomerization* or *enolization*. The tautomers are *not* resonance structures. Each tautomer represents a different compound. They are constitutional isomers. The enol form is difficult to isolate in most cases since the equilibrium is shifted strongly toward the keto form. (The keto/enol ratio for acetone is very large, $1 \times 10^{11}/1$).

QUESTION 22-1

Draw the structure of the enol tautomer of acetaldehyde, CH_3CHO.

ANSWER 22-1

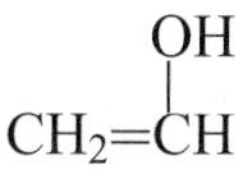

$$CH_2=\overset{\displaystyle OH}{\underset{\displaystyle |}{CH}}$$

RESONANCE STRUCTURES

Ketone 22-1a has two α carbon atoms, one on each side of the carbonyl group. One enol form of this ketone is represented by resonance structures 22-1b and 22-1c. Note that Structure 22-1c has a formal negative charge on the α-carbon atom.

QUESTION 22-2
Draw the resonance structure for the enol tautomer of acetone with a formal charge on the α-carbon atom.

ANSWER 22-2

$$\overset{+}{O}H$$
$$CH_3\overset{\|}{C}-\bar{C}H_2$$

In the presence of a strong base, an *enolate ion* is formed. The enolate anion of ketone 22-1a is represented by resonance structures 22-1d and 22-1e, with a formal negative charge on the α-carbon atom and oxygen atom, respectively. A carbon atom with a formal negative charge is called a *carbanion*. The electron density on the carbon atom in enols and enolate anions explains why electrophiles are attacked by this carbon atom. An enolate ion is more reactive (has a higher electron density) than an enol. An enol is even more nucleophilic than an alkene.

QUESTION 22-3
Draw the resonance form of the enolate anion of acetone with the formal negative charge on the α-carbon atom.

ANSWER 22-3

$$O$$
$$CH_3\overset{\|}{C}\bar{C}H_2$$

REACTIVITY OF ENOLS

Kinetic studies have shown that the enol, not the ketone tautomer, is the reactive species in α-substitution reactions. The rate of a reaction is proportional to the enol concentration, not the keto concentration. Since the enol concentration is usually very low, it is desirable to form additional enol as fast as possible while it is being consumed in an α-substitution reaction. Formation of the enol tautomer is catalyzed by acid or base, as shown in

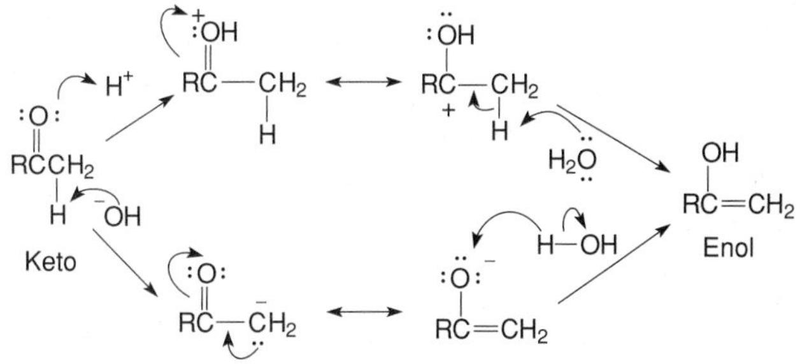

Fig. 22-2. Acid- and base-catalyzed enol formation.

Fig. 22-2. Acid or base is therefore used to increase reaction rates in α-substitution reactions.

Alpha Monohalogenation of Aldehydes and Ketones

Aldehydes and ketones are *monohalogenated* under *acidic conditions* at the α position by chlorine, bromine, or iodine. The rate of halogenation is independent of the halogen concentration. The common mechanism for halogenation is shown for bromine and a methyl ketone in Fig. 22-3, Reaction (a). The rate-controlling step is enol formation. Once the enol is formed, the carbon atom (which has a high electron density as shown in Structure 22-3a) reacts rapidly with the halogen.

ACID CATALYSIS

The first step in the acid-catalyzed halogenation is protonation of the carbonyl oxygen atom. If an electron-withdrawing halogen atom is already bonded to an α-carbon atom, it withdraws electrons, decreasing electron density of the α-carbon atom and the oxygen atom. This decreases the basicity and extent of protonation of the oxygen atom, slows the rate of enol formation, and decreases the rate of reaction with another halogen. The rate of monohalogenation is thus faster than the rate of dihalogenation. Monohalogenation can thus be achieved by reacting 1 equiv. of halogen with 1 equiv. of enol. *The result is a monosubstitution reaction: one hydrogen atom is replaced by one halogen atom.*

QUESTION 22-4
What is the structure of the monochloro α-substitution product of acetone?

ANSWER 22-4

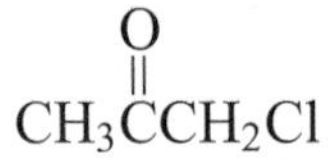

$$CH_3\overset{\displaystyle O}{\overset{\|}{C}}CH_2Cl$$

BASE CATALYSIS

When the reaction is carried out under *basic conditions, polyhalogenation* occurs. α-Monohalogenated compounds are more reactive toward further halogenation under basic conditions than are the nonhalogenated starting carbonyl compounds. This is shown in the reaction of bromine with a bromomethyl ketone (Reaction (b) in Fig. 22-3). The electron-withdrawing halogen atom helps to stabilize (lowers the energy of) the halogenated enolate anion, thus promoting its formation. This stabilization is shown in resonance structure 22-3b. Each halogen that is added speeds up the rate of subsequent halogenations. Thus polyhalogenation tends to occur. Methyl ketones are halogenated to trihalomethyl ketones.

Reaction (c) in Fig. 22-3 shows how the tribromomethyl group is cleaved by hydroxide anion. The tribromomethyl anion is rapidly protonated to give bromoform ($HCBr_3$) and the corresponding carboxylate anion. Chloroform ($HCCl_3$) and iodoform (HCI_3) are formed when chlorine and iodine are used in this reaction. These reactions are *base-promoted*, not base-catalyzed, since the base is consumed in the reaction.

QUESTION 22-5
What product results from the base-promoted iodination of *tert*-butyl methyl ketone using excess iodine?

ANSWER 22-5
2,2-Dimethylpropionate ion and iodoform.

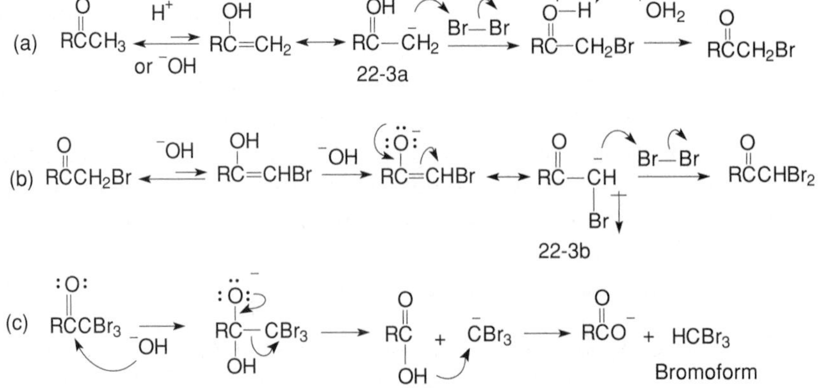

Fig. 22-3. Bromination of methyl ketones.

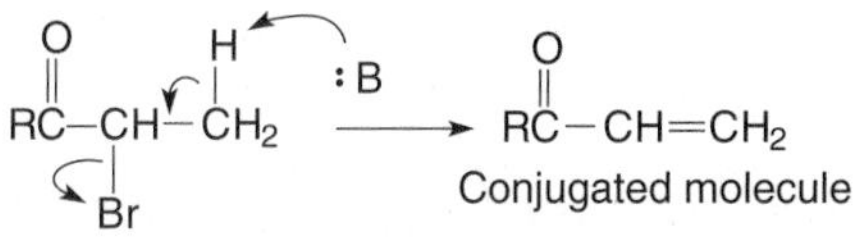

Fig. 22-4.   α,β-Unsaturated ketones.

Alpha-Beta Unsaturated Ketones

α-Haloketones and α-haloaldehydes are dehydrohalogenated when heated with base (e.g., pyridine) to give α,β-unsaturated carbonyl compounds. The mechanism of this reaction is shown in Fig. 22-4. The chemistry of these versatile α,β-unsaturated carbonyl compounds is discussed in Chapter 23.

Alpha-Bromination of Carboxylic Acids

Carboxylic acids cannot be directly α brominated since acids do not form stable enol tautomers. A carboxylic acid can be α brominated by first converting it into a carboxylic acid bromide by reaction with bromine in the presence of PBr_3. The carboxylic acid bromide forms sufficient enol for α bromination to occur. The α-bromocarboxylic acid bromide is subsequently hydrolyzed in aqueous acid to the α-bromocarboxylic acid. This is called a Hell–Volhard–Zelinski reaction, shown in Fig. 22-5.

Acidity of Alpha-Hydrogen Atoms

The Brønsted–Lowry definition of an acid is a compound that can donate a proton. A carbonyl compound with an α-hydrogen atom is capable of donating an α *proton* and acting as a Brønsted–Lowry acid. These carbonyl compounds include aldehydes, ketones, acid halides, and esters. These compounds are very weak acids relative to carboxylic acids (carboxylic acids are also weak acids when compared to inorganic acids like HCl). The pK_a values of a variety of organic compounds are given in Table 22-1. Acidity is inversely related to the pK_a value;

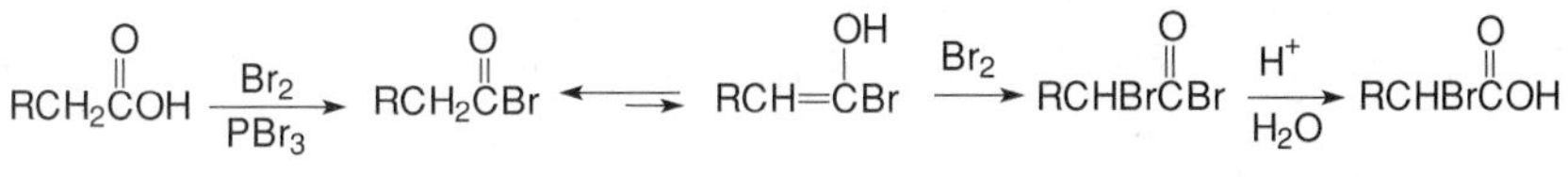

Fig. 22-5.   Bromination of carboxylic acids.

Table 22-1. Acidity (pK_a) values of organic compounds.

Structure	Name	pK_a
CH_3COH (Acetic acid structure, with C=O)	Acetic acid	5
$CH_3CCH_2CCH_3$ (with two C=O)	Acetylacetone	9
$CH_2=CH$ (with OH)	Enol of formaldehyde	10.5
$CH_3C=CH_2$ (with OH)	Enol of Acetone	11
$CH_3CCH_2COC_2H_5$ (with two C=O)	Ethyl acetoacetate	11
$C_2H_5OCCH_2COC_2H_5$ (with two C=O)	Malonic ester	13
H_2O	Water	15.7
CH_3CH_2OH	Ethanol	16
CH_3CH (with C=O)	Acetaldehyde	17
CH_3CCH_3 (with C=O)	Acetone	19
$CH_3COC_2H_5$ (with C=O)	Ethyl acetate	25
$HC\equiv HC$	Acetylene	25
$H_2C=CH_2$	Ethylene	44
CH_3CH_3	Ethane	50

the larger the pK_a value, the weaker the acid. The pK_a values are logarithmic. A difference in 6 pK_a units of two compounds represents a difference of 1,000,000 in their K_a values. (See Chapter 3 for a review of pK_a, K_a, and acid–base chemistry.)

TRENDS IN pK_aVALUES

Several important trends are shown in Table 22-1. Acidity decreases in the order aldehydes > ketones > esters for monocarbonyl compounds. A similar trend is seen for *α,β-dicarbonyl compounds*: β-diketones > β-carbonyl monoesters > β-diesters. The enol tautomer is much more acidic than the keto tautomer for a given compound. For example, the enol of acetone has a pK_a = 11 while acetone has a pK_a = 19. These trends in acidity will be discussed in several of the following reactions.

QUESTION 22-6

Which compound is more acidic, *tert*-butyl methyl ketone or *tert*-butyl methyl ether?

ANSWER 22-6

tert-Butyl methyl ketone. A carbonyl group increases the acidity of an α hydrogen to a greater extent than does an ether oxygen atom.

Strong bases such as sodium hydride (NaH), sodium amide (NaNH$_2$), and lithium diisopropylamide (LDA, Li [N(i-C$_3$H$_7$)$_2$]) are commonly used to form enolate anions by essentially completely removing the α hydrogen of *monocarbonyl* compounds. Hydroxide and alkoxide bases are usually not strong enough to completely remove these α hydrogens. LDA has the advantage of being soluble in the organic solvents used for organic reactions. LDA is a good nucleophile, however its bulkiness prevents it from acting as a nucleophile and attacking the electrophilic carbonyl carbon atom. The acid–base equilibrium expressions and pK_a values for the reaction of acetone with ethoxide (Reaction (a)) and LDA (Reaction (b)) are shown in Fig. 22-6. The equilibrium is shifted toward the weaker acid (survival of the weakest), acetone in Reaction (a) and diisopropyl amine in Reaction (b).

STABILITY OF ENOLATE ANIONS

Anions are stabilized if the negative charge can be delocalized (spread out) throughout the molecule. Enolate anion resonance stabilization is shown in Fig. 22-7. The charge density is greater on the more electronegative oxygen

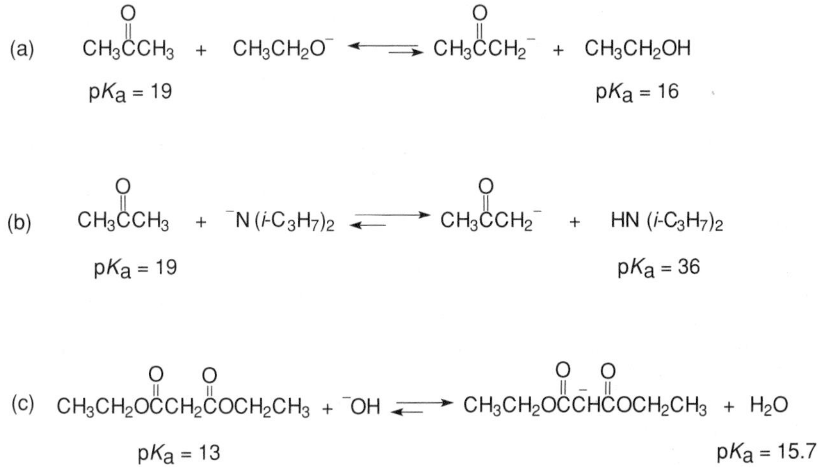

Fig. 22-6. Acid–base reactions of carbonyl compounds.

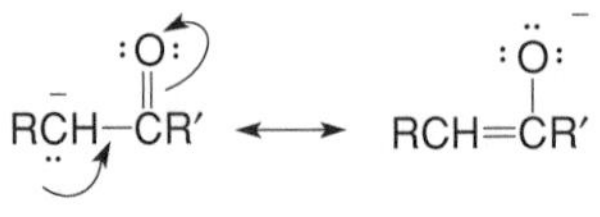

$$RCH-CR' \longleftrightarrow RCH=CR'$$

Fig. 22-7. Resonance-stabilized carbanion.

atom than on the carbon atom. However the carbon atom almost always reacts with an electrophile since a *carbanion* (carbon anion) is a better nucleophile than is an oxygen anion.

1,3-Dicarbonyl compounds

α-Hydrogen atoms that are flanked by two carbonyl groups are more acidic than α-hydrogen atoms adjacent to only one carbonyl group (see Table 22-1). Three common 1,3-dicarbonyl (β-dicarbonyl) compounds are shown in Fig. 22-8. Each compound is known by multiple names. Malonic ester (Structure 22-8a) is one common name for diethyl propanedioate (formal) and diethyl malonate (common). Acetoacetic ester (Structure 22-8b) is a common name for ethyl 3-oxobutanoate (formal) and ethyl acetoacetate (common). Structure 22-8c is called acetylacetone (common) and 2,5-pentanedione (formal).

Enol formation in 1,3-dicarbonyl compounds

The enol tautomer of 1,3-dicarbonyl compounds is present in much greater concentrations than is the enol tautomer of monocarbonyl compounds. The enol/keto ratio for acetylacetone, a 1,3-dicarbonyl compound, is 1/5 in water and increases to 12/1 in hexane. The enol/keto ratio for a monocarbonyl compound, acetone, is the minute $1/1 \times 10^{11}$. The increased amount of enol in 1,3-dicarbonyl compounds is attributed to hydrogen bonding of the OH hydrogen atom with both oxygen atoms, as shown in Fig. 22-9. Intramolecular hydrogen bonding is more important in nonpolar solvents (hexane) than in hydrogen bonding solvents (water), where the solvent competes for hydrogen bonding with the carbonyl group.

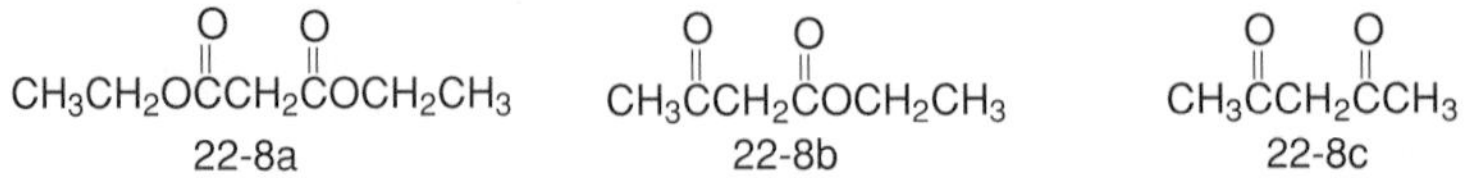

$$CH_3CH_2OCCH_2COCH_2CH_3$$
22-8a

Diethyl propanedioate
or diethyl malonate
or malonic ester

$$CH_3CCH_2COCH_2CH_3$$
22-8b

Ethyl 3-oxobutanoate
or ethyl acetoacetate

$$CH_3CCH_2CCH_3$$
22-8c

2,4-Pentanedione
or acetylacetone

Fig. 22-8. β-Dicarbonyl compounds.

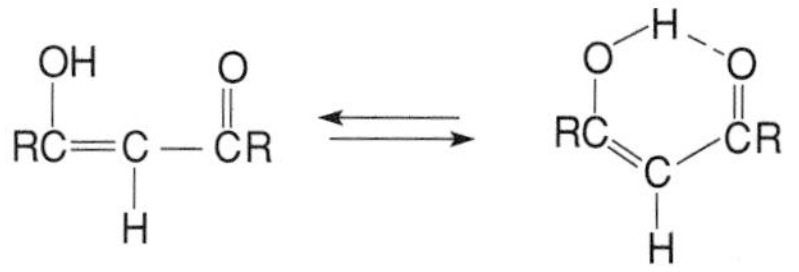

Fig. 22-9. Intramolecular hydrogen bonding.

ACIDITY OF 1,3-DICARBONYL COMPOUNDS

The pK_a values of 1,3-diesters, (β-keto esters) and 1,3-diketones range from about 9 to 13. The increased acidity of these 1,3-dicarbonyl compounds results because the negative charge of the conjugate base is resonance-stabilized over two carbonyl groups, as shown in Fig. 22-10. The α hydrogen can be completely removed in 1,3-dicarbonyl compounds by hydroxide and alkoxide bases. An equilibrium expression for the reaction of malonic ester and hydroxide anion (Reaction (c), Figure 22-6) is shifted strongly toward the weaker acid (water) and the enolate anion. Stronger (and more expensive) bases are not needed to completely remove the α-hydrogen atom.

QUESTION 22-7
Which compound is more acidic, 1,2-cyclohexanedione, 1,3- cyclohexanedione, or 1,4-cylcohexanedione?

ANSWER 22-7
1,3-Cyclohexanedione, because it is a β-diketone.

Malonic Ester Synthesis

Malonic ester (Structure 22-8a) is a β-diester with a methylene (CH_2) group bonded between two carbonyl groups. An acidic (p$K_a = 13$) methylene hydrogen atom can be completely removed with a hydroxide or an alkoxide base. The base used in malonic ester reactions is the ethoxide anion and any transesterification reactions that might occur do not change the ester (ethyl) group. This reaction is shown in Fig. 22-11. The resulting carbanion (22-11a) acts as a nucleophile and undergoes S_N2 substitution reactions with methyl bromide (or primary alkyl, allyl, or benzyl halides), giving an α-alkyl-substituted product 22-11b.

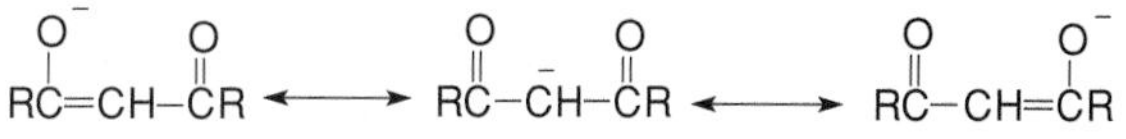

Fig. 22-10. Resonance-stabilized enolate anion.

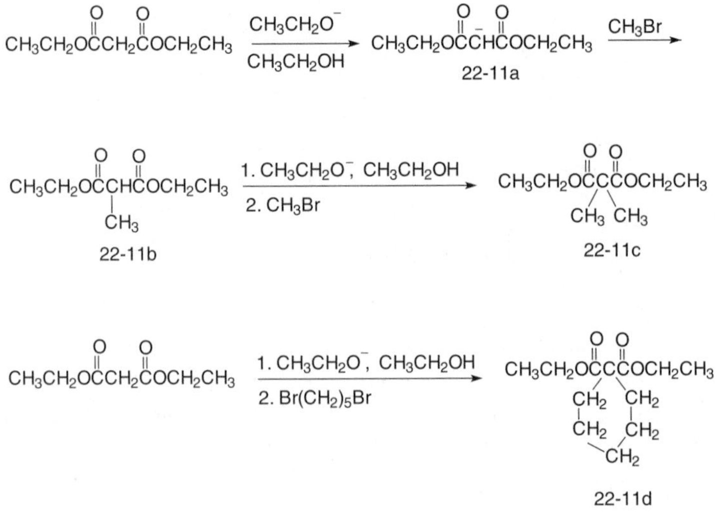

$$CH_3CH_2OCCH_2COCH_2CH_3 \xrightarrow[CH_3CH_2OH]{CH_3CH_2O^-} CH_3CH_2OCCHCOCH_2CH_3 \xrightarrow{CH_3Br}$$

22-11a

$$CH_3CH_2OCCHCOCH_2CH_3 \xrightarrow[\text{2. } CH_3Br]{\text{1. } CH_3CH_2O^-, CH_3CH_2OH} CH_3CH_2OCCCOCH_2CH_3$$

22-11b

22-11c

$$CH_3CH_2OCCH_2COCH_2CH_3 \xrightarrow[\text{2. } Br(CH_2)_5Br]{\text{1. } CH_3CH_2O^-, CH_3CH_2OH} CH_3CH_2OCCCOCH_2CH_3$$

22-11d

Fig. 22-11. Alkylation of malonic ester.

Secondary and tertiary alkyl halides tend to give elimination reactions (alkene formation from the halide) with the malonate anion. The remaining hydrogen atom still bonded to the methylene group can be removed with additional base. The resulting carbanion reacts with another alkyl halide to give a dialkyl-substituted malonic ester (22-11c).

QUESTION 22-8
Draw the product that results from the reaction of malonic ester with 1 equiv. of base and subsequently 1 equiv. of ethyl chloride.

ANSWER 22-8

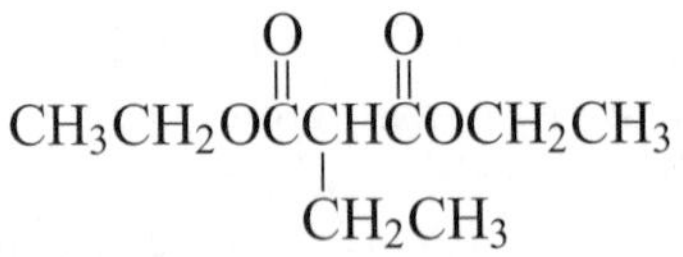

$$CH_3CH_2OCCHCOCH_2CH_3$$
$$CH_2CH_3$$

Cyclic compounds

When malonic ester is reacted with 2 equiv. of base and 1 equiv. of a dihalide, like 1,5-dibromopentane, a six-membered ring compound (Structure 22-11d) is formed.

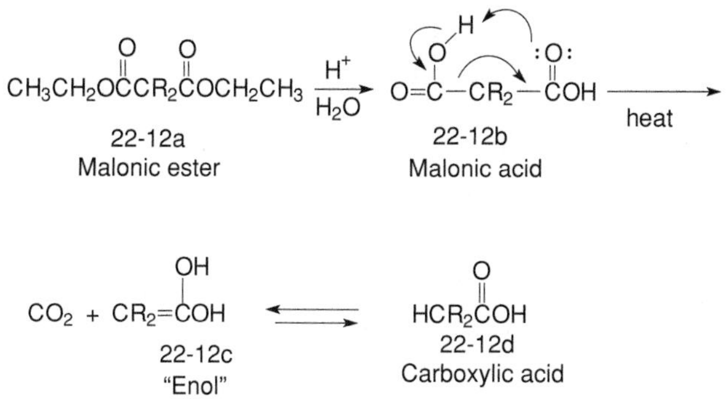

Fig. 22-12. Decarboxylation of malonic ester.

DECARBOXYLATION REACTIONS

Substituted malonic esters can be converted into substituted dicarboxylic acids, as shown in Fig. 22-12. The ester groups on a mono- or disubstituted malonic ester (Structure 22-12a) are hydrolyzed in aqueous acid (or base) to give a substituted malonic acid (Structure 22-12b). Upon heating to $100–150\,°C$, malonic acid undergoes monodecarboxylation to give an "enol" (Structure 22-12c). The "enol" tautomerizes to an acid (Structure 22-12d). Compounds with a carbonyl group β to a carboxylic acid group, that can undergo monodecarboxylation (loss of 1 equiv. of carbon dioxide) upon heating, include various β-diacids and β-keto acids. The mechanism of decarboxylation is analogous to that shown in Fig. 22-12.

QUESTION 22-9
Draw the structure of the product that results from acid hydrolysis and decarboxylation of the substituted malonic ester prepared in Question 22-8.

ANSWER 22-9

$$CH_3CH_2CH_2CO_2H.$$

Acetoacetic Ester Synthesis

Acetoacetic ester synthesis is a method of preparing substituted methyl ketones, as shown in Fig. 22-13. The hydrogen atoms on the methylene carbon atom between the two carbonyl groups are acidic ($pK_a = 11$). Reaction of acetoacetic ester (Structure 22-13a) with 1 or 2 equiv. of base (ethoxide, $CH_3CH_2O^-$, is used with ethyl esters) and a subsequent reaction with a methyl, primary alkyl,

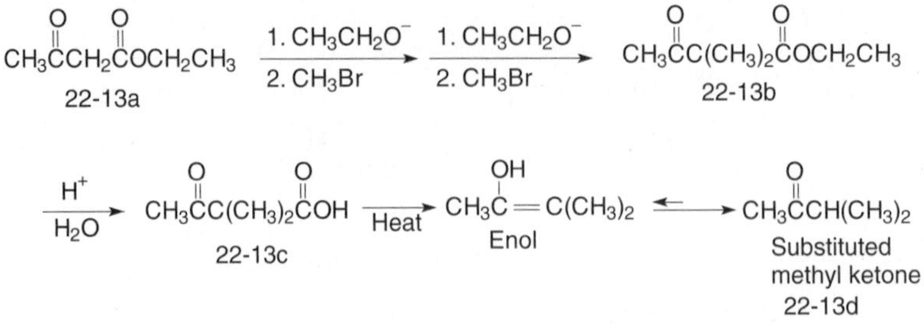

$$\underset{\text{22-13a}}{CH_3\overset{O}{\overset{\|}{C}}CH_2\overset{O}{\overset{\|}{C}}OCH_2CH_3} \xrightarrow[\text{2. } CH_3Br]{\text{1. } CH_3CH_2O^-} \xrightarrow[\text{2. } CH_3Br]{\text{1. } CH_3CH_2O^-} \underset{\text{22-13b}}{CH_3\overset{O}{\overset{\|}{C}}C(CH_3)_2\overset{O}{\overset{\|}{C}}OCH_2CH_3}$$

$$\xrightarrow[H_2O]{H^+} \underset{\text{22-13c}}{CH_3\overset{O}{\overset{\|}{C}}C(CH_3)_2\overset{O}{\overset{\|}{C}}OH} \xrightarrow[\text{Heat}]{} \underset{\underset{\text{Enol}}{}}{CH_3\overset{OH}{\overset{|}{C}}=C(CH_3)_2} \xleftarrow{\quad} \underset{\substack{\text{Substituted}\\\text{methyl ketone}\\\text{22-13d}}}{CH_3\overset{O}{\overset{\|}{C}}CH(CH_3)_2}$$

Fig. 22-13.   Acetoacetic ester synthesis.

allyl, or benzylic halide gives a mono- or disubstituted acetoacetic ester. The reaction using 2 equiv. of methyl bromide gives a dimethylated acetoacetic ester, Structure 22-13b. Subsequent acid-catalyzed hydrolysis of the ester group gives a disubstituted acetoacetic acid, Structure 22-13c, a β-keto carboxylic acid. The β-keto carboxylic acid undergoes decarboxylation upon heating (as explained for malonic acid in Fig. 22-12). The resulting product is a substituted methyl ketone (22-13d). This synthesis can be used to prepare a variety of substituted methyl ketones.

QUESTION 22-10
Draw the structure of the product that results from the reaction of acetoacetic ester with reagents in the following sequence: dimethylation, acid-catalyzed ester hydrolysis, and decarboxylation.

ANSWER 22-10

$$\underset{}{CH_3\overset{O}{\overset{\|}{C}}CH(CH_3)_2}$$

Additional Condensation Reactions

THE KNOEVENAGEL REACTION

Aldehydes and unhindered ketones react with saturated α,β-dicarbonyl compounds to give α,β-unsaturated carbonyl compounds. This reaction is called a Knoevenagel reaction, and an example is shown in Fig. 22-14. If malonic ester is used, the initially formed unsaturated diester (Structure 22-14a) is hydrolyzed to give the corresponding unsaturated diacid. The diacid (Structure 22-14b) can be monodecarboxylated to give a conjugated α,β-unsaturated monocarboxylic

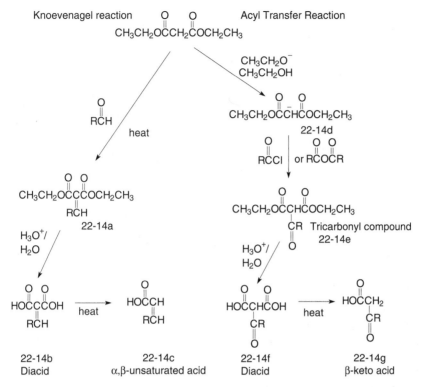

Fig. 22-14.   Additional condensation reactions.

acid (Structure 22-14c). The reaction can also be run with acetoacetic ester to give a conjugated α,β-unsaturated methyl ketone.

ACYL TRANSFER REACTIONS

α,β-Dicarbonyl compounds can undergo acyl transfer reactions. An example using malonic ester is shown in Fig. 22-14. Malonic ester is reacted with base to give the enolate anion (Structure 22-14d). This anion reacts with an acid halide or an anhydride to give the corresponding tricarbonyl compound (Structure 22-14e). Hydrolysis of 22-14e gives the diacid (Structure 22-14f). Subsequent decarboxylation gives the monoacid (Structure 22-14g).

One can now appreciate the huge variety of compounds that can be synthesized by using the few reactions described in this chapter. Organic chemists need to be able to recognize which combinations of reactions are needed to prepare compounds with functional groups in specific positions.

Quiz

1. The term enol is derived form the words
    (a) alkene and nolane
    (b) alkene and alcohol
    (c) ether and alcohol
    (d) ketene and alcohol

2. Tautomers are
    (a) resonance forms
    (b) identical molecules
    (c) constitutional isomers
    (d) derived from tall tomaters

3. Enolate ions have ———— charge.
    (a) a positive
    (b) a negative
    (c) no charge
    (d) a positive and negative

4. The reactive species in the α halogenation of a ketone is
    (a) a ketone
    (b) an aldehyde
    (c) an enol
    (d) a carboxylic acid

5. The intermediate in the base-promoted α halogenation of a ketone is
    (a) an anion
    (b) a cation
    (c) a zwitterion (positive and negative ion)
    (d) a neutral species

6. Carboxylic acids undergo acid-catalyzed α halogenation.
    (a) True
    (b) False

7. The compound containing the smallest pK_a value is
    (a) Acetic acid
    (b) Ethanol
    (c) Acetone
    (d) Ethane

8. The compound containing the smallest pK_a value is
    (a) 3-Pentanone
    (b) 2,3-Pentanedione

(c) 2,4-Pentanedione
(d) 1,5-Pentanedial

9. The reaction of malonic ester with excess ethoxide base and excess
    ethyl bromide gives
    (a) no reaction
    (b) a monosubstituted malonic ester
    (c) a disubstituted malonic ester
    (d) a trisubstituted malonic ester

10. Acid hydrolysis of the product in Problem 9 gives a
    (a) diester
    (b) monoester
    (c) monocarboxylic acid
    (d) dicarboxylic acid

11. Decarboxylation of the product formed in Problem 10 gives
    (a) an acid
    (b) an ester
    (c) a ketone
    (d) an aldehyde

12. Decarboxylation of malonic acid gives
    (a) acetaldehyde
    (b) acetone
    (c) ethyl acetate
    (d) acetic acid

13. Acid hydrolysis and decarboxylation of acetoacetic ester gives
    (a) acetaldehyde
    (b) acetone
    (c) ethyl acetate
    (d) ethyl formate

14. Monomethylation of acetoacetic ester with subsequent acid hydrolysis
    and decarboxylation gives
    (a) ethyl methyl ether
    (b) ethyl acetate
    (c) ethyl methyl ketone
    (d) ethyl methyl alcohol

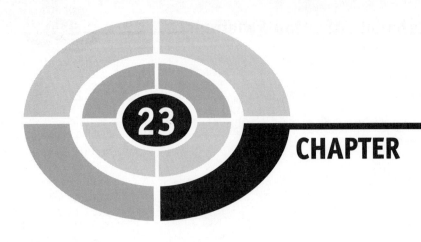

Carbonyl Condensation Reactions

Introduction

Carbonyl compounds can act as nucleophiles or electrophiles. In carbonyl condensation reactions, one carbonyl compound reacts with base to form an enolate anion. This anion acts like a nucleophile attacking an electrophilic carbon atom in the carbonyl group of another molecule. The result is the formation of a new carbon–carbon bond between the carbonyl carbon atom of one molecule and the α-carbon atom of another carbonyl molecule. This is an example of an α-substitution reaction. Various carbonyl compounds can undergo these reactions including aldehydes, ketones, esters, anhydrides, and amides.

CH₃CH $\xrightarrow[\text{CH}_3\text{CH}_2\text{O}^-]{\text{$^-$OH or}}$ CH₂CH + CH₃CH $\rightleftharpoons$ CH₃CHCH₂CH $\xrightarrow{\text{H}_3\text{O}^+}$ CH₃CHCH₂CH

α hydrogen on   23-1a   23-1b   23-1c
an α carbon   Enolate anion   Alkoxyaldehyde   Aldol

β-hydroxyaldehyde

Fig. 23-1. Aldol reaction.

There are numerous biological reactions that involve carbonyl condensation reactions.

Aldol Reactions

Under the appropriate reaction conditions, two aldehyde molecules undergo a rapid and reversible *aldol reaction* to give a β-hydroxyaldehyde. This reaction is shown in Fig. 23-1. A β-hydroxy aldehyde is called an *aldol* (*ald*ehyde and alcoh*ol*). Ketones also undergo this reaction, which is still called an aldol reaction. All steps are reversible.

An aldehyde or ketone will not undergo a base-catalyzed aldol reaction with itself if it does not contain an α-hydrogen atom that can be removed by the base. Aldol reactions are reversible and the equilibrium is shifted strongly toward the starting aldehyde in most reactions employing bulky aldehydes, such as α-substituted aldehydes, and most ketones. The starting materials are apparently favored because of unfavorable steric effects in the aldol product.

BASE CATALYSIS

The aldol reaction of acetaldehyde is shown in Fig. 23-1. Reaction of acetaldehyde with base, CH₃CH₂O⁻ or ⁻OH, forms the enolate anion (23-1a). The enolate anion attacks another molecule of acetaldehyde, giving an alkoxide anion (23-1b). Addition of acid protonates the alkoxide anion giving the aldol product, β-hydroxybutyraldehyde, 23-1c.

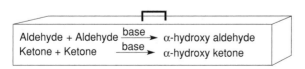

Aldehyde + Aldehyde $\xrightarrow{\text{base}}$ α-hydroxy aldehyde
Ketone + Ketone $\xrightarrow{\text{base}}$ α-hydroxy ketone

Toolbox 23-1.

QUESTION 23-1

What is the product of the aldol reaction of propanal, CH_3CH_3CHO?

ANSWER 23-1

$$
\begin{array}{cc}
\text{OH} & \text{O} \\
| & \| \\
\end{array}
$$
$$CH_3CH_2CHCHCH$$
$$
\begin{array}{c}
| \\
CH_3
\end{array}
$$

Dehydration of Aldol Compounds

The reversible aldol reaction usually takes place under relatively mild conditions. Under more vigorous conditions (heating), the aldol product undergoes dehydration (loss of a molecule of water) to give a conjugated α,β-unsaturated aldehyde. The equilibrium reaction favors formation of the conjugated unsaturated aldehyde. Although the equilibrium reaction for aldol formation is unfavorable, aldol continues to be formed as it undergoes dehydration. Thus good yields of α,β-unsaturated aldehydes are obtained. Ketones undergo an analogous reaction, giving α,β-unsaturated ketones. The dehydration reaction can be catalyzed by acid or base. The mechanism for both catalyzed reactions is shown in Fig. 23-2.

α-hydroxy aldehyde —dehydration→ α,β-unsaturated aldehyde

α-hydroxy ketone —dehydration→ α,β-unsaturated ketone

Toolbox 23-2.

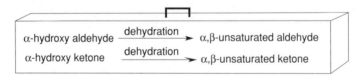

Enol of a β-hydroxyaldehyde

Fig. 23-2. Aldol dehydration reaction.

QUESTION 23-2
What is the dehydration product for the aldol product in Question 23-1?

ANSWER 23-2

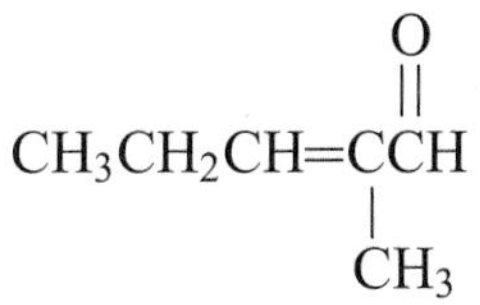

$$\underset{\underset{CH_3}{|}}{CH_3CH_2CH{=}C}CH{\overset{\overset{O}{\|}}{}}$$

Mixed or Crossed Aldol Reactions

The previous discussion was limited to the dimerization of an aldehyde or a ketone. An aldol reaction can take place between two different aldehydes, two different ketones, or an aldehyde and a ketone. If both carbonyl compounds contain the required α-hydrogen atom (for a reaction to occur), there are four possible products. If aldehyde A and ketone B react, the aldol products would be A–A, A–B, B–B, and B–A. This is not a very useful reaction as it would be difficult to separate the four products.

Two different aldehydes, two different ketones, or an aldehyde and a ketone will successfully undergo an aldol reaction if one of the reactants does *not* contain an α-hydrogen atom and the other reagent *does* contain an α-hydrogen atom. This is called a *mixed* or a *crossed* aldol reaction. The reaction of formaldehyde, 23-3a (that contains no α-hydrogen atom), and acetaldehyde, 23-3b (that contains an α-hydrogen atom), is shown in Reaction (a) in Fig. 23-3. Formaldehyde

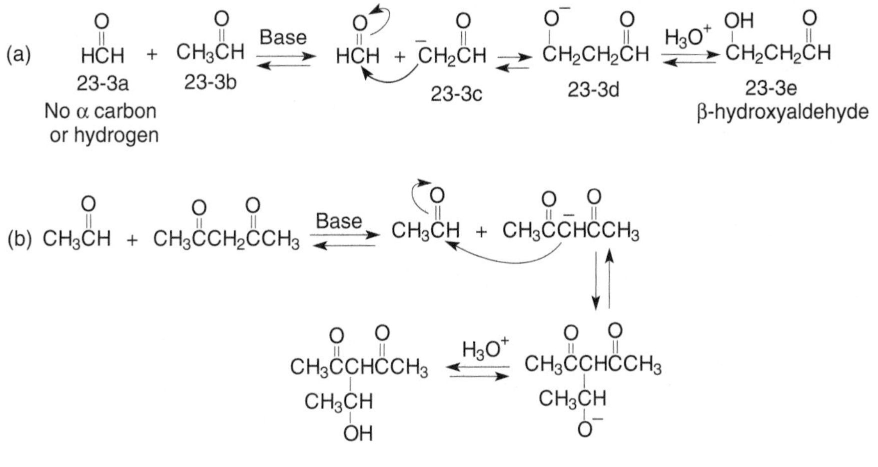

Fig. 23-3. Mixed aldol reaction.

contains a relatively unhindered carbonyl group that is readily susceptible to nucleophilic attack by an enolate anion.

QUESTION 23-3
What is the aldol product from the reaction of formaldehyde with propanal?

ANSWER 23-3

$$\begin{array}{c} \text{OH} \quad \text{O} \\ | \qquad || \\ \text{H}_2\text{CCHCH} \\ | \\ \text{CH}_3 \end{array}$$

DIFFERENT BASE STRENGTHS

One equivalent of base added to one equivalent of acetaldehyde forms the enolate anion 23-3c). This is an equilibrium reaction when (relatively weaker) bases like the hydroxide or alkoxide anions are used. The enolate anion prefers to attack the less hindered carbonyl carbon atom in formaldehyde rather than attacking a more sterically hindered acetaldehyde molecule. Using one equivalent of a much stronger base, such as the amide anion (H_2N^-), converts all the acetaldehyde to the enolate anion. Subsequent addition of formaldehyde gives exclusively the crossed aldol reaction. The initially formed β-alkoxyaldehyde (23-3d) is treated with acid to form the β-hydroxyaldehyde (23-3e).

DIFFERENT ACID STRENGTHS

Mixed aldols can also be prepared if one of the carbonyl compounds is much more acidic than the other reactant. 1,3-Dicarbonyl compounds, such as 2,4-pentanedione ($pK_a = 9.0$), are much more acidic than monocarbonyl compounds such as acetaldehyde ($pK_a = 17$) or acetone ($pK_a = 19.3$). This is a very large difference in acidities. A base selectively removes the most acidic proton to form an enolate anion of the dione. The dione enolate anion then attacks the less acidic aldehyde or ketone. This is shown in Reaction (b) in Fig. 23-3.

Intramolecular Aldol Reactions

1,4- and 1,5-Diketones such as 2,5-hexanedione or 2,6-heptanedione can undergo an *intramolecular cyclization aldol reaction*. The mechanism is identical

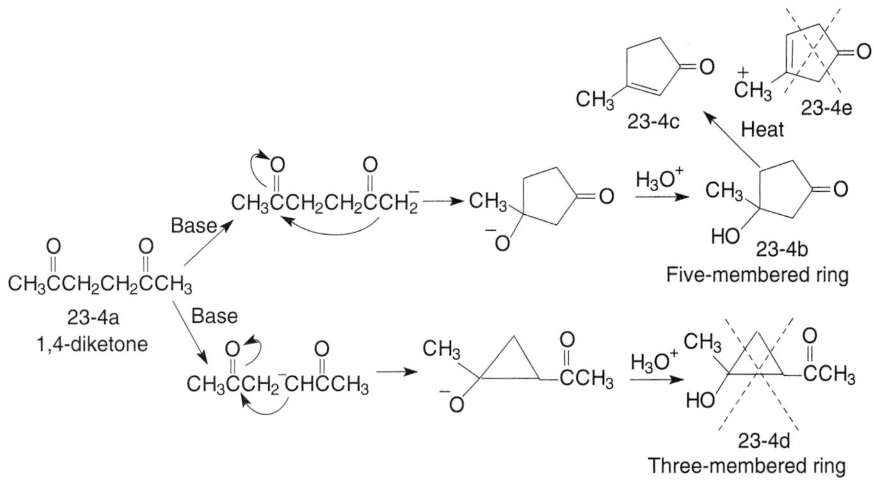

Fig. 23-4.   Intramolecular aldol reaction.

to that shown in Fig. 23-1. 2,5-Hexanedione (Structure 23-4a) could form a five-membered ring (23-4b) or a three-membered ring (23-4d). These reactions are shown in Fig. 23-4. 2,6-Heptanedione could form a six- or a four-membered ring. Because of steric and bond-angle strain, five- and six-membered rings are thermodynamically more stable than larger or smaller rings. Since aldol formation is a reversible reaction, Structure 23-4a will form the more stable five-membered ring (Structure 23-4b) under equilibrium conditions. The intramolecular aldol reaction of 2,6-heptanedione forms a six-membered ring.

If the reaction is carried out under more vigorous basic conditions, dehydration occurs. This equilibrium reaction strongly favors the cyclic conjugated α,β-unsaturated ketone, Structure 23-4c, rather than the β,γ-nonconjugated ketone, Structure 23-4e.

Claisen Condensation Reactions

The mechanism of a *Claisen condensation reaction* is similar to that of an aldol reaction but involves esters rather than aldehydes or ketones. The overall result is the formation of a β-keto ester. The starting ester must contain two α-hydrogen atoms. The reaction between two ethyl acetate esters (23-5a) is shown in Fig. 23-5. An alkoxide base is used to remove an acidic α-hydrogen atom. The alkoxide base corresponds to the ester group so that ester interchange does not change the ester function. For example, ethoxide anion is used with ethyl esters. The reaction is base-promoted, not base-catalyzed, because base is consumed in the

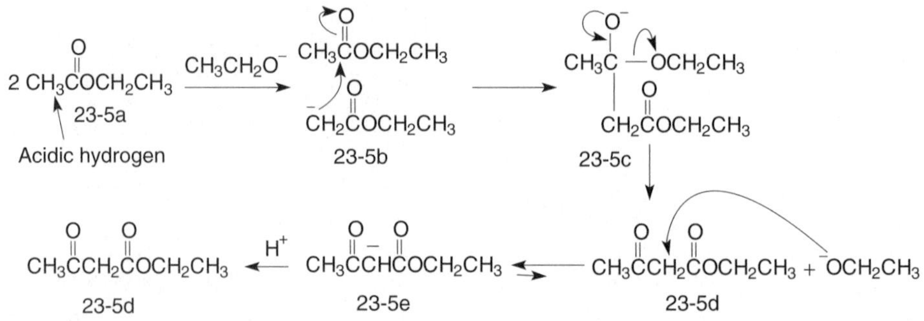

Fig. 23-5. Claisen condensation reaction.

reaction. The resulting enolate anion (23-5b) attacks the carbonyl carbon atom of another ester molecule, forming a tetrahedral alkoxide ion (23-5c).

The fate of the alkoxide anion is different in a Claisen reaction than in an aldol reaction. The alkoxide anion is protonated, forming an alcohol in the aldol reaction. In the Claisen reaction, the ester group leaves as an alkoxy group (an ethoxide anion in this example). The ethoxide anion removes the second acidic α-hydrogen atom from the β-keto ester 23-5d. This shifts the equilibrium toward the condensation product by converting it into an enolate anion 23-5e. Addition of acid protonates enolate 23-5e back to β-keto ester 23-5d. Thus an alcohol molecule (CH_3CH_2OH) is "condensed" from the two reacting ester molecules. An alkene is not formed. This differs from an aldol condensation reaction where a molecule of water is "condensed" from the reaction and an α,β-unsaturated carbonyl compound is formed.

QUESTION 23-4
What is the Claisen condensation product from ethyl propionate?

ANSWER 23-4

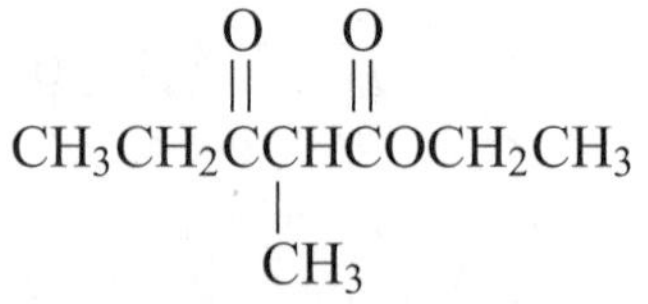

MIXED OR CROSSED CLAISEN REACTIONS

Mixed (or crossed) Claisen reactions involve the reaction between molecules of two different esters. A typical reaction is shown in Fig. 23-6. As in mixed aldol reactions, one of the reactants (an ester in a mixed Claisen reaction)

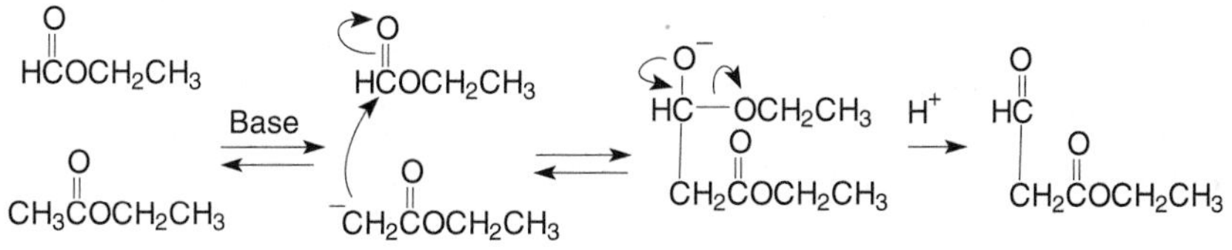

Fig. 23-6.   Mixed Claisen reaction.

must not contain an acidic α-hydrogen atom and must not be able to form an enolate anion. The ester with an acidic α-hydrogen atom reacts with 1 equiv. of an alkoxide base to form the enolate anion. The enolate anion attacks the electrophilic carbonyl carbon of the ester that does not contain an α-hydrogen atom. The resulting product is a β-aldo ester. The alkoxide base used is the same as the ester group (ethoxide anion is used with ethyl esters) so ester interchange does not change the ester group.

QUESTION 23-5
What is the Claisen condensation product from the reaction between ethyl formate and ethyl propionate?

ANSWER 23-5

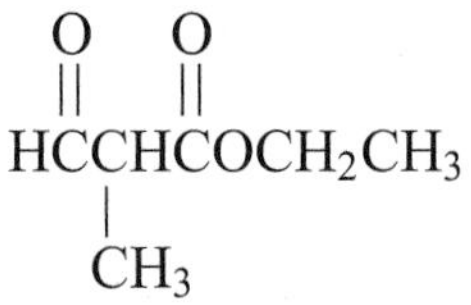

$$\underset{\underset{CH_3}{|}}{\overset{\overset{O}{\parallel}\quad\overset{O}{\parallel}}{HCCHCOCH_2CH_3}}$$

An example of a mixed *Claisen-like* reaction is between an ester and a ketone. If the ester does not contain an α-hydrogen atom, the product is a β-dicarbonyl compound. An example of this reaction is shown in Fig. 23-7. The reaction

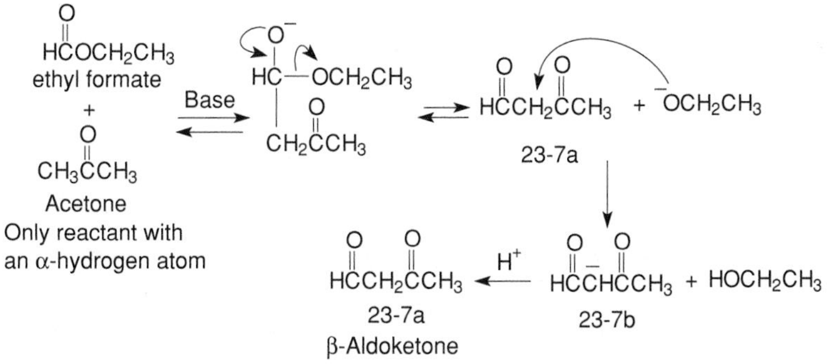

Fig. 23-7.   Mixed Claisen-like reaction.

of acetone with ethyl formate gives a β-aldoketone, Structure 23-7a. A base like NaH or $CH_3CH_2O^-Na^+$ is used to remove the α-hydrogen atom from 23.7a giving enolate anion 23-7b. Acid is added in the final step to protonate the enolate anion reforming 23-7a. The molecule that is condensed from the starting materials in this reaction is ethanol.

INTRAMOLECULAR CLAISEN CONDENSATION REACTIONS

An intramolecular Claisen condensation reaction is also called a *Dieckmann cyclization reaction*. The product of a Dieckmann cyclization reaction is a cyclic β-keto ester. Highest yields are obtained when 1,6- and 1,7-diesters are used as starting materials. These diesters give five- and six-membered cyclic β-keto esters. Rings containing five and six atoms are more thermodymanically stable than larger or smaller rings An example of a cyclization reaction for diethyl 1,7-heptanedioate, Structure 23-8a, is shown in Fig. 23-8. Ethoxide base removes a proton from either of the two α-carbon atoms. Reaction of the enolate anion with the carbonyl group gives cyclic alkoxide 23-8b. An ethoxide anion is eliminated giving β-keto ester 23-8c.

Ester + Ester $\xrightarrow{\text{base}}$ β-keto ester

1,6- or 1,7-Diester $\longrightarrow$ Cyclic β-keto ester

Toolbox 23-3.

QUESTION 23-6
What is the product from the intramolecular Claisen reaction of diethyl 1,6-hexanedioate?

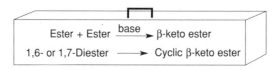

23-8a
1,7-diester

23-8b

23-8c
β-keto ester

Fig. 23-8. Dieckmann cyclization reaction.

ANSWER 23-6

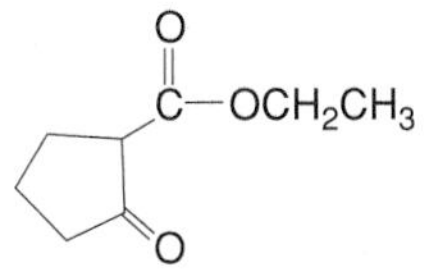

Alpha-Substitution Reactions

ALKYLATION REACTIONS

A β-keto ester that contains an α-hydrogen atom can undergo an additional alkylation reaction. This reaction is shown in Fig. 23-9. Reaction of β-keto ester 23-9a with base gives anion 23-9b. Anion 23-9b undergoes an S_N2 reaction with a methyl, primary alkyl, allyl, or benzyl halide giving the corresponding α-substituted product, 23-9c.

The β-keto ester (before or after alkylation) can be hydrolyzed to a β-keto acid. Upon heating, the β-keto acid can undergo decarboxylation to give a cyclic ketone. (For a review of decarboxylation reactions see the section entitled *Decarboxylation Reactions* in Chapter 22.)

MICHAEL REACTION

A Michael reaction is the nucleophilic addition of an enolate anion to the β-carbon atom of an α,β-unsaturated carbonyl compound. This is another method of making carbon–carbon bonds. The enolate anion can be made from a variety of compounds with acidic α-hydrogen atoms (Michael donors). A variety of α,β-unsaturated carbonyl compounds (Michael acceptors) can also be used. Table 23-1 is a list of Michael donors and acceptors. An example of a Michael reaction between acetylacetone (23-10a) and propenal (23-10c) is shown in Fig. 23-10. Base reacts with diketone 23-10a to form enolate anion 23-10b.

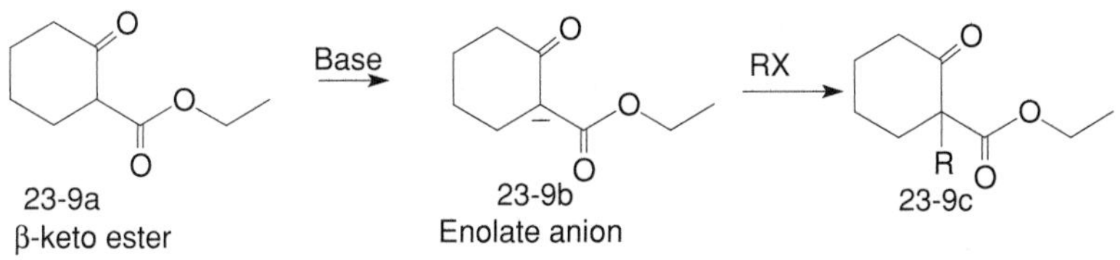

23-9a
β-keto ester

Base →

23-9b
Enolate anion

RX →

23-9c

Fig. 23-9. Alkylation of a β-keto ester.

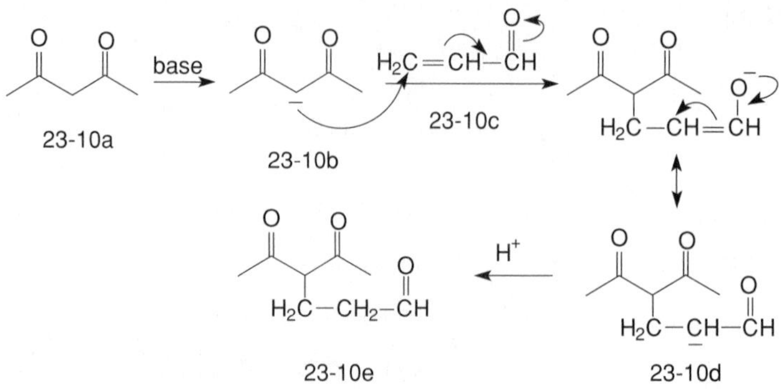

Fig. 23-10. Michael addition reaction.

Anion 23-10b reacts with α,β-unsaturated aldehyde 23-10c to give addition product 23-10d, which is subsequently protonated to give Structure 23-10e.

QUESTION 23-7

What is the Michael addition product for the base-catalyzed reaction between 2,4-pentanedione and ethyl propenoate?

Table 23-1. Michael reaction donors and acceptors.

Donors	Acceptors
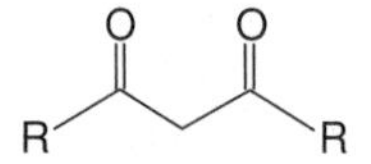	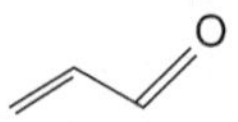
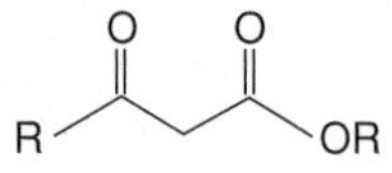	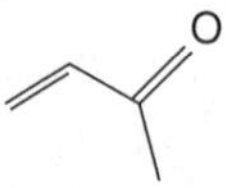
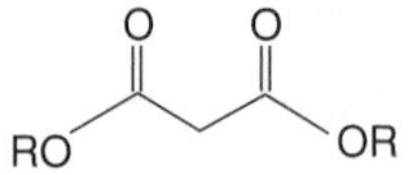	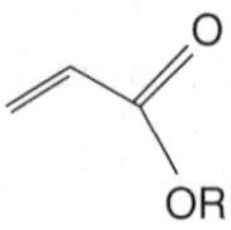
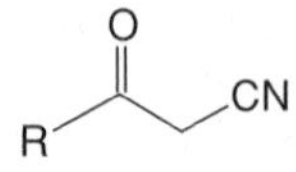	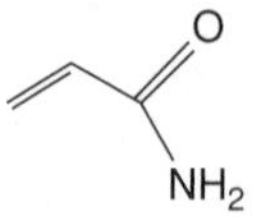
RCH_2NO_2	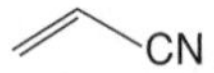

ANSWER 23-7

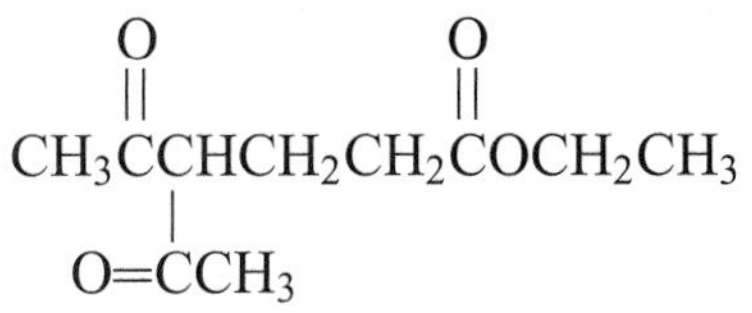

$$CH_3\overset{\underset{\displaystyle \|}{O}}{C}CHCH_2CH_2\overset{\underset{\displaystyle \|}{O}}{C}OCH_2CH_3$$

$$O=CCH_3$$

STORK ENAMINE REACTION

The Stork enamine reaction is one of the better ways of α-alkylating or α-acylating ketones. Three enamine reactions are shown in Fig. 23-11. An enamine (23-11a) is formed by the reaction of a secondary amine with a ketone. (Enamine formation is discussed in the section entitled *Enamines* in Chapter 19.) Enamines are electronically similar to enolate anions. Enamines react with alkyl halides, acyl halides, and α,β-unsaturated carbonyl compounds to give α-substituted enamines. The hydrolysis of enamines gives α-substituted ketones.

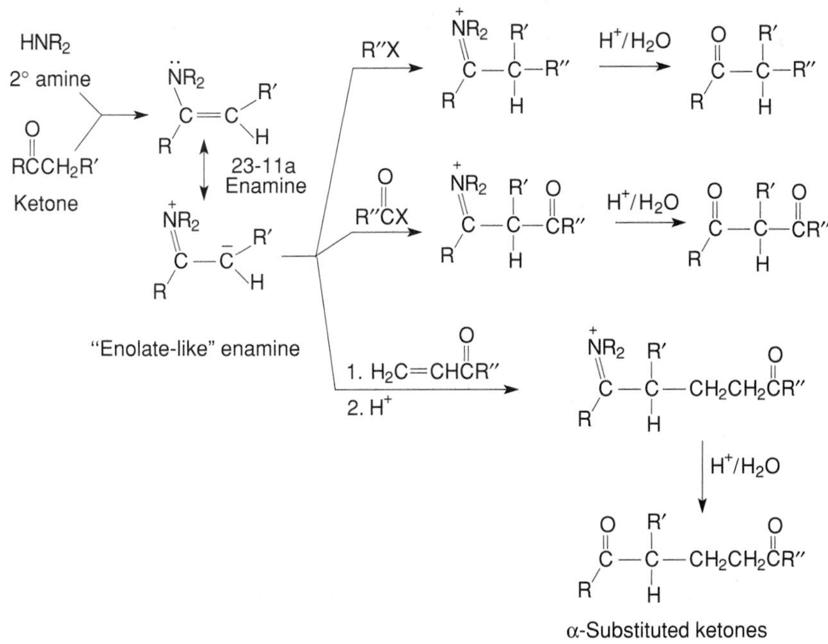

Fig. 23-11. The Stork enamine reactions.

Quiz

1. Aldol is an acronym for
   (a) aliphatic diols
   (b) aldolhydes
   (c) aldehyde and alcohol
   (d) old alcohols

2. The base-catalyzed aldol reaction of two molecules of acetaldehyde gives
   (a) a dialdehyde
   (b) a hydroxyaldehyde
   (c) a dihydroxy compound
   (d) no reaction

3. The aldol reaction between formaldehyde (HCHO) and acetaldehyde (CH_3CHO) gives a
   (a) diol
   (b) dihydroxyketone
   (c) hydroxyaldehyde
   (d) dialdehyde

4. The dehydration of 3-hydroxypropanal, $CH_3CH(OH)CH_2CHO$, gives a/an _____ aldehyde.
   (a) α,β-unsaturated
   (b) β,γ-unsaturated
   (c) α,γ-unsaturated
   (d) β,γ-unsaturated

5. The intramolecular aldol reaction of 1,6-hexanedial, $OHCCH_2CH_2CH_2CH_2CHO$, gives a _____-membered ring.
   (a) four
   (b) five
   (c) six
   (d) seven

6. The Claisen self-condensation reaction of methyl formate, $CH_3C(O)OCH_3$, gives a/an
   (a) diketone
   (b) diester
   (c) keto ester
   (d) aldol

7. The Claisen condensation reaction between ethyl formate and ethyl acetate gives a/an
    (a) diketone
    (b) diester
    (c) keto ester
    (d) aldo ester

8. The intramolecular Claisen condensation of diethyl 1,6-hexanedioate gives a/an
    (a) diketone
    (b) diester
    (c) keto ester
    (d) aldol

9. The Michael reaction between 2,4-pentanedione and methyl vinyl ketone gives a/an
    (a) diketone
    (b) triketone
    (c) aldol
    (d) triol

10. The Stork enamine reaction between an enamine and an alkyl halide, followed by hydrolysis gives a/an
    (a) α-substituted ketone
    (b) β-substituted ketone
    (c) aldol
    (d) keto acid

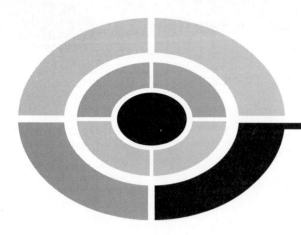

Final Exam

1. The ratio of the diameter of an atom's electron cloud to the diameter of its nucleus is about
   (a) 1/1
   (b) 10/1
   (c) 100/1
   (d) 10,000/1

2. An unidentified element has an atomic number of 7 and a mass number of 15. How many neutrons does it have?
   (a) 7
   (b) 8
   (c) 15
   (d) 22

3. An atom's valence electrons are
   (a) the core electrons
   (b) all the electrons
   (c) the outer shell electrons
   (d) the outer shell electrons minus the core electrons

4. A Lewis structure shows an atom's
    (a) core electrons
    (b) core and outer shell electrons
    (c) outer shell electrons
    (d) outer shell minus the core electrons

5. The maximum number of electrons any orbital can have is
    (a) zero
    (b) one
    (c) two
    (d) eight

6. Molecular orbitals result from
    (a) combing molecules
    (b) hybridizing atomic orbitals
    (c) the aufbau principle
    (d) combining/mixing atomic orbitals

7. VSEPR theory states that
    (a) electrons are very quite
    (b) electron pairs attract each other
    (c) electron pairs repel each other
    (d) electrons go into degenerate orbitals in pairs

8. The electron-domain geometry of an sp^3 hybridized atom is
    (a) tetrahedral
    (b) trigonal planar
    (c) linear
    (d) octahedral

9. Electronegativity refers to
    (a) the attraction of a positive charge for a negative charge
    (b) the attraction between negative charges
    (c) the attraction of an atom for electrons in a bond to that atom
    (d) the negative charge on a atom

10. The intermolecular attraction between hydrogen molecules results from
    (a) ionic-dipole interactions
    (b) dipole-dipole interactions
    (c) hydrogen bonding
    (d) van der Waals interactions

11. Resonance structures represent
    (a) molecules in equilibrium
    (b) the vibrational states of a molecule
    (c) the different arrangements of electron pairs in a molecule
    (d) the interaction of molecules with radiation

12. Brønsted-Lowry acids are defined as
    (a) proton donors
    (b) proton acceptors
    (c) electron pair donors
    (d) electron pair acceptors

13. A conjugate base is
    (a) a species that forms when a conjugate acid donates a proton to a base
    (b) a species that forms when an acid donates a proton to a base
    (c) a complex between two bases
    (d) a base with its valence electrons removed

14. Strong acids have ——— conjugate bases.
    (a) weak
    (b) strong
    (c) two
    (d) no

15. Electron delocalization makes a molecule
    (a) less stable
    (b) ionic
    (c) radioactive
    (d) more stable

16. Acids with large K_a values have
    (a) large pK_a values
    (b) pK_a values equal to their K_a values
    (c) small pK_a values
    (d) no relationship to a pK_a value

17. Lewis acids are
    (a) electron pair acceptors
    (b) electron pair donors
    (c) proton acceptors
    (d) electron pair donors and proton acceptors

18. Nucleophiles
    (a) like electrons

    (b) like carbocations

    (c) dislike electrophiles

    (d) like neutrons

19. Alkanes are ——— compounds.
    (a) saturated
    (b) unsaturated
    (c) conjugated
    (d) aromatic

20. The carbon atoms in alkanes are ——— hybridized.
    (a) sp
    (b) sp^2
    (c) sp^3
    (d) sp^4

21. A alkyl group is
    (a) an alkane
    (b) an alkane without a hydrogen atom
    (c) an alkane with an extra hydrogen atom
    (d) the elements in Group 1A in the periodic table

22. The compound 2-methyl-4-isopropyldecane is
    (a) a branched molecule
    (b) a molecule with two substituents
    (c) a molecule with side groups
    (d) all of the above

23. Molecular conformations refers to
    (a) resonance forms
    (b) cis and trans isomers
    (c) rotation around single bonds
    (d) rotation around double bonds

24. Torsional strain in ethane is a result of
    (a) repulsion between electrons in bonds
    (b) repulsion between nonbonding electrons
    (c) strain in bond angles
    (d) molecules being forced too close together

25. Steric strain in butane is a result of
    (a) repulsion between electrons in bonds
    (b) repulsion between nonbonding electrons

    (c) strain in bond angles
    (d) molecules being forced too close together

26. The most stable conformation about the C2–C3 bond in butane is the
    ——— conformation.
    (a) anti
    (b) eclipsed
    (c) gauche
    (d) chair

27. Cyclohexane is strain free because the ring has a ——— conformation
    (a) planar
    (b) eclipsed
    (c) boat
    (d) chair

28. When a cyclohexane ring flips between chair forms, axial groups
    (a) remain axial
    (b) change from cis to trans
    (c) change from trans to cis
    (d) become equatorial

29. When a cyclohexane ring flips between chair forms, groups that are cis
    (a) remain axial
    (b) change from cis to trans
    (c) remain cis
    (d) remain equatorial

30. *cis*-1,3-Dimethylcyclohexane is most stable when the methyl groups are
    (a) both axial
    (b) both equatorial
    (c) one group is axial and one group is equatorial
    (d) eclipsed

31. A chiral compound
    (a) has a superimposable mirror image
    (b) contains a plane of symmetry
    (c) has a nonsuperimposable mirror image
    (d) contains a mirror plane

32. Molecular configuration refers to
    (a) different orientations resulting from rotation around single bonds

    (b) molecules with the same connectivity but different 3-D orientations

    (c) different conformations of a molecule

    (d) the two chair forms of cyclohexane

33. Enantiomers
    (a) have the opposite configuration at all chirality centers
    (b) have the same configuration at all chirality centers
    (c) have the same configuration at one chirality center but different configurations at other stereocenters
    (d) are mirror images of diastereomers

34. Which of the following molecules is chiral?
    (a) CH_4
    (b) H_3CCl
    (c) H_2CBrCl
    (d) HCBrClF

35. An optically active molecule
    (a) shines in the dark
    (b) rotates the plane of plane-polarized light
    (c) does not rotate the plane of plane-polarized light
    (d) is visible to the naked eye

36. A racemic mixture contains
    (a) equal amounts of diastereomers
    (b) equal amounts of diastereomers and enantiomers
    (c) equal amounts of enantiomers
    (d) an excess of one enantiomer

37. The group with the highest R/S (Cahn-Ingold-Prelog) priority is
    (a) $-H$
    (b) $-CH_3$
    (c) $-CBr_3$
    (d) $-OH$

38. If a stereoisomer has a 2R,3S configuration about C2 and C3, its enantiomer is
    (a) 2R,3R
    (b) 2R,3S
    (c) 2S,3R
    (d) 2S,3S

39. A meso compound
    (a) is optically active

(b) has a plane of symmetry

(c) has a nonsuperimposable mirror image

(d) is chiral

40. *cis*-1,4-Dimethylcyclohexane is

(a) a meso compound

(b) is optically inactive

(c) contains two stereocenters

(d) is optically active

41. Alkenes

(a) are saturated compounds

(b) are unsaturated compounds

(c) contain all carbon-carbon single bonds

(d) contain no carbon-carbon double bonds

42. Rotation around carbon-carbon double bonds in acyclic compounds

(a) occurs at room temperature

(b) never occurs

(c) occurs at low temperatures

(d) occurs at high temperatures

43. E/Z nomenclature refers to

(a) optical activity

(b) giving common names to compounds

(c) the stereochemistry about double bonds

(d) conformation of alkanes

44. The molecular formula C_4H_6NOBr has ―― degrees of unsaturation.

(a) zero

(b) one

(c) two

(d) three

45. The most stable alkyl-substituted alkenes is

(a) tetrasubstituted

(b) trisubstituted

(c) trans disubstituted

(d) monosubstitited

46. A double bond contains

(a) two sigma bonds

(b) two pi bonds

(c) one sigma and one pi bond

(d) one sigma and two pi bonds

47. Thermodymanics is the study of
    (a) how fast reactions occur
    (b) energy and its transformation in equilibration reactions
    (c) the interaction of molecules with light
    (d) rate constants

48. Kinetics is the study of
    (a) how fast reactions occur
    (b) energy and its transformation in equilibration reactions
    (c) the interaction of molecules with light
    (d) exergonic and endergonic reactions

49. A transition state describes the
    (a) starting materials
    (b) reactive intermediates
    (c) products
    (d) unstable complexes that cannot be isolated

50. Carbocations are
    (a) carbon atoms with a negative charge
    (b) carbon atoms with a positive charge
    (c) carbon atoms with no charge
    (d) any atom with a positive charge

51. A ──── carbocation has the greatest stability.
    (a) methyl
    (b) primary
    (c) secondary
    (d) tertiary

52. Carbocations are stabilized by adjacent groups that ──── the carbocation.
    (a) donate electrons to
    (b) withdraw electrons from
    (c) donate protons to
    (d) withdraw protons from

53. The Hammond postulate states that the transition state in exothermic reactions resembles the
    (a) ground state
    (b) excited state
    (c) starting materials
    (d) products

54. The Markovnikov rule states that a hydrogen atom bonds to the carbon atom that already is bonded to ———— hydrogen atoms.
    (a) an equal number of
    (b) the greatest number of
    (c) the least number of
    (d) no

55. Carbocations undergo rearrangement reactions
    (a) to give more stable carbocations
    (b) by hydride shifts
    (c) by methyl shifts
    (d) for all of the above reasons

56. Alkenes undergo an acid-catalyzed addition of water to give
    (a) alkynes
    (b) alkanes
    (c) ethers
    (d) alcohols

57. Alkenes undergo acid-catalyzed addition of water via
    (a) Markovnikov addition
    (b) anti-Markovnikov addition
    (c) syn addition
    (d) anti addition

58. Borane reacts with alkenes
    (a) to give the anti-Markovnikov addition product
    (b) by a concerted mechanism
    (c) to give syn addition
    (d) all of the above

59. Bromine reacts with an alkene in an inert solvent to give
    (a) a monobromide
    (b) a vic dibromide
    (c) a gem dibromide
    (d) a bromohydrin

60. Bromine reacts with an alkene in water to give
    (a) a monobromide
    (b) a vic dibromide
    (c) a gem dibromide
    (d) a bromohydrin

61. Bromine reacts with an alkene in an alcohol to give
    (a) a bromoether

    (b) a vic dibromide

    (c) a gem dibromide

    (d) a bromohydrin

62. Diols are synthesized by reacting an alkene with
    (a) water
    (b) borane
    (c) ozone
    (d) osmium tetroxide

63. Ozone reacts with 2,3-dimethyl-2-butene to give
    (a) one ketone
    (b) two different ketones
    (c) one aldehyde
    (d) one ketone and one aldehyde

64. Alkenes under hydrogenation reactions to give
    (a) alkanes
    (b) alkenes
    (c) alkynes
    (d) conjugated alkenes

65. The carbon atom in a triple bond is ——— hybridized.
    (a) sp
    (b) sp^2
    (c) sp^3
    (d) sp^4

66. A carbon-carbon triple bond is ——— than a carbon-carbon single bond.
    (a) weaker
    (b) shorter
    (c) longer
    (d) less reactive

67. One equivalent of an alkyne reacts with 2 equiv. of HBr to give
    (a) a gem dibromide
    (b) a vic dibromide
    (c) a bromoalkene
    (d) a bromoalkyne

68. An alkyne reacts with excess chlorine to give
    (a) a monochloride
    (b) a dichloride

    (c) a trichloride

    (d) a tetrachloride

69. The acid-catalyzed addition of water to an internal alkyne gives

    (a) a ketone

    (b) a diketone

    (c) an aldehyde

    (d) a dialdehyde

70. The acid-catalyzed addition of water to a terminal alkyne gives, initially,

    (a) an aldehyde

    (b) an enol

    (c) an acid

    (d) an alkane

71. The rapid conversion between an enol and its keto form is called

    (a) resonance

    (b) cationic rearrangement

    (c) tautomerization

    (d) hydrogenation

72. Hydrogenation of an alkyne with Lindlar's catalyst gives

    (a) an alkane

    (b) a cis alkene

    (c) a trans alkene

    (d) a conjugated alkyne

73. An internal alkyne reacts with ozone to give, after reaction with dimethyl sulfide

    (a) alcohols

    (b) aldehydes

    (c) ketones

    (d) acids

74. Terminal alkynes are more acidic than

    (a) alkanes

    (b) alkenes

    (c) cycloalkanes

    (d) all of the above

75. TLC is used to

    (a) separate compounds

    (b) ionize compounds

    (c) induce bond vibrations
    (d) determine the ratio of carbon and hydrogen atoms

76. An R_f value is a measure of
    (a) a compound's melting point
    (b) ionization potential
    (c) molecular weight
    (d) molecular polarity

77. HPLC is used to
    (a) separate compounds
    (b) ionize compounds
    (c) induce bond vibrations
    (d) determine the ratio of carbon and hydrogen atoms

78. Infrared spectroscopy measures
    (a) molecular polarity
    (b) bond vibrations
    (c) ionization potentials
    (d) promotion of electrons into excited states

79. Infrared spectroscopy is used to
    (a) characterize functional groups
    (b) determine molecular weight
    (c) separate compounds
    (d) determine potential energy

80. IR carbonyl absorptions are typically
    (a) strong
    (b) weak
    (c) broad
    (d) in the fingerprint region

81. Ultraviolet spectroscopy measures
    (a) molecular polarity
    (b) bond vibrations
    (c) ionization potentials
    (d) promotion of electrons into excited states

82. Chromophores are functional groups that absorb
    (a) radiowave radiation
    (b) IR radiation
    (c) UV radiation
    (d) X-rays

83. NMR is used to determine
    (a) bond polarity
    (b) molecular structure
    (c) molecular weight
    (d) molecular bond vibrations

84. A proton can exist in ——— spin states.
    (a) one
    (b) two
    (c) three
    (d) four

85. The energy of electromagnetic radiation used in NMR spectrometers is
    (a) greater than that used in IR spectroscopy
    (b) greater than that used in UV spectroscopy
    (c) greater than that used in mass spectroscopy
    (d) less that that used in IR spectroscopy

86. An upfield NMR chemical shift refers to
    (a) a smaller ppm value relative to another chemical shift
    (b) a larger ppm value relative to another chemical shift
    (c) a larger δ value relative to another chemical shift
    (d) shifting a functional group to a higher position

87. An electron-withdrawing group adjacent to a proton of interest causes —— in NMR spectroscopy.
    (a) a downfield shift
    (b) an upfield shift
    (c) proton-proton coupling
    (d) spin-spin splitting

88. Multiplicity in NMR spectra refers to
    (a) splitting patterns
    (b) the number of protons in an absorption
    (c) J values
    (d) chemical shifts

89. The ratio of NMR peak heights in a quartet is
    (a) 1:1:1:1
    (b) 1:2:2:1
    (c) 1:3:3:1
    (d) 1:4:4:1

90. NMR decoupling refers to
    (a) separation compounds
    (b) separate ^1H and ^{13}C spectra
    (c) eliminating splitting patterns
    (d) making all protons chemically equivalent

91. Mass spectroscopy is used to determine
    (a) molecular polarity
    (b) bond vibrations
    (c) molecular ions and fragmentation patterns
    (d) promotion of electrons into excited states

92. Which alkyl halide has the weakest halide bond?
    (a) RF
    (b) RCl
    (c) RBr
    (d) RI

93. The reaction of HBr with t-butyl alcohol gives
    (a) an alkane
    (b) an acid
    (c) an alkyl bromide
    (d) a dibromide

94. Homolytic bond cleavage gives
    (a) cations
    (b) anions
    (c) radicals
    (d) cations and anions

95. The most stable radical is a ——— radical.
    (a) methyl
    (b) primary
    (c) secondary
    (d) tertiary

96. Allylic radicals are stabilized by
    (a) resonance
    (b) interaction with light
    (c) electron-withdrawing atoms
    (d) oxygen atoms

97. Radical chain reactions consist of ——— steps.
    (a) initiation
    (b) propagation

(c) termination

(d) all of the above

98. A Grignard reagent is an
    (a) alkyl halide
    (b) alkylmagnesium halide
    (c) an alkylmanganese halide
    (d) a dialkylcopper compound

99. Nucleophiles are
    (a) electron rich
    (b) electron poor
    (c) nucleophobic
    (d) carbocations

100. Electrophiles are
    (a) electron rich
    (b) electron poor
    (c) electrophobic
    (d) carbanions

101. The best leaving group in S_N1 reactions is
    (a) OH^-
    (b) NH_4^+
    (c) H_2O
    (d) I^-

102. A second-order reaction requires ——— molecule(s) to react in the rate-determining step.
    (a) one
    (b) two
    (c) three
    (d) four

103. A first-order reaction requires ——— molecules to react in the rate-determining step.
    (a) one
    (b) two
    (c) three
    (d) four

104. The components of a nucleophilic substitution reaction include
    (a) a substrate
    (b) a nucleophile

(c) a solvent

(d) all of the above

105. An example of a polar aprotic solvent is
    (a) H_2O
    (b) CH_3CO_2H
    (c) CH_3OCH_3
    (d) CH_3OH

106. An example of a polar protic solvent is
    (a) H_2O
    (b) $CH_3CO_2CH_3$
    (c) CH_3OCH_3
    (d) CH_4

107. An S_N1 reaction shows ——— order kinetics.
    (a) zero
    (b) first
    (c) second
    (d) third

108. ——— alkyl halides react slowest in S_N1 reactions.
    (a) Primary
    (b) Secondary
    (c) Tertiary
    (d) Quarternary

109. Elimination (E1 and E2) reactions of alkyl halides give
    (a) alkanes
    (b) alkenes
    (c) alkynes
    (d) alcohols

110. Zaitsev's rule states the ———substituted alkene tends to be formed in E2 reactions.
    (a) least
    (b) most
    (c) non
    (d) hexa-

111. Alcohols undergo ——— interactions.
    (a) dipole-dipole
    (b) intermolecular hydrogen bonding
    (c) London dispersion force
    (d) all of the above

112. Alcohols can act as
    (a) acids
    (b) bases
    (c) acids and bases
    (d) hydrogenation reagents

113. Primary alcohols react with ——— to give acids.
    (a) H_2O
    (b) chromic acid
    (c) PCC
    (d) osmium tetroxide

114. Primary alcohols react with ——— to give aldehydes.
    (a) water
    (b) chromic acid
    (c) PCC
    (d) osmium tetroxide

115. Aldehydes are reduced to alcohols by
    (a) chromic acid
    (b) PCC
    (c) ozone
    (d) sodium borohydride

116. Formaldehyde reacts with methyllithium to give, after protonation,
    (a) acids
    (b) ketones
    (c) secondary alcohols
    (d) primary alcohols

117. Primary alkyl alcohols react with thionyl chloride (ClS(O)Cl) to give
    (a) primary alkyl halides
    (b) secondary alkyl halides
    (c) acid halides
    (d) no reaction

118. Primary alcohols undergo condensation reactions in strong aqueous acids to give
    (a) anhydrides
    (b) ethers
    (c) acids
    (d) esters

119. Ketones undergo hydrogenation with Raney Ni catalyst to give
    (a) primary alcohols

    (b) secondary alcohols

    (c) tertiary alcohols

    (d) acids

120.   Ethers are good solvents because they are

    (a) very unreactive

    (b) polar

    (c) can form hydrogen bonds with active hydrogen compounds

    (d) all of the above

121.   The oxygen atom in ethers is ——— hybridized.

    (a) sp

    (b) sp^2

    (c) sp^3

    (d) sp^4

122.   The ether formed by the reaction of sodium methoxide ($NaOCH_3$) and methyl bromide (CH_3Br) is

    (a) diethyl ether

    (b) ethyl methyl ether

    (c) dimethyl ether

    (d) methyl ethyl ether

123.   Diethyl ether is hydrolyzed by excess concentrated HI to give

    (a) ethyl alcohol

    (b) ethyl alcohol and ethyl iodide

    (c) ethyl iodide

    (d) ethyl methyl ether

124.   1,2-Propylene oxide reacts with aqueous hydroxide anion to give

    (a) an ether

    (b) a monoalcohol

    (c) a dialcohol

    (d) an ester

125.   1,2-Propylene oxide reacts with aqueous acid to give

    (a) an ether

    (b) a monoalcohol

    (c) a dialcohol

    (d) an ester

126.   The common name for $CH_3S(O)CH_3$ is

    (a) dimethyl sulfide

    (b) dimethyl sulfone

(c) dimethyl sulfoxide
(d) dimethyl ether

127.   Ethanethiol (CH_3CH_2SH) —— ethanol (CH_3CH_2OH).
       (a) is a weaker acid than
       (b) is a stronger acid than
       (c) has the same acidity as
       (d) is a stronger base than

128.   Ethanethiol (CH_3CH_2SH) ——ethanol (CH_3CH_2OH).
       (a) has a higher boiling point than
       (b) has a lower boiling point than
       (c) has an identical boiling point to
       (d) is more soluble in water than

129.   The hydrosulfide anion (HS^-) reacts with ethyl bromide to give ethanethiol by an —— mechanism.
       (a) S_N1
       (b) S_N2
       (c) E1
       (d) E2

130.   Conjugated dienes have
       (a) only single bonds
       (b) only double bonds
       (c) alternating single and double bonds
       (d) double bonds separated by two single bonds

131.   Resonance energy results from
       (a) delocalization of electrons
       (b) overlapping s orbitals
       (c) equilibrium of resonance forms
       (d) overlapping of s and p atomic orbitals

132.   Molecular orbitals are classified as
       (a) bonding
       (b) nonbonding
       (c) antibonding
       (d) all of the above

133.   The highest energy molecular orbitals are
       (a) bonding
       (b) nonbonding
       (c) antibonding
       (d) all are of equal energy

134. The lowest molecular orbitals has ―――― nodes.
    (a) zero
    (b) one
    (c) two
    (d) ten

135. The addition of HCl to 1,3-butadiene gives ―――― constitutional (struc-

on is more stable than

calized on one carbon atom
elocalized on two carbon atoms

method of making

ground state, the HOMO ―――― the LUMO.

ship to
a group being ―――― a reference point.

141. Aromatic compounds are
    (a) linear

(b) acyclic
(c) nonconjugated and cyclic
(d) conjugated and cyclic

142. Meta refers to groups that are ——— to each other.
(a) 1,1-
(b) 1,2-
(c) 1,3-
(d) 1,4-

143. The common name for aminobenzene is
(a) aniline
(b) benzoic acid
(c) phenol
(d) nitrobenzene

144. Annulenes are ——— compounds
(a) conjugated acyclic
(b) nonconjugated cyclic
(c) conjugated cyclic
(d) nonconjugated linear

145. A benzene molecule has ——— bonding molecular orbitals.
(a) one
(b) two
(c) three
(d) four

146. Each carbon atom in a benzene ring is ——— hybridized.
(a) sp
(b) sp^2
(c) sp^3
(d) sp^4

147. Hückel's rule predicts aromaticity for a conjugated ring containing ——— pi electrons.
(a) one
(b) three
(c) six
(d) eight

148. The intermediate in an electrophilic aromatic substitution reaction is
(a) an anion
(b) a tertiary carbocation

(c) a diallyl carbocation

(d) a primary carbocation

149. The electrophilic aromatic substitution of benzene to give bromobenzene involves

(a) Br^-

(b) Br_2

(c) Br_3^+

(d) Br·

150. The methyl group in methylbenzene (toluene) is ––––– directing in electrophilic substitution reactions.

(a) ortho

(b) meta

(c) para

(d) ortha and para

151. The nitro group in nitrobenzene is ––––– in electrophilic substitution reactions.

(a) activating and meta directing

(b) deactivating and meta directing

(c) activating and para directing

(d) deactivating and para directing

152. Friedel-Crafts alkylation reactions give

(a) monoalkylation

(b) polyalkylation

(c) monoacylation

(d) polyacylation

153. Friedel-Crafts acylation reactions give

(a) monoalkylation

(b) polyalkylation

(c) monoacylation

(d) polyacylation

154. Benzyne is a molecule with

(a) only single bonds

(b) only double bonds

(c) single and double bonds

(d) a "triple" bond

155. ––––– is deactivating and ortho/para directing.

(a) $-NO_2$

(b) —Cl
(c) —CH$_3$
(d) —NH$_2$

156. Benzoic acid can be made by oxidizing ———— substituent on benzene.
(a) an ethyl
(b) a *t*-butyl
(c) a hydroxyl
(d) an amine

157. An amine group activates a benzene ring by ———— effect.
(a) an inductive
(b) a resonance
(c) a steric
(d) a hybridization

158. Aniline (aminobenzene) is made by
(a) direct amination
(b) nitration and oxidation
(c) nitration and reduction
(d) sulfonation and reduction

159. Aldehydes are oxidized by chromic acid to give
(a) aldehydes (no reaction)
(b) ketones
(c) acids
(d) alcohols

160. Ketones are oxidized by chromic acid to give
(a) aldehydes
(b) ketones (no reaction)
(c) acids
(d) alcohols

161. A ketone reacts with methylmagnesium chloride (Grignard reagent) to give, after protonation,
(a) an acid
(b) a ketone
(c) an aldehyde
(d) an alcohol

162. The acid-catalyzed addition of excess alcohol to a ketone gives
(a) a hemiacetal
(b) a semiacetal

    (c) a hemiketal
    (d) a ketal

163. Aldehydes react with Wittig reagents ($R_2C–PPh_3$) to give
    (a) acids
    (b) alcohols
    (c) ketones
    (d) alkenes

164. *m*-Chlorophenol is also called
    (a) 1-chlorophenol
    (b) 2-chlorophenol
    (c) 3-chlorophenol
    (d) 4-chlorophenol

165. The strongest acid is
    (a) *o*-nitrobenzoic acid
    (b) *m*-nitrobenzoic acid
    (c) 2,4-dinitrobenzoic acid
    (d) 2,4,6-trinitrobenzoic acid

166. Methylmagnesium bromide reacts with carbon dioxide to give, after protonation,
    (a) an alcohol
    (b) a ketone
    (c) an alkene
    (d) a carboxylic acid

167. Acetic acid is reduced by excess LAH to give
    (a) an alcohol
    (b) a ketone
    (c) an alkene
    (d) a carboxylate anion

168. Which of the following are derivatives of carboxylic acids?
    (a) Anhydrides
    (b) Esters
    (c) Amides
    (d) All of the above

169. Alcohols react with acid chlorides to give
    (a) acids
    (b) anhydrides
    (c) esters
    (d) none of the above

170. The most reactive carboxylic acid derivative is an
    (a) acid chloride
    (b) anhydride
    (c) ester
    (d) amide

171. Anhydrides react with water to give
    (a) acids
    (b) esters
    (c) amides
    (d) none of the above

172. A cyclic ester is called
    (a) an acid
    (b) an anhydride
    (c) a lactam
    (d) a lactone

173. The complete hydrolysis of a nitrile gives
    (a) an acid
    (b) an ester
    (c) an anhydride
    (d) an acid halide

174. An amine reacts with an acid chloride to give
    (a) an acid
    (b) an ester
    (c) a lactone
    (d) an amide

175. An enol results from the reaction of
    (a) an alcohol with an alkene
    (b) an alkene with water
    (c) an alkyne with water
    (d) a ketone with LAH

176. Tautomers are
    (a) resonance structures
    (b) enol and keto structures
    (c) mirror images
    (d) enantiomers

177. The alpha carbon atom is
    (a) the carbonyl carbon atom
    (b) bonded directly to the carbonyl atom

    (c) not related to the carbonyl position

    (d) in the para position in phenol

178. The compound with the most acidic hydrogen atom is
    (a) 3-hexanone
    (b) 2,3-hexanedione
    (c) 2,4-hexanedione
    (d) 1,6-hexanedial

179. Acid-catalyzed hydrolysis of malonic ester gives
    (a) a monoacid
    (b) a diester
    (c) a diacid
    (d) a carboxylate salt

180. The reaction of malonic ester with 1 equiv. of sodium ethoxide and 1 equiv. of ethyl bromide gives
    (a) diethyl ethylmalonate
    (b) ethylmalonic acid
    (c) disodium ethylmalonate
    (d) butanoic acid

181. The acid-catalyzed hydrolysis of diethyl ethylmalonate gives
    (a) dimethyl ethylmalonate
    (b) ethylmalonic acid
    (c) disodium ethylmalonate
    (d) butanoic acid

182. Ethylmalonic acid is decarboxylated by heating to give
    (a) diethyl ethylmalonate
    (b) ethylmalonic acid
    (c) disodium ethylmalonate
    (d) butanoic acid

183. Monoethylation of acetoacetic ester followed by acid-catalyzed hydrolysis and decarboxyation gives
    (a) methyl propyl ether
    (b) methyl propionate
    (c) methyl propyl ketone
    (d) methyl propyl alcohol

184. The base-catalyzed aldol reaction between two molecules of acetaldehyde gives
    (a) a dialdehyde
    (b) an α-hydroxyaldehyde

(c) a β-hydroxyaldehyde

(d) an α,β-dihydroxyaldehyde

185. The dehydration of a β-hydroxyaldehyde gives
    (a) an α,β-unsaturated aldehyde
    (b) an α,γ-unsaturated aldehyde
    (c) a β,γ-unsaturated aldehyde
    (d) an α,β-unsaturated ketone

186. The intramolecular aldol reaction of 1,7-heptanedial gives a _____ -membered ring.
    (a) four
    (b) five
    (c) six
    (d) seven

187. A Claisen condensation reaction of methyl acetate gives a
    (a) diketone
    (b) diester
    (c) ketoester
    (d) aldoester

188. The Claisen condensation reaction between methyl formate and methyl acetate gives a
    (a) diketone
    (b) diester
    (c) ketoester
    (d) aldoester

189. Aldehydes are oxidized by PCC to give
    (a) aldehydes (no reaction)
    (b) ketones
    (c) acids
    (d) alcohols

190. 1-Methylcyclohexene is oxidized by ozone followed by reduction with $(CH_3)_2S$ to give
    (a) two ketone groups
    (b) two aldehyde groups
    (c) one ketone and one aldehyde group
    (d) one ketone and one acid group

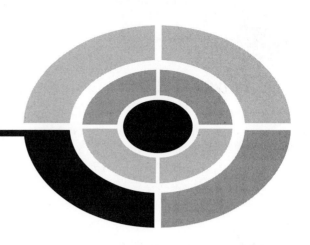

Quiz and Exam Solutions

CHAPTER 1 3/10

1. c	2. b	3. a	4. b	5. b
6. c	7. b	8. c	9. d	10. a
11. b	12. a	13. d	14. a	15. b
16. c	17. b	18. c	19. b	20. c
21. b	22. c	23. b	24. a	

CHAPTER 3 3/10

1. b	2. a	3. c	4. d	5. d
6. b	7. a	8. d	9. d	10. c
11. b	12. b	13. a		

3/4 **CHAPTER 4**

1. b	2. b	3. b	4. b	5. a
6. d	7. b	8. a	9. c	10. b
11. a	12. a	13. b	14. c	15. c
16. b	17. a	18. b	19. a	20. b
21. a	22. c	23. d		

3/12 **CHAPTER 5**

1. d	2. b	3. a	4. c	5. d
6. c	7. c	8. d	9. b	10. d
11. a	12. c	13. c	14. c	15. c
16. a				

3/13 **CHAPTER 6**

1. b	2. b	3. c	4. b	5. b
6. a	7. a	8. b	9. b	10. a
11. a	12. a	13. b		

3/30 **CHAPTER 7**

1. c	2. d	3. b	4. c	5. a
6. b	7. a	8. b	9. c	10. c
11. a	12. a			

CHAPTER 8

1. c	2. b	3. d	4. b	5. b
6. c	7. b	8. c	9. b	10. c
11. b	12. c	13. c	14. b	15. c
16. d				

CHAPTER 9

1. c	2. b	3. c	4. c	5. b
6. b	7. c	8. b	9. c	10. b
11. b	12. b	13. b	14. b	15. a
16. d				

CHAPTER 10

1. b	2. c	3. b	4. b	5. a
6. c	7. b	8. a	9. b	10. b
11. b	12. d	13. c	14. d	15. d
16. c	17. b	18. b	19. d	20. d
21. b	22. c	23. d	24. d	25. b
26. b	27. c	28. c	29. c	30. a
31. a	32. b	33. b		

CHAPTER 11

1. a	2. a	3. a	4. b	5. a
6. b	7. c	8. c	9. c	10. a
11. b				

CHAPTER 12

1. b	2. d	3. c	4. b	5. c
6. c	7. a	8. d	9. d	10. b
11. b	12. b	13. c	14. b	15. a
16. c	17. d			

CHAPTER 13

1. b	2. b	3. b	4. a	5. a
6. b	7. a	8. a	9. c	10. b
11. d	12. b	13. b	14. b	15. a
16. d	17. c	18. c	19. a	20. c

CHAPTER 14

1. c	2. c	3. a	4. b	5. b
6. c	7. a	8. b	9. a	10. c
11. a				

CHAPTER 15

1. b	2. b	3. b	4. b	5. d
6. d	7. c			

CHAPTER 16

1. c	2. a	3. d	4. c	5. a
6. d	7. c	8. a	9. b	10. c
11. d	12. c	13. c	14. b	15. c
16. a	17. c	18. c	19. b	20. b

CHAPTER 17

1. d	2. b	3. c	4. a	5. c
6. b	7. c	8. c	9. c	10. c
11. a	12. b	13. d	14. b	15. b
16. c				

CHAPTER 18

1. d	2. c	3. c	4. b	5. d
6. c	7. a	8. d	9. c	10. d
11. a	12. d	13. c	14. a	15. d
16. b	17. d	18. b	19. b	20. d

CHAPTER 19

1. c	2. d	3. b	4. c	5. a
6. d	7. c	8. b	9. c	10. a

| 11. b | 12. d | 13. a | 14. b | 15. d |
| 16. c | 17. c | 18. a | 19. b | 20. d |

CHAPTER 20

1. c	2. b	3. b	4. a	5. b
6. a	7. c	8. a	9. a	10. d
11. d				

CHAPTER 21

1. d	2. b	3. a	4. d	5. b
6. b	7. d	8. b	9. c	10. b
11. d	12. a	13. a	14. b	15. d
16. a	17. a	18. d	19. c	

CHAPTER 22

1. b	2. c	3. b	4. c	5. a
6. b	7. a	8. c	9. c	10. d
11. a	12. d	13. b	14. c	

CHAPTER 23

| 1. c | 2. b | 3. c | 4. a | 5. b |
| 6. c | 7. d | 8. c | 9. b | 10. a |

FINAL EXAM

1. d	2. b	3. c	4. c	5. c
6. d	7. c	8. a	9. c	10. d
11. c	12. a	13. b	14. a	15. d
16. c	17. a	18. b	19. a	20. c
21. b	22. d	23. c	24. a	25. b
26. a	27. d	28. d	29. c	30. b
31. c	32. b	33. a	34. d	35. b

36. c	37. d	38. c	39. b	40. b
41. b	42. d	43. c	44. c	45. a
46. c	47. b	48. a	49. d	50. b
51. d	52. a	53. c	54. b	55. d
56. d	57. a	58. d	59. b	60. d
61. a	62. d	63. a	64. a	65. a
66. b	67. a	68. d	69. a	70. b
71. c	72. b	73. d	74. d	75. a
76. d	77. a	78. b	79. a	80. a
81. d	82. c	83. b	84. b	85. d
86. a	87. a	88. a	89. c	90. c
91. c	92. d	93. c	94. c	95. d
96. a	97. d	98. b	99. a	100. b
101. d	102. b	103. a	104. d	105. c
106. a	107. b	108. a	109. b	110. b
111. d	112. c	113. b	114. c	115. d
116. d	117. a	118. b	119. b	120. d
121. c	122. c	123. c	124. c	125. c
126. c	127. b	128. b	129. b	130. c
131. a	132. d	133. c	134. a	135. b
136. d	137. b	138. a	139. a	140. c
141. d	142. c	143. a	144. c	145. c
146. b	147. c	148. c	149. b	150. d
151. b	152. b	153. c	154. d	155. b
156. a	157. b	158. c	159. c	160. b
161. d	162. d	163. d	164. c	165. d
166. d	167. a	168. d	169. c	170. a
171. a	172. d	173. a	174. d	175. c
176. b	177. b	178. c	179. c	180. a
181. b	182. d	183. c	184. c	185. a
186. c	187. c	188. d	189. a	190. c

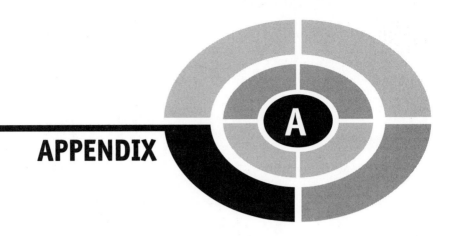

PERIODIC TABLE OF THE ELEMENTS

Periodic Table of the Elements

	I A	II A	III A	IV A	V A	VI A	VII A	VIII A	VIII A	VIII A	I B	II B	III B	IV B	V B	VI B	VII B	VIII B
1	1 H 1.0079																	2 He 4.003
2	3 Li 6.94	4 Be 9.0121											5 B 10.81	6 C 12.011	7 N 14.006	8 O 15.999	9 F 18.998	10 Ne 20.17
3	11 Na 22.989	12 Mg 24.035											13 Al 26.981	14 Si 28.085	15 P 30.973	16 S 32.06	17 Cl 35.453	18 Ar 39.948
4	19 K 39.098	20 Ca 40.08	21 Sc 44.955	22 Ti 47.90	23 V 50.941	24 Cr 51.996	25 Mn 54.938	26 Fe 55.847	27 Co 58.933	28 Ni 58.71	29 Cu 63.546	30 Zn 65.38	31 Ga 69.735	32 Ge 72.59	33 As 74.921	34 Se 78.96	35 Br 79.904	36 Kr 83.80
5	37 Rb 85.467	38 Sr 87.62	39 Y 88.905	40 Zr 91.22	41 Nb 92.906	42 Mo 95.94	43 Tc 98.906	44 Ru 101.07	45 Rh 102.90	46 Pd 106.4	47 Ag 107.86	48 Cd 112.41	49 In 114.82	50 Sn 118.69	51 Sb 121.75	52 Te 127.60	53 I 126.90	54 Xe 131.30
6	55 Cs 132.90	56 Ba 137.33	57 La 138.90	72 Hf 178.49	73 Ta 180.94	74 W 183.85	75 Re 186.20	76 Os 190.2	77 Ir 192.22	78 Pt 195.09	79 Au 196.96	80 Hg 200.59	81 Tl 204.37	82 Pb 207.2	83 Bi 208.98	84 Po (209)	85 At (210)	86 Rn (222)
7	87 Fr (223)	88 Ra 226.02	89 Ac (227)	104 Unq (261)	105 Unp (262)	106 Unh (263)	107 Uns (262)	108 Uno (265)	109 Une (266)	110 Unn (272)								

Lanthanide Series

58 Ce 140.12	59 Pr 140.90	60 Nd 144.24	61 Pm (145)	62 Sm 150.4	63 Eu 151.96	64 Gd 157.25	65 Tb 158.92	66 Dy 162.5	67 Ho 164.93	68 Er 167.26	69 Tm 168.93	70 Yb 173.04	71 Lu 174.96

Actinide Series

90 Th 232.03	91 Pa 231.03	92 U 238.02	93 Np 237.04	94 Pu (244)	95 Am (243)	96 Cm (247)	97 Bk (247)	98 Cf (251)	99 Es (254)	100 Fm (257)	101 Md (258)	102 No (259)	103 Lr (260)

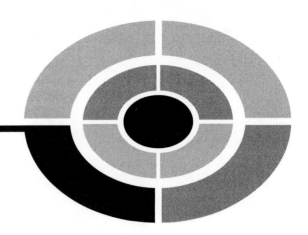

Bibliography

There are many excellent organic texts available that can be used to explain in greater detail the topics presented in *Organic Chemistry Demystified*. Six of these are mentioned here.

Books

Brown, W.H., and Foote, C.S., *Organic Chemistry*, 3rd ed, Brooks/Cole, New York, 2002.

Bruice, P.Y., *Organic Chemistry*, 4th ed, Prentice Hall, New Jersey, 2004.

McMurry, J., *Organic Chemistry*, 5th ed, Brooks/Cole, New York, 2000.

Schmid, G.H., *Organic Chemistry*, Mosby, St. Louis, 1996.

Solomons, T.W.G., *Organic Chemistry*, 6th ed, John Wiley, New York, 1996.

Wade, L.G., Jr., *Organic Chemistry*, 6th ed, Prentice Hall, New Jersey, 2006.

INDEX

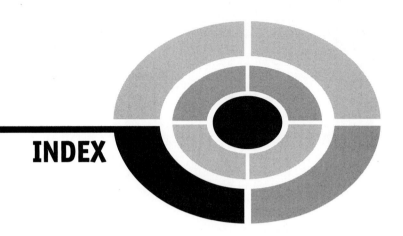

Index

Index

Index

Index

Index

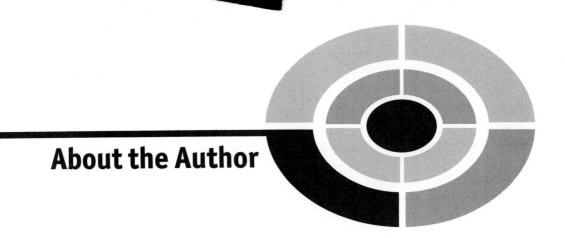

About the Author

Daniel Bloch received a PhD in organic chemistry from the University of Illinois. He spent 30 years in the industrial world in scientific and management positions at SC Johnson and Aldrich Chemical. Dr. Bloch has held visiting professorships in chemistry at several universities, teaching at the undergraduate and graduate levels. He has over 30 publications, patents, and technical presentations. Dr. Bloch has lectured nationally and internationally. Currently he is president of Lakeshore Research, LLC, a contract consulting firm.